SCIENTIFIC WORDS

SCIENTIFIC WORDS

Their Structure and Meaning

by

W. E. FLOOD

GREENWOOD PRESS, PUBLISHERS
WESTPORT, CONNECTICUT

Library of Congress Cataloging in Publication Data

Flood, Walter Edgar.
Scientific words: their structure and meaning.

Reprint of the ed. published by Duell, Sloan, and Pearce, New York.
1. Science--Dictionaries. I. Title.
[Q123.F57 1974] 503 74-6707
ISBN 0-8371-7541-0

Originally published in 1960 by Duell, Sloan and Pearce, New York

Reprinted with the permission of Hawthorn Books, Inc. Publishers

Reprinted by Greenwood Press, Inc.

First Greenwood reprinting 1974
Second Greenwood reprinting 1977

Library of Congress catalog card number 74-6707

ISBN 0-8371-7541-0

Printed in the United States of America

ACKNOWLEDGEMENTS

My thanks are due to many friends and colleagues for the help and advice which they have given me in the preparation of this book, notably to Mr. A. C. Townsend, M.A., Librarian, Natural History Museum, who read the first draft and gave special attention to the classical origins of the elements and the explanations of terms in the field of Natural History.

W. E. F.

INTRODUCTION

A ZOOLOGIST has described Man as:

> 'metazoan, triploblastic, chordate, vertebrate, pentadactyle, mammalian, eutherian, primate'.

A chemist has written that:

> 'in the formation of mono-substitution products of benzoic acid the halogen takes up the meta-position with respect to the carboxyl'.

From a technical dictionary we read that a carbuncle is:

> 'a circumscribed staphylococcal infection of the subcutaneous tissues'.

Such passages as these are probably unintelligible to the non-scientist and might even puzzle a scientist if he specialised in a totally different field. They are, however, sensible statements of certain scientific facts. Their difficulty lies in the concepts which are involved—concepts with which the reader may not be familiar—and also in the technical terms which are used. A reader who does not understand such terms should not be tempted to dismiss the passages as mere technical jargon. If he is thoughtful he might well ask a number of questions about the specialised words of science.

Why do scientists use such unfamiliar, and apparently difficult, words? Why do they need a special vocabulary of their own?

What is the nature of the specialised words of science? What are their origins? Are they just fanciful inventions or have they been sensibly and logically constructed?

Are these words necessarily unintelligible to all but the scientific expert or can an ordinary educated person, who knows a little science, make some sense of them and gain at least a general idea of their meanings?

The Purpose and Nature of Scientific Words

The development of an appropriate vocabulary is essential to the development of any subject. Words are the elements of language; language is the vehicle of ideas. By silent language thoughts are developed in the mind, and by written or spoken language thoughts are communicated to others.

It is obvious that a scientist must have names by which to

identify and refer to the various chemical substances, minerals, plants, animals, structural units, instruments, etc., with which he deals. He must have suitable adjectives for describing these things and suitable verbs for defining their behaviour.

He also needs suitable names by which to identify the various abstractions with which he deals—processes, states, qualities, relationships, and so on. Thus, after Faraday had investigated the passage of electric currents through different solutions and noted the resulting liberation of chemical substances, the term *electrolysis* was invented. This one word was a kind of shorthand symbol for the process; it 'pinned down' the process and conveniently embraced its many aspects. From then on it was possible to think about the process and to talk about it to others. Similarly, the single term *symbiosis* conveniently summarises a biological state; *diathermancy* identifies a physical quality.

Many scientific words are of this kind. Without the name (or technical term) a concept remains vague and ill-defined; the scientist is hindered in his mental processes, in his recording of what he thinks and does, and in his communication with others.

In his communication a scientist is mainly concerned with the exact and logical expression of that which he wishes to pass on to another. His purpose is to inform (as clearly as possible), not to excite emotion. It follows that each of his words must have a precise meaning, and one meaning only, so that there is no risk of confusion or ambiguity. Of course he must know himself what his words mean and he must assume that the person with whom he communicates attaches the same meanings to them. (If he is communicating with a person who is unlikely to understand his specialised terms he must take care not to use them, even if that may mean some loss of precision or elegance. A number of 'popular' science talks fail because the speaker, often an expert scientist, thoughtlessly uses words which the ordinary person does not understand.)

The meanings of many ordinary words of our language are not single and precise. Although the original, basic meanings may be clear, the words have acquired a range of meanings over the years. Thus the familiar word *fair* has somewhat different meanings when used to describe the weather, a person's hair, an action or decision, or a boy's performance at school; some words (e.g. *rude*) suffer a significant change in meaning. Hence a scientist avoids the ordinary words of the language; he prefers his own words. These words can then be rigorously defined and given the necessary precision of meaning.

The use of words which are 'set apart' from everyday life also enables the scientist to avoid evoking irrelevant and distorting associations. Some ordinary words convey more than their literal meanings; they evoke further images, emotions and reactions on the part of the hearer or reader. (Thus *red*, basically a word denoting a certain colour, may conjure up thoughts and feelings relating to danger, to blood, or to a particular political outlook.) The specialised words of science, if used in their proper contexts, are largely free from distorting associations. It is interesting to note that when a scientific term, originally well-defined, becomes a word of ordinary speech, it usually suffers a widening of meaning and acquires a number of associations. Thus criticism (as well as sulphuric acid) may be *vitriolic*, a man may be *electrified* into action, and people may claim to be *allergic* to all sorts of things and conditions. The word *atomic*, whose meaning is quite clear to the scientist, may conjure up in the public mind a picture of widespread destruction or of unlimited power.

In addition to precision of meaning and freedom from associations, most scientific words have a third quality: by their form and structure they reveal something of their meanings. Many scientific words are logically built up from simpler word-elements (usually of Greek or Latin origin) and the general meaning of the whole can be inferred from an understanding of the parts. Some terms, in fact, are self-explanatory if the Latin and Greek roots are known; they have only to be 'translated' for their meanings to become apparent.

Thus a *quadrilateral* is clearly a four-sided figure, *entomology* is the study of insects, *gastrectomy* is the cutting out of the stomach (or part of it). In the case of a large number of words the full or precise meaning may not be directly disclosed but the general meaning is apparent and the word is seen to 'make sense'. Thus *cyanosis* indicates a state (possibly a morbid state) of blueness; it is a sensible word to use to denote the blue condition of the skin which results from insufficient oxygen in the blood. A *xerophyte* (literally "a dry plant") is one which is adapted for living in very dry conditions; a *hydrophyte* is one which lives on the surface of, or submerged in, water. A *polymer* consists of "many parts"; the term is an appropriate one for a giant molecule which is built up from a large number of simple units.

In a similar way, many chemical names are essentially descriptions of the compounds which they denote. Thus whereas the common name *aniline* for a certain oil discloses nothing about the nature of the compound (except, perhaps, that it is vaguely related to indigo), the chemical name *aminobenzene* immediately indicates the molecular composition and structure.

Scientific language, to be efficient, must be universally intelligible. The classical languages, Latin and Greek, are so fundamental to the civilised world that words constructed from elements of these languages are readily understood the world over. (Even if scientists know little of the classical languages, they can easily learn to 'translate' the scientific terms which they may meet.) Most scientific terms are effectively international.

Sources of Scientific Words

Scientific words in English may conveniently be divided, from the standpoint of their origins, into three groups:

(*a*) those taken from the ordinary English vocabulary;
(*b*) those taken virtually unchanged from another language;
(*c*) those which have been invented.

The third group is by far the largest.

Just as the cricketer has taken certain everyday words, such as *run*, *over*, *maiden*, from the general English vocabulary and given them specialised meanings within the context of his game, so the scientist has occasionally taken ordinary English words and endowed them with specialised meanings. *Energy*, *work*, *power*, *salt*, *base*, *fruit* are examples of such words. They are unsatisfactory as scientific terms because they lack the essential qualities which we have described. Although the scientist may give them precise meanings, they are liable to be interpreted more loosely (or even differently) by the non-scientist. They are not free from irrelevant associations; they reveal little of their meanings by their forms; and usually, they are not understood outside the English-speaking countries. There are not many words of this kind but, unfortunately, most of them stand for concepts of fundamental importance.

The English language contains a number of words which have been taken from another language with little or no change of spelling. Amongst them are *café*, *morgue*, *souvenir*, *trek*, *marmalade* and *agenda*. Practically all the scientific words of this kind have been taken from Latin or Greek. As examples of Latin words we may note *axis*, *fulcrum*, *larva*, *radius*, *locus*, *nimbus*, *cortex*. Many parts of the human body, e.g. *cerebrum*, *pelvis*, *cornea*, have Latin names. There are fewer unaltered Greek words—*thorax*, *stigma*, *iris*, *helix* are examples—but it should be noted that many terms adopted in Latin form, e.g. *trachea*, *bronchus*, *phylum*, were themselves based on Greek. Many of the Greek or Latin terms have retained their original meanings but in some cases the meanings have been restricted and rendered more precise.

The largest group of scientific words are those which have been invented. The advance of science during the last few centuries has been so rapid and so extensive that no language has been capable of providing, ready-made, all the words which were required. Further, the classical languages do not contain words appropriate to modern discoveries, inventions and concepts. (There is no Latin word, for example, for photography!) Hence the scientist has had to invent new words for his own purposes.

It is very rare for a scientist to make up a word 'out of his head'; the term *ester* for a compound formed by the interaction of an alcohol and an organic acid was perhaps such an invention. A small but interesting group of terms comprises those based on proper names. In the naming of the chemical elements recourse has been made to the names of places (as in *polonium*, *ytterbium*), of gods and goddesses (as in *thorium*, *vanadium*), of planets and asteroids (as in *uranium*, *cerium*), and of scientists themselves (as in *curium*, *gadolinium*). Scientists' names have also been used to provide the names of units (e.g. *watt*, *volt*, *gauss*, *joule*) and hence the names of measuring instruments (e.g. *voltmeter*). Among the other terms based on the names of scientists are *daltonism*, *nicotine*, *bakelite* and *mendelism*. A number of plants, e.g. *fuchsia*, *dahlia* are named after botanists.

In his task of inventing new terms, however, the scientist has usually turned to the classical languages for his raw material. He has taken 'bits and pieces'—roots, prefixes, suffixes—from these languages and joined them together to form the terms he needed. Thus, when he needed a general name for animals such as snails and slugs which apparently walk on their stomachs, he took the Greek roots *gast(e)ro-* (stomach) and *-pod* (foot) and formed the new word *gastropod*. When he wanted a word to describe a speed greater than that of sound he took the Latin prefix *super-* (above, beyond) and the Latin root *son-* (sound) and coined the adjective *supersonic*. Thousands of scientific words have been built up from classical word-elements in this way.

It may be asked why the scientist should have turned to the classical languages for the words and word-elements which he needed. By turning to a language other than his own he was certainly able to find words and elements which were distinct from those of ordinary speech but he turned to the classical languages for an important historical reason. The fifteenth and sixteenth centuries witnessed that great revival of classical learning which is commonly called the Renaissance. Latin was regarded as the universal language of scholarship; it was the 'perfect' language of philosophy, theology and science. This classical tradition persisted into the seventeenth century—both Harvey and Newton wrote their great works in Latin—and it was not until towards the end of that century that English was fully accepted as an adequate and suitable language for a scholarly exposition of science.

During this period many Latin words were taken into the scientific vocabulary and many new words were constructed (chiefly in the form of Latin words) from classical elements. The tradition of using the classical languages as a source of scientific words remains.

Greek was not used in the same way as a medium of expression but it was held in respect as the language of the people who at one time led the world in art, science and philosophy. Moreover, it provided a particularly suitable basis for scientific language. It had been developed by a long line of philosophers as a medium for accurate expression and its elements were such that derivatives and compounds were readily formed. The scientists therefore mainly went to the Greek for the new terms which they needed (though, as has been pointed out, the terms were at one time often framed in Latin form). Greek is still the source of most of the new terms of science and more than half of the words of the great vocabulary of science are ultimately of Greek origin.

The Formation of Scientific Words from Classical Word-elements

Despite the enormous size of the modern vocabulary of science, the basic elements from which the words have been constructed are comparatively few. This book lists about 1,150 word-elements. They have produced, and will go on producing, tens (or probably hundreds) of thousands of words. A large proportion of the words have been built up from a much smaller number of elements. (The greater part of the vocabulary of medicine and anatomy—perhaps 30,000 words—has been constructed by the use of only about 150 standard word-elements and the names of the parts of the body.) Many elements appear in a range of words distributed among a number of different sciences. Thus the element *pter-* (Gk. *pteron*, wing) appears in the names of many sub-classes of insects (e.g. *Diptera*, *Lepidoptera*), of certain types of aircraft (e.g. *helicopter*) of a group of chemical substances (e.g. *methopterin*) and in the name of a mesozoic flying reptile (*Pterodactyl*).

As will be seen from the Glossary, the word-elements are generally used in forms which are specially adapted to word-building. Thus the Greek noun *nephros* (kidney) is used in the combining-form *nephro-* (or *nephr-* before a vowel). Let us take this root and look at the range of words which have been built up from it. We may suffer from *nephropathy* (disease of the kidney), *nephralgia* (pain in the kidney), *nephritis* (inflammation of the kidney) or *nephroptosis* (a dropping of the kidney). We may undergo the surgical operations of *nephrotomy* (a cutting of the kidney), *nephrectomy* (a cutting out), *nephrorrhaphy* (a sewing up) or *nephropexy* (a fixing in place). Yet more terms will be found in a medical dictionary. We might like to invent a few more terms ourselves. The kidney can suffer the processes of *nephrothermolysis* (being cooked) and *nephrophagy* (being eaten)! The root has also been used in forming the names of excretory structures in certain lower animals. In an Earthworm, for example, each normal segment contains a pair of excretory organs which have been called the *nephridia* (literally, the little kidneys).

Prefixes which indicate degree, position or number are of particular value in word-building. Thus we may suffer from *hyperpiesis* (high blood pressure) or from *hypopiesis* (low blood pressure), the two terms being formed by the addition of contrasting prefixes to the same root. Similarly, the *ectoplasm* is the thin protoplasm near the outside of a cell and the *endoplasm* is the denser protoplasm well within the cell. The *Apoda* have no legs, the *Decapoda* have ten, and the *Myriapoda* have many. Radio valves may be classified as *diodes* (two electrodes), *triodes* (three electrodes) . . . *pentodes* (five electrodes) . . . *octodes* (eight electrodes), and so on.

Often both Greek and Latin elements with the same meaning are available. Thus a flesh-eating animal may be described as *sarcophagous* (Gk.) or *carnivorous* (L.); both *hypodermic* (Gk.) and *subcutaneous* (L.) mean under the skin. Occasionally slight differences of meaning have been arbitrarily assigned to corresponding words of different origins. There is no rule as to whether Greek or Latin elements should be used in word-building though often the Greek elements fit together more easily.

Sometimes both Greek and Latin elements are combined in the same word. *Television* is a well-known example; the prefix *tele-* (from afar) is Greek and the root *vis-* (seeing) is Latin. (The 'all-Greek' word *teleorama* would have been more satisfying to the purists but it is unlikely to be adopted.) The formation of 'hybrid' words of this kind may be considered objectionable if 'pure' alternatives are readily available and equally convenient. Thus the term *odoriphore** is a needless hybrid; the 'all-Greek' term *osmophore* would serve just as well. There appears to be no justification for the invention of the hybrid word *pluviometer* (rain gauge) when two all-Greek terms, *hyetometer* and *ombrometer*, are available. And chemists still seem not to have made up their minds whether to use Latin or Greek prefixes of number before the Latin root *-valent*.

Undoubtedly some hybrids have been formed because of thoughtlessness or ignorance, but many have been formed because certain prefixes and suffixes have become well known and have been found to be convenient. Thus the familiar Greek root *-meter* (measurer) has been added to all sorts of stems, e.g. to a Latin stem in *audiometer* and to an English stem in *weatherometer*. (Note the insertion of the *o* before *-meter*; in all-Greek terms an *o* normally arises as the ending of the stem.) The Greek element *-logy* (often regarded as *-ology*) is now freely added to stems of various kinds and origins; the three common medical elements *-itis* (inflammation), *-oma* (growth, tumour) and *-osis* (morbid state) are not infrequently added to Latin stems (e.g. as in *gingivitis*, *fibroma*, and *silicosis*). Certain prefixes of classical origin, e.g. *re-*, *pre-*, *micro-*, *sub-*, *tele-*, are still 'living' and are freely used in combination with words of any origin, e.g. in *re-oxidise*, *pre-Cambrian*, *microfilm*, *substandard* and *telecommunication*.

The process of word-building has certainly resulted in some peculiar-looking words, e.g. *heterochlamydeous*, *otorhinolaryngology* and *postzygapophysis* (in which one prefix of Latin origin and two of Greek have been added to the Greek word *physis*), but many of them readily break down into their component parts and reveal their meanings. Some of the ugliest words, perhaps, are found in the field of medicine but the longest words are the names given to certain chemical compounds. *Tetrahydronaphthylamine*, with twenty-three letters, is a very humble example; some names contain over sixty letters. These long names, however, are easily understood by a chemist, for they are logically constructed and provide detailed descriptions of the compounds to which they are given.

It is not easy to explain the nature of these chemical names without presupposing a fair knowledge of chemistry. Perhaps one simple example will help the reader to appreciate the way these

* *Odori'phore*—"odour bearer"—a group of atoms which confer a particular smell on a chemical compound.

names are constructed. The molecule of benzene (C_6H_6) consists of a ring of six carbon atoms to each of which is joined a hydrogen atom. If two (*di-*) of the hydrogen atoms are replaced by chlorine atoms, the compound is conveniently called *dichlorbenzene*. There are, however, three forms of dichlorbenzene depending upon the relative positions of the two chlorine atoms. The forms can be distinguished by the use of appropriate prefixes. Thus one form, a substance sometimes used for protecting clothes from moths, is known as *para-dichlorbenzene*.

The Analysis and Interpretation of Scientific Words

Not many people are in the position of needing to invent new scientific words. A scientist may need to do so occasionally, particularly if he is researching in a new field. Sometimes a manufacturer invents a pseudo-scientific name (often a verbal monstrosity) for his products, apparently to make them seem more attractive. The layman is never called upon to invent scientific words.

All kinds of people, however, may find themselves needing to interpret the meanings of scientific words. The scientist may meet new terms invented by other scientists; he may meet words which are unfamiliar to him because they are in specialised fields outside his own. The student frequently meets words which are strange to him but which he must learn and understand in order to progress in his studies. And, in these modern times, the layman meets scientific words in his newspapers, in advertisements, and through television.

It must be recognised at the outset that a reader (or listener) cannot understand a discourse on a subject if he lacks the necessary background knowledge; he must be able to meet the author 'part way'. This is true of all kinds of reading. One cannot fully understand a passage of Shakespeare if one lacks the background which he presupposed when fashioning his metaphors; one cannot follow an account of the working of a synchrotron if one knows nothing about electric fields and particles. Similarly, one cannot interpret the meaning of a word if one has an inadequate understanding of the subject to which it relates. The term *melanosporous* must remain unintelligible if one does not know what spores are, and *stereo-isomerism* cannot be understood by one who knows nothing about molecular structure. (This does not mean, however, that it is impossible to understand that which is outside one's immediate knowledge and experience. One's knowledge can be extended by building up from that which is known. The concept of *stereo-isomerism*, for example, could be explained to a layman if care were taken to build up his knowledge step by step.)

Words which are pure Latin or pure Greek, and which cannot be broken down into simpler parts, do not readily disclose their meanings; one either knows the meanings or one does not. Thus one cannot infer the meaning of *tibia*, *thallus*, or *soma* merely from the spelling. It has been shown, however, that the majority of scientific words have been constructed from simpler word-elements and thus, from an understanding of the parts, one may deduce the meaning (or at least the general sense) of the whole. This is, indeed, one of the virtues of scientific words.

The criticism is sometimes made that deduction of meaning on the basis of etymology may be misleading. It is true that some scientific terms are misnomers; they were coined in the light of knowledge which is now known to be inaccurate. Thus *vitamins* are not amines, the *maria* of the Moon are not seas, and *oxygen* is not necessarily a producer of acids. Many minerals have been misnamed. With the great majority of scientific terms, however, etymology can be of great value in the deduction of meaning.

As has already been pointed out, the meanings of a large number of scientific words are directly revealed by simple translation. *Conchology* is obviously the study of shells, a *lignicolous* fungus is clearly one which lives on wood, and what else can *hypodermic* mean than under (or below) the skin? *Antiseptic*, *microphyllous*, *anemometer*, *centripetal*, *pentadactyl*, *hyperglycaemia* are among the thousands of scientific words whose meanings may be readily deduced by simple analysis. It is possible that by simple translation one might occasionally miss some subtle shade of meaning or of application but one would nevertheless gain a useful idea of what the words denote.

There are thousands of other words, of course, whose full meanings cannot be determined by simple deduction. Thus *pericardium* clearly means "round the heart", but we cannot deduce exactly what it is; an *electrometer* is apparently an instrument for measuring electricity but we cannot tell what property of electricity it measures. The word *isotope* tells us no more than that 'it' is in the same place as something else. The translation of the names given to plants and animals is often of no help in identification; we cannot recognise *Myosotis* by translating the name as "mouse ear" nor do we know what *Oligochaeta* are even if we deduce that they have "few bristles".

Even if a scientific word does not reveal its full meaning on simple analysis, it is seen to 'make sense' when its full meaning has been explained. It is not an unintelligible assembly of letters. It is seen to fit in with its meaning; it is more easily recognised on another occasion, it is more easily remembered; its relation to other similar words will be appreciated. An understanding of structure and derivation converts an unintelligible word into one which makes sense.

The main purpose of this book is to provide an explanatory list of the more important word-elements which enter into the formation of scientific terms. By the use of this list, and with the help of the illustrative examples, the reader should be able to break down and interpret many of the scientific terms which he meets and to 'make sense' of thousands of others. Let us take a few words in illustration.

The word *photometer* readily breaks down into the elements *photo-* (light) and *-meter* (a measurer); it is evidently the name of an instrument for measuring some quality (e.g. intensity) of light. The word *geomorphology* breaks down into the elements *geo-* (Earth), *morpho-* (form, shape) and *-logy* (which may be interpreted as 'the study of'); geomorphology is thus the study of the shape of the Earth (actually of the origin and nature of its surface shape and features). The term *gastromyotomy* breaks down into its elements *gastro-* (stomach), *myo-* (muscle) and *-tomy* (cutting); we deduce that gastromyotomy is the surgical cutting of the muscles of the stomach. Similarly, we deduce that *nephro'ptosis* is a dropping of the kidney and that *arterio'sclerosis* is a hardening of the arteries. We understand why lines on a map passing through places which have the same temperature are called *isotherms* (*iso-*, equal, *therm-*, heat) and so deduce the meanings of *isobar*, *isoneph*, *isohyet*, and similar terms. Having learned that the element *cyto-* indicates a cell, we can make sense of such terms as *cyto'logy*, *cyto'genesis* and *cyto'lysis*.

Let us take one example to illustrate how a long chemical name may be interpreted. What can be made of the name *polytetrafluoroethylene*? As is often useful when analysing chemical names, we work from right to left. We start with *ethylene*, the name of a well-known hydrocarbon (hydrogen-carbon compound) with the chemical formula C_2H_4. *Tetra-fluoro-* indicates that four fluorine atoms are taking the place of four (in this case all) hydrogen atoms in the molecule. So we reach a compound which may be represented by the formula C_2F_4. The prefix *poly-* in chemical names indicates that a giant molecule, as in a 'plastic', has been built up by the joining together of a large number of simple units.

Polytetrafluoroethylene is, in fact, a 'plastic' substance built up from C_2F_4 units, known commercially as P.T.F.E. or Teflon.

An understanding of the structure and origin of a word is not only a guide to its meaning; it is often a guide to its spelling. No schoolboy who understands the origin of the word *bicycle* should ever spell it wrongly. Nor should he wonder whether to put one *s* or two in such words as *disappear*, *disappoint* and *dissolve*. There must clearly be two *n*'s in *innocuous*—one from the prefix and one from the root—but only one in *inoculate*. The word *desiccate* should never be a notorious spelling difficulty. As the science student or layman learns the commoner word-elements and recognises their presence in the words he meets, he is also learning how to spell the words. The spelling of such words as *anaesthetic*, *diarrhoea*, *dysentery*, *haemorrhage*, *paraffin*, *parallel*, *psycho-neurosis*, *rhododendron* should present no difficulties if their origins are understood.

The criticism may be made that in these days few scientists, and few laymen, are acquainted with the classical languages and hence they cannot analyse words in the ways we have described. In former times a scientist was often a man who, having received an education in the classics, subsequently devoted himself to scientific studies. He was well able to invent the new words which he needed and to interpret words invented by others. Nowadays, however, a scientist usually knows little or nothing of the classics (and, let it not be overlooked, the classicist usually knows nothing of the sciences).

We live in a scientific age; an understanding of science is at least as necessary to the make-up of an educated man as a knowledge of the arts. More and more people need to understand the words of science. This does not mean that traditional courses of Latin and Greek should therefore form a part of everyone's education but it indicates the desirability of teaching the more important roots which enter into the formation of English, and especially scientific, words. A study of 'Words and their Origins', with a bias towards scientific words, should form a part of the normal work of all our secondary schools.

NOTES ON THE GLOSSARY

THE glossary lists about 1,150 word-elements (roots, prefixes, suffixes) which enter into the formation of scientific terms. (Very common elements, e.g. *un-*, *-ation*, *-able*, which are sure to be known to the reader, are not included.)

The meaning of each element is given and also its origin (usually Latin or Greek). It should be noted, however, that many words and elements whose origins are given as Greek passed into Latin before becoming part of the English language. Greek words have been written with the corresponding English letters; θ, υ, φ, χ, ψ, and the aspirated ϱ are shown as *th*, *y*, *ph*, *ch*, *ps* and *rh* respectively; γ is shown as *n* in those words (e.g. *enkephalos*, *planktos*) in which it effectively has the sound of *n*.

Wherever it is thought helpful or interesting, attention is drawn to the occurrence of an element in a familiar word of ordinary speech.

The use of each element in word-building is illustrated by a selection of scientific terms which incorporate the element. The meaning of each term is given. The terms have been selected to show:

(*a*) the various forms which the element may take;

(*b*) the use of the element in building terms in different sciences.

The glossary is not intended to be a complete scientific dictionary—it does not give all the terms which incorporate each element—but it does provide, in fact, simple explanations of several thousands of terms.

The sign ˈ is used to break up a word into its component parts in order to demonstrate the structure of the word. The sign is *not* an indication of stress nor is it necessarily a guide to pronunciation.

Double inverted commas (" ") are used to show literal meanings, i.e. direct 'translations'.

Chemical formulae are given wherever they serve a useful purpose. In some cases a formula is an aid to the explanation; in many cases a formula is given to help in the identification of the substance named.

GLOSSARY OF SCIENTIFIC WORD-ELEMENTS

A

α-, ALPHA-

α, the first letter of the Greek alphabet, is sometimes added before the name of a series or group of things to denote the first member of the series or group. (The succeeding letters β, γ, . . . are used for other members.)

α-rays, alpha-rays—one of the three types of radiation given off by radioactive substances, consisting of a stream of positively charged particles (**α-particles**).

α-brass, alpha-brass—a form of brass (a solid solution of zinc in copper) containing up to about 38 per cent. zinc.

In naming organic compounds, α is sometimes used to show that a certain group of atoms is in the first of two (or more) possible positions in the molecule.

α-hydroxy'propionic acid — the acid (lactic acid) whose molecule is represented by the formula $CH_3.CH(OH).COOH$. The hydroxy-group (-OH) takes the place of one of the hydrogen atoms in propionic acid $CH_3.CH_2.COOH$; counting back from the characteristic acid group -COOH, the hydroxy-group replaces one of the hydrogen atoms joined to the first carbon atom. (The acid represented by $CH_2(OH).CH_2.COOH$, in which the hydroxy-group is joined to the second carbon atom, is β-hydroxypropionic acid.)

Similarly, in naphthol $C_{10}H_7OH$, an -OH group takes the place of one hydrogen atom of naphthalene $C_{10}H_8$. There are two different positions which the -OH group could occupy. Hence there are two forms of naphthol: **α-naphthol** and **β-naphthol.**

A- An alternative form of AB- (q.v.).

A-, AN-

not, without, lacking (Gk. *a-*, *an-*). (This prefix is used in the form AN- before *h* or a vowel.)

a'cephalic—without a head.

a'symmetric(al) — not symmetric(al), not divisible by a line or plane (or lines or planes) into two (or more) parts exactly similar in size, shape and position.

a'phasia—"without speech"—a disorder of speech due to disease or brain injury.

a'sthenia—"lack of strength", **my'asthenia**—weakness of the muscles.

a'sphyxia — "without pulse" — suffocation.

a'morphous — "without shape" — not having a definite shape; (in chemistry) not having a crystalline form.

a'neroid—"not wet"—a form of barometer which does not contain a liquid.

a'vitamin'osis—the state of lacking, or being deficient in, vitamins; a disease caused by such a deficiency.

an'aemia—"lack of blood"—lack of red cells (or of the red pigment haemoglobin) in the blood.

an'aesthesia — "lack of feeling" — a state of being made unconscious (e.g. by chloroform).

an'aerobic—"without air living"—(organism) which lives without air.

An'opheles—"not helpful, i.e. hurtful" —kinds of mosquito, especially that which is responsible for malaria.

an'hydrous — "without water" — e.g. anhydrous copper sulphate is a white powder; copper sulphate crystals contain some water and are blue.

-A

A large number of scientific words, taken virtually unchanged from Latin or

Greek, or built up as if they were Latin or Greek, end in -A. This termination arises in a variety of ways of which the more common are given below.

(1) Latin and Greek feminine singular nouns, e.g. **larva, nebula, Alga, Hydra**. (Latin words of this type usually take the Latin plural -AE, e.g. *larvae*.)

(2) Abstract nouns ending in -IA (including a large number of medical terms), e.g. **mania, anaesthesia, myopia, hysteria, hydrophobia, neuralgia**.

(3) Nouns formed by the addition of -MA to Greek roots (see -M), e.g. **exanthema, sarcoma**.

(Note. Greek neuter nouns ending in -MA, e.g. *stoma, exanthema*, have stems ending in -MAT-, so that the plurals end in -ATA, e.g. **stomata, exanthemata**.)

(4) The plural of Latin neuter nouns which end in -UM, e.g. **bacterium/bacteria, cilium/cilia, sporangium/sporangia, stratum/strata**.

(5) The plural of Greek neuter nouns ending in -ON, e.g. **phenomenon/phenomena, tetrahedron / tetrahedra, ganglion/ganglia**. (Words of modern application, e.g. *electron*, or of modern construction, e.g. *cyclotron*, take English plurals.)

(6) Neuter plural nouns or nouns constructed as if they were neuter plurals are commonly used as the names of groups and classes of plants and animals.

Spermato'phyta—the great class of plants which reproduce by means of seeds.

Pterido'phyta—the class of plants which includes the Ferns, Club-mosses, etc.

Annel'ida—the group (phylum) of "ringed" (segmented) animals which includes the Earthworms.

Arthro'poda—the great group (phylum) of animals with "jointed feet", i.e. Insects, Crabs, Spiders, etc.

Coleo'ptera—the order of insects with "sheathed wings", i.e. the Beetles.

Cest'oda—the cestode ("ribbon-like") animals, i.e. the Tapeworms.

Vertebr'ata—the vertebrate animals, i.e. those with a backbone (vertebral column).

(7) In the names of some earths and minerals, the termination -A indicates the oxide of the corresponding element whose name ends in -IUM (or -ON), e.g. **alumina**—oxide of aluminium, **strontia**—oxide of strontium, **silica**—oxide of silicon.

In most of these cases the naming of the oxide preceded the naming of the element. Some of the names (e.g. **lithia, alumina**) were formed after the type **soda** (which, however, is not an oxide). **Cadmia** is derived from *kadmia* (*gē*), i.e. Cadmean earth, and **magnesia** from *Magnēs* (*lithos*), i.e. the Magnesian stone.

AB-, ABS-, A-

away, away from, from (L. *a, a-, ab, ab-, abs-*).

(This prefix is used in the form ABS- before *c* or *t*.)

This common prefix is seen in **absent** ("being away"), **abstract** ("to draw out from"), and **aversion** ("a turning away").

ab'rasive—a substance (e.g. emery, sand) used for rubbing away rough particles on wood, stone, etc.

ab'original—"from the beginning"—primitive.

ab'sorb—"to drink away from"—to take liquid into itself as blotting-paper takes in ink; so also, to take in gas, heat, light, etc.

ab'scission—"a cutting away". **abscission layer**—the layer of plant cells where the stem of a fruit or leaf breaks off from the branch.

abs'cess—"a going (retreating) away"—a collection of pus in infected tissue and separated by a wall from healthy tissue.

AC- See AD-.

ACANTH-, ACANTHO-

a thorn, a spine (Gk. *akanthos*).

Acanthus is a genus of prickly plants including, especially, the Bear's Breech or Brank Ursine. A representation of Acanthus leaves was used ornamentally in Greek architecture.

acanth'ous—spiny, thorny.

acantho'cladous—having spiny branches. **acantho'carpous** — having spiny fruit.

trag'acanth—"goat thorn"—a spiny shrub found in Western Asia; a whitish or reddish gum obtained from this shrub.

acantho'sis—a condition in which coloured warty growths appear on the surface of the body.

hex'acanth—"having six hooks"—a stage in the life-history of some Tapeworms.

-ACEA

A Latin suffix used in forming the names of some orders and classes of animals.

Cet'acea—the order which includes the Whales and Dolphins.

Crust'acea—the class which includes Crabs and Lobsters.

The suffix is really an adjectival ending. (Compare the English forms *cetaceous*, *crustaceous*.) The full Latin name of, e.g., the first order given above is *Animalia cetacea*, i.e. the whale-like animals, but the word *animalia* is omitted as understood.

-ACEAE

A Latin suffix used in forming the names of many of the families (Natural Orders) of plants.

Ros'aceae—the family which includes the Rose, Apple, Strawberry, etc.

Papaver'aceae—the Poppy family.

The suffix is really an adjectival ending. (Compare the English forms *rosaceous*, *herbaceous*.) The full Latin name of, e.g., the first family given above, is *Plantae rosaceae*, i.e. the rose-like plants, but the word *plantae* is omitted as understood. There are a few well-known flower families (e.g. Compositae, Umbelliferae) whose names do not end in -ACEAE. Proposals have been made to rename these families to bring them into line but the proposals have not been generally accepted.

ACET-, ACETO-

These elements are derived from the Latin *acetum*, vinegar. An **acetabulum** is "a little receptacle (cup) for vinegar". The word is used in zoology for a variety of cup-shaped cavities, e.g. the sucker of a Cuttlefish, a socket for an insect's leg or for the thigh-bone of a vertebrate animal.

The elements are more usually met, however, in the field of chemistry. **Acetic acid**, $CH_3.COOH$, is the acid of vinegar and the elements usually mean related to, or derived from, acetic acid.

acet'ates—salts of acetic acid, e.g. sodium acetate CH_3COONa.

acet'yl—the characteristic group of atoms CH_3CO- of acetic acid, e.g. acetyl chloride CH_3COCl.

acet'amide—the amide (q.v.) corresponding to acetic acid, CH_3CONH_2.

acet'one — the ketone (q.v.) CH_3COCH_3, a colourless liquid valuable as a solvent.

aceto'phenone — $CH_3COC_6H_5$- the ketone which contains a methyl group CH_3- (compare acetic acid) and a phenyl group C_6H_5-.

ACIDI-, ACIDO-

acid (L. *acidus*, sour, acid).

acidi'fy—to make or become sour, to make into an acid.

acidi'metry—the measurement of the strengths of acids.

acid'osis—the state of having too much acid in the blood.

ACR-, ACRID, ACRYL

sharp, sour, pungent (L. *acer*, *acr* -).

This root is seen in **acrid** (bitter, pungent, irritating) and **acri'mony** (bitterness of temper or manner).

acrid'ine—a compound found in crude anthracene. It has a very irritating action on the skin.

acr'ole'in — allyl aldehyde CH_2: CH.CHO. Formed by the action of heat on glycerol. The unpleasant, acrid smell of burnt fat is due to this substance.

acr'yl'ic acid — the acid (CH_2:CH.COOH) formed by the oxidation of acrolein.

ACRO-
at the tip, topmost, terminal (Gk. *akros*). An **acrobat** is one who "goes on his tips", i.e. walks and acts on tiptoes.

acro'carpous—bearing fruits at the ends of the branches. (Contrast pleurocarpous.)

acro'megaly—"extremity largeness"—a disease which causes the head, hands and feet to become very large.

acro'petal — "tip seeking" — (buds, leaves, etc.) developing in turn from below and gradually upwards towards the tip of the branch, etc.

ACTIN-, ACTINO-
ray, sun-ray, and hence ray- (i.e. star-) shaped. (Gk. *aktis*, *aktin(o)-*).

actin'ic rays—those rays of sunlight which affect a photographic film.

actin'ium — one of the chemical elements which, being radioactive, sends out rays.

actino'meter — an instrument for measuring the strength of sunlight.

actino'therapy—the medical treatment of diseases by light rays.

actino'morphic—star-shaped (as, e.g. are Starfish and Sea-anemones).

Actino'mycetes—the star-shaped Fungi. (They cause the damp smell of underground rooms.)

ACU-, ACULE-, ACUMIN-, ACUTI-
sharp, pointed (like a needle), with a needle (L. *acus*; *aculeus*; *acumen*, *acumin-*; *acutus*).
Acumen is sharpness in discerning things; an **acute angle** is a sharp angle, i.e. one less than a right-angle.

acumin'ate — tapering to a point, having a tapering point.

acuti'foliate—having sharp, pointed leaves.

acu'puncture—the putting of needles into the sciatic nerve to ease pain.

aculeus—"a little needle"—the sting of an animal; a prickle of a plant. **aculeate** — having a sting; bearing prickles.

AD-, (AC-, AF-, etc.)
to, towards, at, near to (L. *ad*, *ad-*). This very common prefix takes a form which depends upon the initial letter of the stem to which it is attached, i.e. to which letter it is assimilated, as follows:

- AC- before *c* and *q*. **ac'cumulate**—to "heap together"; **ac'quire**—to "seek to (one's self)", hence to gain.
- AF- before *f*. **af'ferent**—"carrying towards"—descriptive of blood vessels which carry blood towards an organ, also of nerves which carry impulses to the central nervous system.
- AG- before *g*. **ag'glutination**—a "sticking together", e.g. of small particles, bacteria, etc. into a larger mass.
- AL- before *l*. **al'luvium**—that which is "washed to"—a deposit of sand, etc., left by rivers and floods.
- AN- before *n*. **an'nihilate**—to bring "to nothing", to blot out of existence.
- AP- before *p*. **ap'pendage**—that which "hangs to"—a part which is added to, is extra to, or grows out from another part.
- AR- before *r*. **ar'rogate**—to "ask to", to claim (unduly).
- AS- before *s*. **as'sociate**—to join together, to become partners.
- AT- before *t*. **at'tract**—to "draw towards (one's self)".
- A- before *sc*, *sp*, *st*. **a'scend**—to "climb to".
- AD- before all other letters. **ad'here**—to "stick to"; **ad'renal**—near to the kidney, e.g. the adrenal gland.

-AD
(1) A Greek noun-forming suffix. Commonly used to form collective numerals.

gon'ad—"a reproductive thing"—a sex gland.

mon'ad—a thing which is single, e.g. a one-celled organism.

tri'ad—a triple thing, a group of three things.

tetr'ad—a group of four things, e.g. of spores.

myri'ad—a group containing a very large, indefinite number of things.

(2) A suffix (from L. *ad*) invented to form adjectives and adverbs with the sense of 'to, towards' the part indicated by the main element of the word.

dextr'ad—to or towards the right side of the body.

caud'ad—towards the tail.

dors'ad—towards the back.

mes'ad, mesi'ad — towards the median line of a body.

ADEN-, ADENO-

a gland (Gk. *aden*, *aden*(*o*)-).

aden'oids—"gland-like" things—soft masses (of lymph tissue) at the back of the nose and throat.

aden'itis—inflammation of a gland.

aden'oma—a tumour with a gland-like structure.

adeno'phyllous — having gland-like leaves.

ADIP-

fat (L. *adeps*, *adip*-).

adip'ose—pertaining to fat, e.g. adipose tissue in the body.

adiposis dolorosa—"a sorrowful state of fattiness"—a condition in which painful masses of fat develop under the skin.

adip'ic acid — a crystalline acid, $(CH_2)_4(COOH)_2$, formed by the oxidation of various fats.

-Æ-, -AE-

The ligature ("tied letters") Æ is passing into disuse; it is more usual to 'untie' the letters and write them separately -AE-. This combination occurs in a number of elements given in this book, e.g. ARCHAEO-, GYNAE-, HAEM-, PAED-, PALAEO-.

When the ligature is thus untied, the *a* then appears to be superfluous and there is now a tendency, particularly in the U.S.A., to omit all such *a*'s. Thus the spellings **hemorrhage, medieval, peritoneum** (and others) are not uncommon. (The *a* is never dropped when the AE indicates a Latin plural—see below.)

-AE

-AE forms the plural of Latin feminine nouns ending in -A (and of corresponding adjectives). Among the many examples in science are **larva/larvae, nebula/nebulae, formula/formulae, vertebra/vertebrae, Plantae labiatae** — the plants with labiate (lipped) flowers.

Non-scientific words of this kind often take English plurals (e.g. *arenas*) and there is a modern tendency, particularly in the U.S.A., to give certain scientific words English plurals. Thus *formulas* is not uncommon.

-AEMIA, (-EMIA)

blood (Gk. (*h*)*aima*, (*h*)*aimat*-).

This root is aspirated (see HAEM-) when it occurs at the beginning of a word.

an'aemia—"lack of blood"—lack of red cells (or of the red pigment haemoglobin) in the blood.

septic'aemia—blood-poisoning.

ur'aemia—"urine in the blood"—a diseased condition due to inadequate working of the kidneys.

hyper'glyc'aemia—the presence of too much sugar in the blood.

AEOLI-, (AEOLO-)

the wind.

Aeolus (L.), *Aiolos* (Gk.) was the god of the winds. An **Aeolian harp** is a stringed instrument which produces musical sounds when the wind blows on it.

aeoli'an rocks—rocks formed by the accumulation of sand grains blown by the wind.

aeoli'pyle—a "wind gate"—an instrument for showing the force of steam, particularly in causing a turning movement, as it issues from a small opening.

AER-, AERI-, AERO-

air, of air, by air (L. *aer*, *aer*(*i*)-; Gk. *aēr*, *aero*-).

An **aeroplane** is a "wanderer through the air". Some English writers, apparently copying the fashion of the U.S.A., use the spelling *airplane*. *Air* is essentially an English word; there is no objection to adding it to *craft* or *worthy* but it seems pointless to substitute a hybrid word for an all-Greek word especially as the latter has already been accepted into the language.

aer'ate—to force air into (e.g.) a liquid.

aeri'al—a wire put up (in the air) to receive or transmit radio waves.

aero'be — "air liver" — an organism which can only live in air. (An **an'aerobic** organism lives in an almost complete absence of oxygen.)

aero'lite—"air stone"—a meteorite of a stony composition.

aero'dynam'ics—"air force study"—the study of gases which are in motion and particularly of the forces which the air exerts on a body which is moving in the air.

aero'tropism — "air-turning" — the growth and movement of a part of a plant towards a supply of air.

AESTHESIA, AESTHET-, AESTHO-
perception, feeling (Gk. *aisthēsis, aisthēt-*).
An **aesthete** is one who perceives, and is sensitive towards, things of beauty.

an'aesthesia — "lack of feeling" — a state of being made unconscious as by an **anaesthetic** (e.g. chloroform).

hemi'an'aesthesia—"half lacking feeling"—loss of the sensation of touch (and often of pain and temperature) on one side of the body.

par'aesthesia—"disordered feeling"—abnormal sensations, e.g. tingling, tickling.

kin'aesthetic sensations—"movement feelings"—sensations (from muscles, etc.) which tell of the movement or position of a part of the body.

aestho'physiology—the study of the organs of sensation.

AF- See AD-.

AG- See AD-.

-AGOGUE, (-AGOG)
that which leads, guides, draws out, attracts (Gk. *agōgos*).
A **synagogue** is a place in which Jews are drawn together.

ped'agogue—one who guides and draws out children, i.e. a school-teacher. **Pedagogy, pedagogics**—the science of teaching.

lact'agogue—a substance which causes the production of milk.

chol'agogue—a drug which stimulates a flow of bile.

sial'agogue—a drug which stimulates a flow of saliva. (The spelling **sialogogue,** apparently based on the mistaken idea that the structure is *sialo'gogue*, is now so common that it has become accepted.)

AGRI-, AGRO-
field, land (L. *ager, agr(i)-*; Gk. *agros*).

agri'culture—the art of cultivating fields.

agro'nomy, agro'nomics — "the management of the land"—the management of farms.

AL- See AD-.

AL-
the (Arabic *al*).
Both the mathematicians and the chemists are indebted to Arabic for some of their words though the modern meanings of the words may seem rather remote from the meanings of the original Arabic.

The words *algebra, cipher* and *zero* may be traced back to Arabic. **Algebra** is derived from *al jebr* which means "the reunion of broken parts". The word was originally used to indicate the surgical treatment of broken bones. But during the early growth of mathematics the Arabs used the phrase *al jebr wa'l maqābalah* for a form of calculation. It may be translated "the reunion (or integration) and the comparison (or equation)". It is from this phrase that the word algebra, in its mathematical sense, is derived. It has been in use in England since about 1550.

Some of the oldest words of chemistry are of Arabic origin. The old name **alchemy**, from which the more modern name *chemistry* is derived, is an example. There is some doubt as to the exact origin of the Arabic *al kimia*. It probably corresponds to the Greek *chemia*, the land of Chem, an ancient name for Egypt. It was perhaps in the dark lands of ancient Egypt that the art of alchemy originated and was first practised.

al'kali—*al qalīy*—"the roasted ashes". The name was originally given to a soda-ash obtained from the calcined ashes of marine plants.

al'embic—*al ambig*—"the still (or cup)". An apparatus formerly used for distilling, a retort.

al'izarin—*al açarah*—"the extract". The red colouring-matter of Madder.

alcohol. See -OL.

The term **alkahest** was applied to the alchemist's universal solvent. The word, however, is sham Arabic and was probably invented by Paracelsus (1493–1541). He was among those who attacked the science and philosophy of Aristotle and the Arabs.

AL-, ALA, ALI-

a wing (L. *ala*).

alar—pertaining to wings.

al'ate — having wings or wing-like parts.

ali'form—wing-shaped.

Note. The term *aliphatic* (as applied to a certain class of organic compounds) comes from the Greek *aleiphar*, oil.

The word *alimony* (maintenance, an allowance for maintenance) comes from the Latin *alimonia* (nourishment) from *alo, alere*, to nourish.

ALB-, ALBI-

white (L. *albus*).

An **album** is "a white, blank tablet", now (of course) a blank book.

Lamium album—the White Dead-nettle.

alb'escent—becoming white, tending towards whiteness.

albi'florous—bearing white flowers.

alb'ite—white soda feldspar.

albedo—the "whiteness" of a planet's surface as measured by the proportion of the light which it reflects.

albino—a human being with a marked lack of colouring pigment in, e.g. skin, hair, eyes. Similarly, an animal or plant lacking normal colouring. The condition is called **albinism.**

ALBUMEN, ALBUMIN-

Albumen is the white of an egg (L. *albumen, albumin-*) and hence, chemically, the essential substance in the white of an egg. It is typical of the **albumins**—a group of simple proteins, widely distributed in plant and animal tissues, soluble in water and salt solution, easily coagulated (made solid) by heat.

albumin'oid—similar to, of the same nature as, albumen. (Chemically, **albuminoids** are simple proteins containing sulphur, insoluble, found in strengthening tissues.)

albumin'ous—(seed) containing food-material (endosperm) for the embryo. **ex'albuminous**—(seed) which lacks such food material because, as in the Bean, the material is within the embryo itself.

albumin'uria—the presence of albumins in the urine.

ALDEHYDE

An **aldehyde** is formed by the oxidation of an alcohol. Thus ordinary ethyl alcohol $CH_3.CH_2.OH$ is converted into **acetaldehyde** $CH_3.CHO$. The group of atoms -CHO is characteristic of the aldehydes.

In the process of oxidation, hydrogen is removed. The name *aldehyde* is a contraction of *al*(*cohol*) *dehyd*(*rogenatum*), i.e. de-hydrogenated alcohol.

Oxidation of an aldehyde produces an acid. Thus acetaldehyde is converted into acetic acid. The aldehydes are named according to the acids they produce.

form'aldehyde — a colourless gas, H.CHO; its solution is called formalin and is used as an antiseptic and in the production of Bakelite. (Named from formic acid.)

acet'aldehyde — colourless liquid, $CH_3.CHO$, produced by the oxidation of ethyl alcohol. (Named from acetic acid.)

met'aldehyde—a solid form of acetaldehyde with a molecule consisting of three acetaldehyde molecules.

benz'aldehyde—oil of bitter almonds, $C_6H_5.CHO$. (Named from benzoic acid.)

-ALES

A Latin adjectival plural used as a suffix in the names of orders or groups of related families of plants.

Lili'ales—the group comprising the Lily, Snowdrop and Iris families.

ALG-, -ALGIA, -ALGESIA

pain (Gk. *algos*, *algēsis*).

neur'algia—pain in a nerve.

gastr'algia—pain in the stomach.

an'algesia—absence of pain, loss of sensitivity to pain. **analgesic, analgetic**—(substance) which lessens or prevents pain.

hyper'algesia — increased, excessive sensitivity to pain.

Note. The name *Algae* of certain lower plants has no connection with this root. *Alga* is Latin for seaweed.

ALIMENT-

food, nourishment (L. *alimentum*).

aliment'ary canal—the passage from the mouth, through the body, to the anus (outlet) in which food is received, digested and absorbed.

ALLEL-, ALLELO-

one another (Gk. *allēlōn*).

Parallel lines are "at the side of one another".

allelo'morph—"one another's form"—one of a pair of contrasted inheritable characteristics, e.g. tall/dwarf (peas), blue/brown (eyes).

ALLO-, ALL-

other, another (Gk. *allos*).

allo'morphous—"having other forms"—having the same chemical composition but different physical forms.

allo'tropes—substances with "other manners"—different physical forms of the same chemical element, e.g. diamond, graphite and lamp-black are **allotropes** of carbon. Such a condition of existing in several different physical forms is called **allotropy**.

allo'gamy—"other marrying"—cross-fertilization.

allo'pathy — "other suffering" — the treatment of one disease by inducing another, different disease. (Contrast homoeopathy.)

all'ergy—"other energies (reactions)"—an abnormal reaction and unusual sensitiveness (often resulting in inflammation and destruction of tissue) to particular foods, to pollen, to insect bites, etc.

ALLYL, ALL-

The characteristic hydrocarbon group of oil of garlic (L. *allium*, garlic).

Oil of garlic is allyl sulphide, $(CH_2{:}CH.CH_2)_2S$. The group of atoms $CH_2{:}CH.CH_2$- is the **allyl group**.

allyl bromide—$CH_2{:}CH.CH_2Br$.

allyl alcohol—$CH_2{:}CH.CH_2OH$.

allyl aldehyde—$CH_2{:}CH.CHO$ (also called acrolein.)

allyl *iso*-thiocyanate—oil of mustard, $CH_2{:}CH.CH_2NCS$.

Note that the corresponding hydrocarbon, $CH_2{:}CH.CH_3$, is usually called propylene. The related hydrocarbon with two double joins, $CH_2{:}C{:}CH_2$, is called **allene**.

ALPHA- See α-.

ALTI-, ALTO-

high (L. *altus*).

The **altitude** of, e.g., a mountain or a triangle is its height. **Alto** is (or perhaps was) the highest male singing part; the female voice of similar range is properly called **contralto**.

alti'meter—an instrument for measuring height (e.g. in an aeroplane).

alto-cumulus—rounded, heaped clouds, in groups or lines, at heights of 10,000–25,000 feet.

ALUM, ALUMIN-, ALUMINO-

Ordinary **alum** (L. *alumen*, *alumin-*), a crystalline mineral, is a double sulphate of potassium and **aluminium**. Other alums contain sodium, ammonium (or other monovalent element) in place of potassium, and chromium, iron (or other trivalent element) in place of aluminium.

alumina—oxide of aluminium, occurring as the mineral corundum.

alumin'ates—salts (e.g. sodium aluminate) formed by the combination of aluminium oxide with bases.

alumino'thermic process—the reduction of a metal oxide to a metal by the use of finely divided aluminium powder. Much heat is given out in the process.

AMBI-

on both sides, in both ways (L. *ambo*, *amb*(*i*)-).

amb'iguous—"acting (driving) in both ways"—of doubtful meaning or classification.

ambi'dextrous—able to use both hands equally well.

amb'ient air—the surrounding air. (L. *amb'ire*, to go around.)

AMBLY-

blunt, hence dull, faint (Gk. *amblys*).

ambly'gon—obtuse-angled, i.e. with a "blunt angle" (greater than a right-angle).

Ambly'stoma — "blunt mouthed" animal—a tailed, gill-less amphibian, the mature form is the Axolotl.

ambly'opia — weak, impaired eyesight.

AMBUL-

to walk (L. *ambulo*).

An **ambulance** (formerly an *hôpital ambulant*) was a moving hospital which attended an army as it moved; the term is now more often used for a carriage for conveying sick or injured people.

ambul'ant, ambul'atory—able to walk about; used for walking about.

somn'ambulist—one who walks in his sleep.

AMIDE, AMINE, AMINO-, IMIDE, IMINE

The term *ammonia* may be traced back, through Latin, to Greek and means "pertaining to Ammon (the Egyptian god Amun)". *Sal ammoniac* ("salt of Ammon") was supposed to have been prepared from the dung of camels near the shrine of Jupiter Ammon. The substance is now known chemically as ammonium chloride.

It is from this term *ammoniac* that the name of the gas ammonia (NH_3) is derived and hence the names of various compounds of ammonia.

An **amide** was originally a compound in which one of the hydrogen atoms of ammonia is replaced by a metal or a group of atoms. Thus **sodamide**, $NaNH_2$, is a compound in which one of the hydrogen atoms of ammonia is replaced by a sodium atom. (Compare the names *sod*(*ium*) *amide* and *sodium chloride*.) But in modern chemical naming the use of the term amide is restricted to compounds in which one of the hydrogen atoms of ammonia is replaced by the characteristic 'stem' of an organic acid.

acet'amide—$CH_3CO.NH_2$; the acetyl group (of acetic acid $CH_3CO.OH$) here replaces one hydrogen atom of ammonia.

cyan'amide—$NC.NH_2$, in which the stem of cyanic acid ($NC.OH$) replaces a hydrogen atom.

ox'amide—$NH_2.CO.CO.NH_2$, a kind of 'double' amide in which the stem of oxalic acid ($HO.CO.CO.OH$) joins with two amide groups.

The replacement of two of the hydrogen atoms of ammonia by acid groups results in an **imide**.

The replacement of one (or more) hydrogen atoms of ammonia by other organic groups (especially hydrocarbon groups) of atoms results in an **amine**.

methyl'amine—$CH_3.NH_2$; the methyl group CH_3- (derived from the hydrocarbon methane CH_4) here replaces one hydrogen atom of ammonia.

phenyl'amine—$C_6H_5.NH_2$. In this compound, more commonly known as aniline, the phenyl group C_6H_5- replaces one hydrogen atom of ammonia.

The term **vitamins** ("life-amines") is a misnomer. It was invented in 1912, as a shorter alternative to 'accessory food factors', when it was thought that these substances were amines. In spite of more exact knowledge about their natures, the name has been retained.

Secondary amines, in which two of the hydrogen atoms of ammonia are replaced, are sometimes called **imines**.

The group of atoms NH_2-, characteristic of the amines, is called the **amino-group**.

amino'acetic acid—$CH_2(NH_2)COOH$; here the amino-group replaces one of the hydrogen atoms of acetic acid CH_3COOH.

amino-acids—fatty acids in which the amino-group takes the place of a hydrogen atom. They are important in the building of proteins within the body.

AMPHI-, AMPHOTER-
about, around, on both sides, in both ways (Gk. *amphi*, *amph-*; *amphoteros*).

amphi'bious—"living in both ways"—(animal) able to live both on land and in water.

Amphi'oxus—a small water creature (also called Lancelet) which is pointed at both ends.

amphi'stomous — having mouth-like openings at both ends (as have some parasitic worms).

amphi'genous—growing all round, e.g. as a fungus which grows all round a leaf.

amphi'podous—having both walking and swimming legs.

amphoter'ic—having both acidic and basic ('alkaline') properties, e.g. aluminium oxide can, as an acid, form salts (aluminates) and also, as a base, form metal compounds (e.g. aluminium chloride).

AMPLEXI-
embracing (L. *amplexus*).

amplexi'caudate—"with the tail embraced"—having the tail enveloped in the membrane between the legs.

amplexi'caul—"embracing the stem"—said of a stalkless leaf whose base wraps round the stem.

AMPLI-
large, wide (L. *amplus*).
This root is seen in **ample**—of sufficient size.

ampli'fy—"to make wider"—to make sound waves, electric currents, etc., stronger.

ampli'tude—the "width" of waves, i.e. the distance between the top of a wave and the middle position. Similarly, of an oscillation (e.g. of a pendulum), the distance between the extreme and middle positions.

AMYGDAL-
almond (Gk. *amygdalē*, Almond-tree).

amygdal'oid—almond-shaped.

amygdal'in—a substance (a glucoside) found in Bitter Almonds, Cherry and Peach nuts.

AMYL, AMYL-, AMYLO-
(1) starch (fine meal) (L. *amylum*, Gk. *amylon*).

amyl'aceous—starchy, of starch.

amyl'oid—like starch; in chemistry, a starch-like cellulose compound.

amylo'genesis—the building up of starch by a plant.

amylo'lysis—the breaking down of starch into sugars.

amyl'ases — enzymes (ferments), such as ptyalin in the saliva of the mouth, which bring about the breaking down of starch.

(2) In the fermentation of sugars and starch, a proportion of higher alcohols (fusel oil) may be obtained. The chief of these alcohols is called **amyl alcohol** (though the connection with starch is rather remote and the name is not a good one). Normal amyl alcohol has the formula $CH_3CH_2CH_2CH_2CH_2OH$, i.e. $C_5H_{11}OH$; the group of atoms C_5H_{11}- is the **amyl group** (corresponding to the hydrocarbon pentane C_5H_{12}).

amyl acetate — $CH_3CO.OC_5H_{11}$, a liquid with a pear-like smell, used as a fruit essence and as a solvent for nitrocellulose (e.g. in nail varnish).

amyl'ene—a hydrocarbon, C_5H_{10}, corresponding to pentane but with a double join between two of the carbon atoms. (See -ENE.)

AN- If a prefix, see A- or ANA-, or AD- if followed by *n*.

-AN

A chemical suffix used in forming the names of classes of polysaccharides (carbohydrates).

dextr'an—a polysaccharide built up from dextrose units. (The joins are made by the elimination of water molecules between parts of the sugar molecules; *dextran*=*dextr*(*ose*) *an*(*hydride*).)

galact'an—a polysaccharide built up from galactose units.

pent'an, pentos'an—a polysaccharide, e.g. a gummy carbohydrate in wheat bran and woods, built up from pentose units.

ANA-, AN-

up, up again, throughout (Gk. *ana*, *ana-*).

ana'batic — "up going" — (winds) caused by the upward movement of warm air.

ana'bolism—"upward throwing (together)"—the building up of complex substances from simple substances inside a living organism.

ana'leptic—"picking up again"—restoring, strengthening; a drug which restores and strengthens.

ana'lysis—"loosening up again"—the breaking up of a material (or sentence, problem, etc.) into simpler parts so as to examine its composition.

ana'mnesis—the bringing up to the memory again (e.g. of all matters relating to a person's illness).

an'ode—"up way"—the electrode by which an electric current enters an electrolytic cell, discharge tube, etc.

ana'stomo'sis—"a state of joining up by mouths"—the joining together of two blood vessels, nerves, etc.; the artificial joining together of two parts of the alimentary (food) canal.

ana'tomy—"cutting up"—the cutting up of bodies for scientific study; the parts which are revealed by such cutting.

ANCHYL-, ANKYL-, ANKYLO-

bent, crooked, hence joined together to become crooked (Gk. *ankylos*).

anchyl'osis, ankylosis—the joining together of bones at a joint so that the joint is stiff and crooked.

ankylo'stomiasis — "crooked mouth disease"—hookworm; a disease caused by a small organism (*Ankylostoma duodenale*) in the bowel and resulting in general ill-health.

ANDR-, ANDRO-

male (Gk. *anēr*, *andr*(*o*)-).

andr'oecium—the male part (stamens) in a flower; the group of male organs in a moss.

andro'gynous—"male and female"—having both male and female parts (i.e. stamens and pistil) in the same flower.

heter'androus—having stamens which are not all of the same size.

prot'androus—"male first"—(flower) in which the male part ripens before the female part.

andro'logy — the study of diseases which are peculiar to males.

andro'gens—sex hormones (stimulating substances) which are responsible for the development of the male sex organs and the masculine characteristics.

-ANE

A chemical suffix denoting a paraffin hydrocarbon. (The molecule consists of a chain of carbon atoms (to which hydrogen atoms are attached) without double joins between them.) The suffix was devised by Hofmann to give a series with -ENE and -INE (q.v.).

meth'ane—the simplest hydrocarbon, CH_4. Also called marsh gas.

but'ane—a paraffin hydrocarbon with four carbon atoms, C_4H_{10}. (Supplied compressed in steel containers for domestic heating purposes.)

oct'ane—a paraffin hydrocarbon with eight carbon atoms, C_8H_{18}.

ANEMO-

the wind (Gk. *anemos*).

Anemone—"the daughter of the wind" —a genus of flowering plants (of the Crowfoot family) the best known of which is commonly called the Wind-flower. (Possibly so called because some species like exposed, windy situations.)

anemo'meter — an instrument for measuring the speed (or power) of the wind.

anemo'philous — "wind loving" — (flowers) pollinated by means of the wind.

ANGI-, ANGIO-, -ANGEA, -ANGIUM

a vessel, a container, hence a blood vessel (Gk. *angos*, *angei(o)-*).

Hydr'angea—"water containers"—a well-known flowering plant, so called because of the cup-like form of the seed capsules.

Angio'sperms—the great group of plants whose seeds develop and ripen inside a container (the ovary), i.e. all ordinary flowering plants.

spor'angium—the case in which spores are produced (e.g. in a Moss).

gamet'angium—the organ in which gametes (reproductive cells) are formed.

angi'oma—a tumour (growth) composed mainly of blood vessels.

tel'angi'ectasis — "end blood-vessel extension"—a diseased enlargement of the small blood vessels, resulting in small red or purple tumours in the skin.

Note. *Angina pectoris* (L.) means "quinsy of the breast". It is marked by spasms in the chest as a result of over-exertion when the heart is diseased.

ANIM(A)

breath, life (L. *anima*).

This root, which is seen in **animal, animate** (having life) and **inanimate** (not having, and never having had, life), needs no elaboration.

ANISO-

not equal (Gk. *anisos*). Used especially to form negatives of words beginning with ISO- (q.v.).

aniso'dactylous — having "unequal fingers"—(birds) having three toes turned forwards and one turned backwards when perching.

aniso'gametes—gametes (reproductive cells) of two different kinds. (Unlike kinds unite in pairs.)

aniso'phylly—the state of having more than one kind of leaf on the same shoot. (Also called heterophylly.)

aniso'tropic—having different properties (qualities) in different directions, e.g. (crystal) having different electrical resistances in two directions at right-angles.

Note. The terms *Anise* (the plant), *aniseed* (its seed) and *anisole* (chemical compound) are not, of course, in any way related to this prefix.

ANKYL- See ANCHYL-.

ANNEL-, ANNUL-

a ring (L. *an(n)ellus*; *annulus*).

annul'ar eclipse—an eclipse of the Sun in which the Moon does not wholly obscure the Sun but leaves a ring of light all round.

annulus—(*Maths.*) the ring-like area between two concentric circles. (*Botany*) a ring of thickened cells in the spore-container of a Fern. (It helps in discharging the spores when the container is dry.)

Annel'ida—the "small ringed" animals —the class of animals, including Earthworms and Bristle worms, whose bodies consist of a number of ring-like segments.

ANOMAL(O)-, ANOMO-

irregular, not even, lawless (Gk. *anomalos*, *anomos*).

This root is seen in **anomalous**—irregular, not in agreement with the general order.

anomo'phyllous—having leaves irregularly placed.

anom'odont—having irregular or no teeth. (Applied to certain extinct reptiles.)

ANSER-
goose (hence duck) (L. *anser*).

anser'ine—of, or like a goose or duck. (Hence, stupid.)

Anser'iformes—the bird family which includes Geese, Ducks and Swans.

ANT- See ANTI-.

ANTE-
before (in position or in time) (L. *ante*, *ante-*).
An **ante-room** is a small room before, and leading to, a main room. To **antedate** is to date before the true time.

anterior—more to the front; the front end of an animal. (Contrast posterior.)

ante'cubital—before the elbow.

ante'pectoral—in front of the breast.

ante'natal clinic—a place for the examination and treatment of women before the birth of a child.

Note. This prefix should not be confused with ANTI- (against). The English word *anticipate* ("to take before") has an *i* where strictly it should have an *e*.

ANTH-, ANTHE-, ANTHO-, ANTHER(O)-
a flower (Gk. *anthos*, a flower; *anthēros*, flowering, flowery).
A **Chrysanthemum** is a "golden flower"; a **Polyanthus** has "many flowers". An **anthology** was originally a collection of flowers; it is now a collection of chosen literary items.

peri'anth—the ring (or rings) of coloured parts round the reproductive parts of a flower. (Used especially when the petals and sepals cannot be distinguished.)

nyct'anth'ous—having flowers which open only at night.

ex'anthe'ma—an "out-flowering"—an eruption on the surface of the skin.

antho'bian—an animal (e.g. a beetle) which lives in, or feeds on, flowers.

antho'cyanins — chemical substances which produce the colours of many plants and flowers.

antho'lites—certain fossil plants resembling flowers.

The Latin *anthera*, from the Greek, denoted a medicine extracted from flowers. It has given rise to **anther**—the pollen-container at the top of a stamen. Because an anther contains male cells, the word-element ANTHER(O)- has come to mean "male".

anther'idium—a "little male part"—the male reproductive organ of lower plants.

anthero'zoid—a free-moving male cell, a sperm.

Note. The root ANTHROP- (Man) has no connection with the root ANTHERO-.

ANTHRAC-, ANTHRAX
charcoal, and hence coal (Gk. *anthrax*, *anthrak-*).

anthrac'ite — a hard, slow-burning form of coal which leaves little ash.

anthrac'ene—a crystalline hydrocarbon ($C_{14}H_{10}$) obtained from coal-tar and used as a raw material for certain dyestuffs.

anthrax—wool-sorters' disease (caused by the anthrax bacillus) which causes black spots on the skin.

ANTHROP-, ANTHROPO-
Man (human being) (Gk. *anthrōpos*).
A **phil'anthrop'ist** ("lover of Man") is one who, by his kindness, generosity and acts, exerts himself for the well-being of his fellow-men.

anthrop'oid—man-like, e.g. an anthropoid ape.

anthropo'logy—the scientific study of Man, as an animal, his place in nature, his ways of living, etc.

anthropo'metry—the measurement of the human body (e.g. height, weight, etc.), especially for the comparison of different races of Man.

anthropo'geny—the investigation of the origin of Man.

ANTI-, ANT-
against, acting against, opposite to (Gk. *anti*, *anti-*).
(The form ANT- is used before *h* or a vowel.)
This prefix is well-known in such words as **anti-clockwise, antidote** (for counteracting a poison), **anti-slavery**, etc. As a living prefix it is used with words of all origins.

anti'septic—(substance) used to prevent sepsis (the development of bacteria in wounds).

anti'podes — "opposite feet" — the opposite point on the Earth, e.g. Australia from England.

anti'pyretic — (substance) which counteracts fever.

anti'body—a substance set free in the blood to act against bacteria and other substances which are harmful.

anti'gen—any poison, bacterium, etc., which, when put into the blood, causes the production of an antibody.

anti'cathode—a plate in an X-ray tube which is opposite to the cathode (negative plate).

anti'cline, anticlinal fold—a large fold in the rock layers of the Earth's crust in which the layers 'lean against' each other, i.e. the fold is arch-shaped.

ant'acid—any substance which neutralises or destroys an acid; a substance which prevents acidity (especially in the stomach).

ant'helminthic — (substance) which destroys worms in the bowel.

AP- See AD-.

API-
bee (L. *apis*).

api'ary— a place where bees are kept.

api'culture—bee-keeping.

Note. *Apical* (pertaining to the apex) is not derived from this root.

APO-
from, away, away from; in modern scientific terms, separate, detached (Gk. *apo*, *apo-*).
To **apologise** is to "speak away" that which might have caused offence.

apo'geny—loss or absence of the power to reproduce.

apo'plexy — "striking completely away"—unconsciousness and paralysis as a result of flowing out of blood in the brain.

apo'gee—the point on the Moon's (or satellite's) orbit which is farthest away from the Earth (or central planet).

ap'helion—the point on a planet's orbit which is farthest from the Sun.

apo'carpous—bearing two or more carpels (ovule containers) which are distinct and separate from one another.

APOPHYSIS
An **apo'physis** (Gk. *apophysis*) is a "growth away from", an outgrowth. It is used in many contexts, e.g. a growth from a bone (usually for the attachment of a muscle), the swollen end of the seta (stalk) of a moss capsule, an enlargement of the outer end of a scale of a pine cone, a side-branch or offshoot of a larger vein of igneous rock.

neur'apophysis—either of the two processes (outgrowths) of a vertebra which form the neural arch (enclosing the spinal cord).

pleur'apophysis—a sideways process (outgrowth) of a vertebra.

hyp'apophysis—an apophysis on the lower side of a vertebra.

par'apophysis—an apophysis (of a vertebra) in a ventro-lateral ("bellywards and sideways") position.

APSE, APSIS (plural **APSIDES**)
a fitting together, a loop, a network, an arch, a vault, an orbit (L. *apsis*, *apsid-*; Gk. *hapsis*, *hapsid-*).

apse line, line of apsides—the line which joins the points of greatest and least distance of a planet from the Sun.

syn'apse—the looping together of one nerve cell and another by the interlacing of the branch-like endings.

AQUA, AQUE-
water (L. *aqua*).

aquarium—a container of water for keeping living fish and plants.

aque'ous—watery, e.g. an aqueous solution (a substance dissolved in water), aqueous humour (watery liquid in front of the lens of the eye).

aque'duct—an artificial channel or pipe for conveying water, especially over long distances (e.g. from a reservoir to a town).

aqua'marine — "sea water" — a blue-green form of beryl used as a gem.

aqua regia—"kingly water"—a mixture of one part nitric acid and two to four parts hydrochloric acid which dissolves the "royal" metal gold.

ARACHN-, ARACHNO-
a spider (Gk. *arachnē*).

arachno'logy—the study of spiders.

Arachn'ida — the class of animals which includes Spiders, Scorpions and Mites, all of which have four pairs of walking legs.

arachn'oid—spider-like; like a cobweb; covered with long, cobweb-like hairs.

arachn'idium—the spinning organ and silk glands of a spider.

ARBOR-, ARBORI-
a tree (L. *arbor*).

arbor'escent—tree-like in growth or appearance.

arbori'culture—the cultivation of trees and shrubs.

arboretum—a garden for the scientific study of trees.

ARCHAEO-, (ARCHEO-)
ancient (Gk. *archaios*).

archaeo'logy—the study of the early history of Man as shown by the buildings, carvings, pots, etc., he has left.

Archaeo'zoic era—the oldest of the five eras of the Earth's history (several thousand million years ago). The name, probably constructed so as to fall in line with the names of the other eras, means "ancient life" but no undisputed remains of living things have been found in the rocks. The term Archaean is to be preferred.

Archaean rocks — rocks which are older than those of the Cambrian times, especially those of the Archaean era.

Archaeo'pteryx—"ancient wing"—the oldest fossil bird, forming a link between reptiles and modern birds.

ARCHE-, ARCHI-, ARCH-
first, original, primitive, hence chief (Gk. *archos*, *arche-*). An **archbishop** is a "chief bishop"; an **architect** is a "chief builder".

arche'gonium—a "primitive" form of female organ as in Mosses and Ferns.

archi'cerebrum—the "primitive brain" of invertebrate animals.

arche'type, archi'type—the original type or model (e.g. of an aeroplane) from which others are copied.

archi'trave—the main beam across the tops of pillars.

AREN-, ARENA, ARENI-
sand (L. *arena*).
An **arena** was originally a sand-strewn place where combats took place before spectators.

aren'aceous—sandy, sand-like.

areni'colous (or **arenaceous**) **plants**—plants which grow best in sandy places.

ARGENT-, ARGENTI-
silver (L. *argentum*).
Ag, the chemical symbol for silver, is derived from *argentum*.

argenti'ferous—(rocks) which bear or yield silver.

argent'ite—an important ore of silver (silver sulphide).
See the note under LITH- concerning **litharge**.

ARGILL-, ARGILLI-
clay (L. *argilla*; Gk. *argillos*).

argill'aceous rocks—rocks consisting largely of clay.

argilli'colous—(plants) which flourish in clayey soils.

ARGYR-, ARGYRO-
silver (Gk. *argyros*).

argyr'anthous—having silvery flowers.

argyro'phyllous—having silvery leaves.

argyria—silver poisoning.

cer'argyr'ite — horn silver — mineral silver chloride.

hydr'argyrum—"water (liquid) silver" —the Latinised form of the Greek name for mercury (quicksilver), hence the chemical symbol **Hg** for mercury. **hydrargyrism**—poisoning by mercury or its compounds.

-ARIUM

A suffix meaning 'a thing connected with or used for' or (especially) 'a place for or container of'. (Originally the neuter ending of Latin adjectives ending in *-arius*.)

aqu'arium—"a thing containing water" —an artificial pond or tank in which aquatic plants and animals are kept for observation and study.

herb'arium—a book, case, or room, in which dried plants are collected.

viv'arium — an artificially prepared place or container for keeping live animals.

formic'arium—a container in which ants are kept for observation and study. The Latin form -ARIUM is represented in English by -ARY.

api'ary—a place in which bees are kept.

avi'ary—a place in which birds are kept.

ARTERI-, ARTERIO-

an artery (a main blood vessel which carries blood outwards from the heart) (Gk. *artēria*, wind-pipe).

The word **artery** is derived through Latin from the Greek. The Greek word may be related to a verb meaning 'to raise, to carry' or to the word *aēr* (air). The ancients certainly regarded the arteries as air-ducts connected with the wind-pipe.

arterio'tomy—the cutting of an artery for letting out blood.

arterio'sclerosis—a state of hardening of the walls of the arteries (especially with old age).

arteri'ole—a small artery.

ARTHR-, ARTHRO-

a joint (of the body) (Gk. *arthron*).

arthr'itis—inflammation of, or pain in, the joints.

arthro'tomy—the surgical cutting into a joint.

arthro'pathy—painful suffering in the joints.

Arthro'poda—the large group (phylum) of animals which have "jointed legs". It includes Crabs, Spiders, Mites, Centipedes, Insects.

-ARY See -ARIUM.

AS- See AD-.

ASC-, ASCO-

a bag (a wine skin) (Gk. *askos*).

ascus—a bag-like organ in which spores are formed. It is characteristic of the large group of Fungi called the **Asco'mycetes**; spores formed in an ascus are called **asco'spores.**

asc'idium—"a little bag"—(1) a bag-shaped or pitcher-shaped leaf. (2) a genus of the Ascidiacea.

Asc'idiacea, Ascidia—the family of Sea-squirts. (The adult is encased in a bag-like covering.)

-ASE

A suffix used in forming the names of enzymes (substances, of a protein nature, produced by living cells, which bring about chemical changes in the body). The suffix was probably taken from **diastase** (see -STASIS).

diastase—an enzyme (or group of enzymes) which brings about the change of starch into sugar. Found in seeds (e.g. barley) during germination.

amyl'ases — the general name for enzymes which cause the breakdown of starch into simpler substances.

prote'ases—enzymes (e.g. pepsin in the stomach juices) which break down proteins.

lip'ases—enzymes which break down fats.

invert'ase—an enzyme which breaks down sucrose (cane sugar). So called because a solution of sucrose twists the plane of polarised light to the right and the solution resulting from the breakdown twists it to the left, i.e. the twisting is "inverted".

Note that enzymes may be named by reference to the kind of substance in which they bring about the chemical change (e.g. amylase) or to the result of their action (e.g. invertase, oxidase).

-ASIS

The addition of -SIS (q.v.) to certain Greek roots, notably verb forms, produces nouns (e.g. of state or condition). Thus the stem *sta-* (stand) is converted into *stasis* (a standing, a state of standing or of being put in a position).

The suffixes -ASIS and, particularly, -OSIS (q.v.), which have taken in with themselves an *a* or *o* of the combining stem and are 'artificial', are freely used in medicine to denote a *diseased*, *unhealthy* or *damaged* condition. -ASIS is normally preceded by *i*.

psori'asis—a disease in which red patches and scales appear on the skin.

filari'asis—a general name for diseases caused by thread-like parasitic worms (*Filaria*), e.g. **elephantiasis**—a disease in which there is enlargement of the limbs and a thickening of the skin.

ASPID-, ASPIDO-

a shield (Gk. *aspis*, *aspid(o)-*).
This root is seen in the name of the well-known plant **Aspidistra.**

Aspid'ium—a "little shield"—a fern with a shield-shaped cover to the spores.

Aspido'branchia — animals with "shielded gills"—the group of sea creatures which includes the Limpets.

ASTER, ASTRO-

a star (L. *aster*, *astr-*; Gk. *astēr*, *astr(o)-*).
An **Aster** is a well-known star-shaped flower; an **asterisk** is a star (*) used in printing. A **disaster** is an unfortunate happening which was supposed to come when the stars were unfavourable.

astro'logy—"the study of the stars"—originally, the study of the stars for practical purposes; now confined to the non-scientific 'art' of judging the effects of the stars on human affairs and so foretelling the future.

astro'nomy—"the arranging (in order, law) of the stars"—the scientific study of the stars and other heavenly bodies.

aster'oids—small planets (in fact not stars) whose paths round the Sun lie between those of Mars and Jupiter.

astr'oid—a star-shaped curve.

astro'lithology—"star stone study"—the study of meteorites and similar bodies which reach the Earth from outer space.

astro'physics—the branch of astronomy which deals with the material nature of the stars, e.g. their compositions, temperatures, etc.

astro-navigation—the steering of a ship or aeroplane by the stars.

astro'cyte—a much branched ("star-shaped") fibrous cell in the supporting tissue of the brain and spinal cord.

ASTHEN-, ASTHENIA

lack of strength (Gk. *asthenēs*).

asthenia—lack of strength; weakness, debility.

my'asthenia—weakness of the muscles.

asthen'opia—weakness of the eye or of the eye muscles.

AT- See AD-.

-ATA See -A.

-ATE

This common suffix of Latin origin appears in certain nouns (e.g. **mandate,** a **vertebrate,** i.e. a vertebrate animal), a larger number of verbs (e.g. **calculate, saturate, permeate**), and many adjectives.
Some adjectives have the meaning of 'like'—

hastate — (leaf) like a spear, spear-shaped.

palmate — (leaf) like the palm of a hand.

Other adjectives have the meaning of 'possessing'—

nucleate—possessing a nucleus.

septate — possessing septa (cross-walls).

This suffix needs no further elaboration here other than its use in the naming of chemical substances.
(*Chemistry*) -ATE is used in forming the names of salts.

acet'ate—a salt of acetic acid, e.g. sodium acetate $CH_3CO.ONa$.

carbon'ate—a salt of carbonic acid, e.g. calcium carbonate $CaCO_3$.

When there are two similar salts (derived from two similar acids), -ATE denotes the salt with the higher proportion of oxygen.

nitr'ate—a salt of nitric acid, e.g. sodium nitrate $NaNO_3$. (Contrast sodium nitrite $NaNO_2$.)

sulphate—a salt of sulphuric acid, e.g. sodium sulphate Na_2SO_4. (Contrast sodium sulphite Na_2SO_3.)

ATEL-, ATELE-, ATELO-

"without an end", unfinished, imperfect, undeveloped (Gk. *atelēs*).

atel'iosis—"a state of imperfection"—imperfect growth resulting in a dwarf with correct body proportions.

atel'ectasis—"a state of imperfect opening"—failure of the lungs to expand at birth.

atelo'glossia—imperfect development of the tongue.

AUDI-, AUDIO-

to hear (L. *audio*).
This root is seen in **audience** (the group of people who hear, e.g., a lecture or play) and **audible** (able to be heard).

audio'meter—an instrument for testing the power of hearing.

audio'frequencies—frequencies of oscillation (of waves, of electric currents) which give rise to, or correspond to, audible sound.

AUR-, AURI- (1)

the ear (L. *auris*).

aur'al—pertaining to the ear.

auri'scope—an instrument for examining the ear and the Eustachian passage.

auri'cle—"a little ear"—(1) the external ear of animals (formerly the lower lobe of the human ear); (2) an ear-shaped lobe at the base of a leaf; (3) either of the two upper chambers of the heart.

auri'culate—having ear-shaped attachments.

Auricula—a kind of Primula (also called Bear's Ear), so called from the shape of its leaves.

AUR-, AURI- (2)

gold (L. *aurum*).
An **aureole** is a golden disc shown surrounding the head of a saintly person in a picture. (It is also a kind of solar halo.) **Au,** the chemical symbol for gold, is derived from *aurum*.

aur'ic—pertaining to gold, e.g. **auric chloride**—gold chloride $AuCl_3$.

auri'ferous rock—rock which bears or yields gold.

AUT-, AUTO-

self, by itself (hence independently) (Gk. *autos*).
This word is well-known in **automatic** (working by itself)—hence the mongrel word **automation**—in **automobile** (a vehicle which moves by its own power) and in **autograph** ("a self writing").

auto'gamic—"self marrying"—fertilizing itself.

auto'plasty — "self moulding" — the repair of wounds by tissue taken (surgically) from some other part of the same individual.

auto'phagy—"self eating"—the eating of a part of its own body. The adjective **autophagous** also has a more pleasant meaning—capable of feeding itself from birth (as some birds).

auto'lysis — "self loosening" — the breaking down of a cell or an organ by chemical actions produced within itself.

aut'opsy—"seeing for one's self"—the inspection of a dead person to learn the cause of death.

aut'oxidation—the slow oxidation of certain substances when exposed to air.

AUX-, AUXAN(O)-

to increase, to grow (Gk. *auxanō*, *auxō*).

auxano'meter — an instrument for

measuring the rate of growth of a part of a plant.

aux'ins—substances formed in a plant which stimulate growth. (Plant hormones.)

aux'etic—inducing or stimulating cell division.

AVI-

a bird (L. *avis*), hence flight.

avi'ary—a place for keeping birds.

avi'fauna—the bird population of a district or country.

avi'ation—the science of flying aircraft.

AZO-, AZ-

Azoic (A+ZO-, Gk. *azōos*) means "without life". The term is sometimes applied to the pre-Cambrian geological times (the Proterozoic and Archaean eras).

When Lavoisier found that the gas which remained when oxygen was removed from air would not support life, he proposed the name **azote** for it. Chaptal renamed the gas (or, strictly, the essential part of it) nitrogen. Although the old term azote has passed into disuse, the element AZO- is still used to denote the presence of nitrogen in a chemical compound.

azo'benzene—the compound $C_6H_5.N:N.C_6H_5$, i.e. a compound derived from nitrogen and benzene.

azo dyes—yellow, red and brown dyes containing the azo-group of atoms -N:N- in the molecule.

hydr'az'ine — a fuming liquid $H_2N.NH_2$ composed of hydrogen and nitrogen.

B

β-, BETA-

β, the second letter of the Greek alphabet, is sometimes added before the name of a series or group of things to denote the second member of the series or group. (The first member is denoted by α.)

β-rays, beta-rays—one of the three types of radiation given off by radioactive substances, consisting of a stream of electrons (negatively charged particles). Such particles are called **β-particles.** A **betatron** is an instrument which produces high speed electrons.

β-brass, beta-brass—a form of brass containing 40–49 per cent. of zinc.

In naming organic compounds, *β* is sometimes used to show that a certain group of atoms is in the second of two (or more) possible positions in the molecule. For examples, see α-.

BACILL-, BACILLI-

a little rod (L. *bacillum*; *bacillus*).

bacilli'form—shaped like a little rod.

bacillus — a rod-shaped bacterium. (Plural **bacilli.**)

bacill'ary—(tissue, membrane, etc.) having the structure of small rods; pertaining to bacilli.

bacill'aemia—the presence of bacilli in the blood.

BACTERIO-

Pertaining to bacteria.

Bacteria (sing. **bacterium**) are microscopic, unicellular organisms, without chlorophyll (green colouring matter) or well-defined nuclei. They occur almost everywhere; some kinds cause disease. They received their name (Gk. *baktērion*, a little rod, a staff) from the appearance of those which were first detected.

bacterio'logy—the study of bacteria.

bacterio'lysis—the process of breaking down and destroying bacteria.

bacterio'phage—a body (possibly a virus) which "eats" (destroys) bacteria.

bacterio'rrhiza—a beneficial partnership between bacteria and a plant root.

BALLIST-

the "throwing" of projectiles (L. *ballista*, from Gk. *ballō*, to throw).

A *ballista* was an ancient military machine used for hurling stones.

ballist'ics—the scientific study of projectiles and gunfire.

ballistic pendulum—a heavy block, suspended on strings, at which bullets are fired in order to measure the speed of the bullets.

BALNE(O)-
bath, hence bathing (L. *balneum*).

balne'ology—the scientific study of bathing and of medicinal springs.

balne'o'therapy—the treatment of diseases by bathing, especially in medicinal springs.

BAR-, BARO-, BARY-
weight, heaviness (Gk. *baros* (n.); *barys* (adj.)).

bar'ium—the metal element obtained from the heavy mineral **barytes**.

bary'sphere—the supposed heavy central mass of the Earth.

The root is more commonly met with reference to the weight (i.e. the pressure) of the atmosphere.

baro'meter — an instrument for measuring the pressure of the atmosphere.

baro'thermo'graph — "air-pressure temperature writer" — an instrument which continuously records the pressure and the temperature of the air.

iso'bars—lines drawn on a map to pass through places which have the same atmospheric pressure.

milli'bar—a unit for measuring atmospheric pressure. (1,000 millibars = 1 **bar** = 75 cms. (approx.) of mercury in an ordinary barometer.)

BARB
a beard, a tuft (L. *barba*).
The root is well-known in the word **barber**.

barb—a beard-like growth; a side growth from the shaft of a feather.

barb'ate—having hairy tufts.

barbel—a fleshy, beard-like attachment hanging from jaws of some fish.

barb'ule—"a little barb"—a filament branching from the barb of a feather.

BASI-, BASO-
(1) of, or at, or belonging to, the base (bottom) (L. from Gk. *basis*, a foundation, a base).

basi'cranial—related to, or situated at, the base of the skull.

basi'vertebral—related to, or situated at, or coming from, the lower (hinder-most) side of a vertebra.

(2) Pertaining to a chemical base—a substance which reacts with an acid, without the evolution of a gas, to form a salt. (The term includes alkalis but is of wider application.)

basic carbonate—a carbonate (e.g. of lead) which also contains a proportion of the hydroxide (a base).

basic dyes—'alkaline' dyes which are fixed by an acid.

basi'phil, baso'phil—a cell which reacts readily with basic dyes.

baso'philia—an increase of the basophil cells in the blood.

BASIDIO-
A **basidium** (L. *basidium*, a little base) is a rounded or club-shaped cell which bears spores on its top. It is characteristic of the **Basidio'mycetes**—a large group of the higher Fungi (including Toadstools). The spores borne on a basidium are called **basidio'spores**.

BAT-
one that goes (Gk. *batēs*, from *bainō*, to go, walk, step).
An **acrobat** is "one who walks on his tips", i.e. on tiptoes. **Aerobatics** is expert performance of movements in an aeroplane.

ana'batic—(winds) caused by the upward motion of warm air.

a'dia'batic—"not going through"—said of a process (e.g. the expansion of a gas) which takes place under such conditions that heat does not enter or leave.

BATHO-, BATHY-
depth (or height), especially the depth of the sea (Gk. *bathos* (n.), *bathys* (adj.)).

batho'meter, bathy'meter—an instrument for measuring the depth of the sea.

batho'philous — "depth loving" — (plants, animals) suited to life at great depths.

bathy'sphere — a spherical diving-apparatus (large enough to contain two men and instruments) used for investigating life in deep waters.

bathy'scaphe—"depth boat"—a kind of underwater boat, used for deep-sea studies, consisting of a buoyant container filled with petrol supporting an observation chamber underneath.

batho'lith, bathy'lith—"deep stone"—a large mass of igneous rock (often granite), intruded in the Earth's crust, with steep sides and no known base.

eury'bathic, eury'bathytic—"of broad depths"—able to live in a wide range of depths.

-BE See BI-, BIO-.

BENTH-, BENTHO-, BENTHON

the depth of the sea, the sea-bottom (Gk. *benthos*).

benthon, benthos—the plants and animals living on the bottom of the sea.

phyto'benthon—plants living at the bottom of water.

bentho'potamous—living on the bottoms of rivers and streams.

epi'benth'ous — "above the sea-bottom"—(plants and animals) living below low-tide level but not deeper than 100 fathoms.

archi'benthal—pertaining to, or living on, the steep slopes which lead down from a continent to the main ocean bed.

BENZ-, BENZO-

Benzoin is a fragrant resin obtained from a Javanese tree. In the sixteenth century the name was *benjoin*. This name may be traced back through Italian and other Romance languages to the Arabic *luban jawi*, meaning "frankincense of Java". (The *lu* was dropped in the Romance languages on the supposition that it stood for *the*.) The resinous substance is sometimes called gum-benzoin to distinguish it from the chemical compound benzoin which is its main constituent.

The term **benzoic** basically means "of, or derived from, benzoin". It is particularly applied to the acid with the formula C_6H_5COOH.

Distillation of benzoic acid with lime yields a hydrocarbon to which the name **benzene** was given. (The hydrocarbon had been discovered by Faraday in 1825 in the liquid obtained by the destructive distillation of oils and fats; it is now obtained from coal-tar.)

benz'ene—a colourless liquid, the simplest cyclic (ring) hydrocarbon, C_6H_6.

benz'aldehyde—oil of bitter almonds, C_6H_5CHO.

benzo'ic acid—See above. The salts are called **benzo'ates**.

benzo'yl—the characteristic 'stem' of benzoic acid, C_6H_5CO-, e.g. **benzoyl chloride**—$C_6H_5CO.Cl$.

benz'yl — the group of atoms $C_6H_5CH_2$-, e.g. **benzyl alcohol**—$C_6H_5CH_2OH$. (The benzyl group is so named because benzyl alcohol, on oxidation, gives, first, benzaldehyde and then benzoic acid. Note that the group is *not* C_6H_5-, i.e. benzene less one hydrogen atom.)

BETA- See β-.

BI-, BINI-

two, twice, having two (L. *bis*); two by two, in twos (L. *bini*).

Note. The form BIN- is sometimes used instead of BI- before vowels (e.g. **bin'oxalate**); more often it represents the Latin *bini*.

This common prefix is seen in **bicycle** (which has two wheels) and in **bigamy** (having two wives or husbands).

bi'ped—an animal (e.g. Man) which has two legs.

bi'valve — a shellfish (e.g. Oyster) which has a double shell.

bi'parous—giving birth to two offspring at a time.

bi'ennial—a plant which lives for two years (producing seed in the second year). (Note that such words as **bi-weekly** may mean occurring every two (weeks) or twice in one (week).)

bi'ceps—a muscle (e.g. that in the upper arm) which has "two heads", i.e. two points of attachment.

bi'metallic strip—a strip made by fastening two strips of different metals together side by side. (It bends when heated because of the unequal expansions.)

bi'refringence—"double bending back" —the formation by some mineral (e.g. calcite) of two refracted rays.

bi'nomial—consisting of two terms, e.g. **binomial nomenclature**—the use of a two-word name (one for the genus and one for the species) for an animal or plant (e.g. *Bellis perennis*=Daisy), **binomial expression**—an algebraic expression consisting of two terms (e.g. $x+y$).

bi-stable — (arrangement, system) which can come to rest in either of two positions, e.g. a coin can come to rest with either 'heads' or 'tails' uppermost.

bin'ary—consisting of, or involving, two parts, e.g. a binary star (=double star).

bin'aural—pertaining to, or for the use with, two ears, e.g. a binaural stethoscope.

bin'ocular—pertaining to two eyes, e.g. binocular vision. **binoculars**—field-glasses (or opera glasses) for use with both eyes.

In the naming of chemical compounds, BI- denotes a double amount of the element, group, etc. named. When used rigidly it denotes an acid salt, i.e. one in which only a part of the hydrogen of the acid has been replaced by a metal.

bi'carbonate—a salt containing the group $-HCO_3$, e.g. sodium bicarbonate $NaHCO_3$. (Contrast sodium carbonate Na_2CO_3. The bicarbonate contains twice as much of the carbonate group, in relation to the sodium, as the carbonate.)

bi'sulphate—an acid salt containing the group $-HSO_4$, e.g. sodium bisulphate $NaHSO_4$. (Contrast sodium sulphate Na_2SO_4.)

The prefix BI- is also used more loosely as an alternative to DI-, e.g. bisulphide (=disulphide), bichromate (=dichromate). The use of DI- is to be preferred.

BI-, BIO-, -BE

life (Gk. *bios*).

bio'logy—the study of living things (plants and animals).

bio'chemistry — the study of the chemical changes which take place in living things.

bio'luminescence—the production of light by living things (e.g. by Glow-worms).

amphi'bi'ous—able to live in "both" ways, i.e. on land and in water.

sym'bio'sis—"a state of living together"—a partnership between two living organisms, e.g. the association of an Alga and a Fungus in a lichen.

anti'biotic — "against life" — a substance (e.g. penicillin) which destroys or injures living organisms (especially bacteria).

micro'be—"a very small living being" (plant or animal)—a somewhat loose term for micro-organisms, especially for bacteria which cause disease.

an'aero'be—"without air liver"—an organism which lives in an almost complete absence of oxygen.

BLAST-, BLASTO-

a sprout, a bud, a germ (Gk. *blastos*). Used especially with reference to the early stages of a developing embryo.

blast'ula—"a little bud"—the hollow ball, consisting of a single layer of cells, into which the fertilized egg first develops.

blasto'coele—the hollow cavity of a blastula.

blasto'mere—"a part of the bud"—any cell formed during the early stages of the division of the egg cell.

triplo'blastic—having three primary layers in the developing embryo.

hypo'blast—the innermost layer of the developing embryo.

plano'blast—"a wandering bud"—a free-swimming reproductive unit of a Coelenterate (e.g. Obelia).

BLENNO-

slime, mucus (watery jelly as from the nose) (Gk. *blennos*).

blenno'rrhoea—a flow of mucus from the eyes, bowel, etc.

blenno'genous—producing mucus.

BLEPHAR-, BLEPHARO-
eye-lid (Gk. *blepharon*).

blephar'itis—inflammation of the eye-lid.

blepharo'ptosis—a dropping of the upper eye-lid due to paralysis of the muscle.

blepharo'spasms—uncontrolled twitchings of the eye-lid muscles.

-BOLA, -BOLISM
These word elements are derived from the Greek *bolē*, *bolos*, a throw, from *ballō*, to throw. Perhaps the following term is the nearest to the idea of throwing.

em'bolism—a thing "thrown in"—an obstruction of an artery, etc., by a clot carried from another part of the blood system.

In most terms the elements are used in somewhat indirect senses.

A **parabola** is a thing "thrown at the side, i.e. a thing for comparison". (Compare the English word **parable.**) The mathematical curve called a parabola is formed by the cutting of a cone by a plane which is parallel to a slanting side of the cone. The hyperbola and ellipse may be compared with this curve.

A **hyperbola** is "an over-throw, an excess". The curve is formed by the cutting of a cone by a plane which is more steeply inclined to the base than a side. (It is also, in non-mathematical terms, relatively wider than a parabola.)

An **ellipse** is a "defect, a coming short" (Gk. *elleipsis*). The curve is formed by the cutting of a cone by a plane which is less steeply inclined to the base than a side. (It is also relatively less wide than a parabola.)

-BOLISM frequently refers to the "throwing" together or apart of chemical substances within a living body.

ana'bolism—the building up of complex substances from simple substances within a living body.

kata'bolism—the breaking down of complex substances into simple substances within a living body.

meta'bolism—the total of the chemical and physical changes constantly taking place within a living body.

BOTHR-, BOTHRIO-
a pit, a trench, a depression (Gk. *bothros*; *bothrion*).

bothrium—a groove-shaped sucker in a Tapeworm.

bothr'enchyma—tissue consisting of pitted cells.

Bothrio'lepis—"pitted scale"—a fish-like creature found as a fossil.

BOTRY-, BOTRYO-
a bunch of grapes (Gk. *botrys*).

botry'oid, botry'ose, botry'tic—shaped like a bunch of grapes; (*Botany*) bearing flowers in clusters; (*Geology*) made up of grains resembling a bunch of grapes.

botryo'mycosis—a disease of horses in which there is an excessive growth of fibrous tissue.

botryoidal tissue—tissue consisting of pigmented cells, end to end, containing canals filled with red liquid, as in Leeches.

BRACHI-, BRACHIO-
an arm (L. *brachium*, from Gk. *brachion*). This root may be traced in **brace, bracelet,** and **embrace** (to fold in the arms).

brachi'al—pertaining to an arm, e.g. the brachial artery.

brachi'ate—possessing an arm; with branches out at right-angles.

brachium—the part of the fore-limb of a vertebrate animal which is nearer the body, i.e. the upper arm in Man; any arm-like structure.

Brachio'poda—"arm-footed" animals —a phylum of two-shelled marine animals (Lamp shells) which have arm-like projections near the mouth for wafting in food and dissolved oxygen.

BRACHY-
short (Gk. *brachys*).

brachy'cephalic — having a "short

head", i.e. the length of the skull is not more than 1¼ times the breadth.

brachy'dactyly—unusual shortness of the fingers and toes.

brachy'pterous—having short wings (wings which when folded do not reach the base of the tail).

BRADY-
slow (Gk. *bradys*).

brady'cardia—slowness of the heart-beat.

brady'pepsy—slowness of digestion.

brady'phrenia—slowness of the mental processes.

Brady'pus—"slow foot"—the three-toed Sloth, a sluggish arboreal animal of tropical America.

BRANCHI-, BRANCHIO-
gills (breathing organs of fish and other water creatures) (L. *branchiae*, Gk. *branchia*).

branchiae—gills. **branchi'al**—pertaining to, or of the nature of, gills.

branchi'ate—possessing gills.

branchio'stegal—pertaining to the gill-covers.

Branchio'poda—the "gill-footed" animals—a class of sea creatures (which includes Shrimps) whose flattened swimming feet also serve as breathing organs.

Fili'branchia—animals with "thread-like gills"—an order (group) of shellfish which includes Sea Mussels, Oysters and Scallops.

BREVI-
short (L. *brevis*).
This root is well-known in **brevity** and **abbreviate** (to make shorter). Although a **breve** is now the longest note in music, in early notations it stood for a short note.

brevi'lingual—having a short tongue.

brevi'pennate—having short wings.

brevi'rostrate—having a short beak.

BROM-, BROMO-
a stench, an offensive smell (Gk. *brōmos*).

brom'idrosis—a disorder of the sweat glands in which the sweat has an offensive smell.

The root more usually refers to the chemical element **bromine**—a dark red, heavy liquid which has an irritating smell.

bromide—a compound of bromine and another element (or group of atoms), e.g. silver bromide AgBr, methyl bromide CH_3Br.

bromo'benzene—a compound in which a bromine atom has replaced a hydrogen atom of benzene, C_6H_5Br.

bromo'form — a colourless liquid, $CHBr_3$, similar to chloroform.

BRONCH-, BRONCHI-, BRONCHO-
The **bronchi** (sing. **bronchus**) are the two main branches of the wind-pipe, leading to the lungs; the **bronchia** are branches of a bronchus. (L. *bronchus*, from Gk. *bronchos*, wind-pipe.)

bronchi'al—pertaining to the bronchi or bronchia.

bronch'itis—inflammation of a bronchial membrane.

bronchi'ole—one of the small end-branches of the bronchia.

bronchi'ectasis—enlargement of the bronchi as a result of weakening of the bronchial walls.

broncho-pulmonary—pertaining to the bronchi and the lungs.

broncho'scope—an instrument for inspecting the inside of the bronchi.

BRONTO-
thunder (Gk. *brontē*).

bronto'logy—the branch of meteorology which is concerned with thunder.

Bronto'saurus — "thunder lizard" — a genus of huge prehistoric reptiles.

BRYO-
moss (Gk. *bryon*).

bryo'logy—the study of Mosses.

Bryo'phyta—"the moss plants"—the large class of non-flowering plants which comprises the Mosses and Liverworts.

Bryo'phyllum—"mossy leaf"—a tropical plant, sometimes cultivated indoors in this country, whose leaves develop roots and shoots at the edges when detached from the plant (or sometimes before detachment).

Note. The root EMBRYO- has a very different meaning but it is related to this root through the Greek verb *bryō*, to swell.

BUCC-, BUCCO-
the cheek (L. *bucca*).

bucc'al cavity—the space inside the mouth between the cheeks.

bucc'inator—the flat, thin cheek muscle.

bucco-lingual — pertaining to the cheeks and the tongue.

BURS-, BURSA
hide, skin; hence a bag, a pouch. (Compare **purse**.) (L. *bursa*, from Gk. *byrsa*.)

bursi'form—shaped like a bag or pouch.

bursa—any bag-like cavity, especially a sac containing lubricating liquid at places of friction in the body (e.g. between muscles, tendons, skin, bony projections).

burs'itis—inflammation of a bursa.

Bursa pastoris—the common plant Shepherd's Purse, so called from the form of the fruits.

BUTYR-, BUT-, BUTYL
butter (L. *butyrum*; Gk. *boutyron*).

butyr'aceous—of the nature of butter.

butyr'ic acid—a thick liquid with a rancid smell, formed in bad butter, C_3H_7COOH.

These elements are more commonly used in forming the names of chemical compounds which are derived from, or related to, butyric acid.

but'ane—a paraffin hydrocarbon with four carbon atoms, C_4H_{10}.

butyl'ene—the corresponding hydrocarbon which has a double join between two of the carbon atoms, C_4H_8.

butyl—the hydrocarbon group of atoms C_4H_9-, e.g. butyl alcohol C_4H_9OH.

BYSS-
a mass of fine threads (like cotton or flax) (Gk. *byssos*).

byss'aceous, byss'oid—consisting of a mass of fine threads.

byssus—the 'beard' (thread-like tuft) of some bivalve molluscs (e.g. Oyster, Mussel) used for attachment; a thread-like tuft of some Fungi.

Note. *A'byss* is of different derivation; it means the bottomless deep.

C

CAC(O)-
bad, evil, unhealthy (Gk. *kakos*).

cac'odorous—bad smelling.

cac'hexia, cachexy—"unhealthy habit (condition)"—an unhealthy state of the body, e.g. weakness, wastage of tissue, grey complexion, as of a person suffering from cancer.

cac'ophthalmia—"bad eyesight"—an eye disease.

cac'od'yl—"evil smelling substance" —the group of atoms $As(CH_3)_2$, i.e. arsenic plus two methyl groups; its compounds, e.g. **cacodyl** $(CH_3)_2As.As.(CH_3)_2$, **cacodyl chloride** $As(Ch_3)_2Cl$, have intensely unpleasant smells.

CAINO-, (KAINO-)
new, recent (Gk. *kainos*).

Caino'zoic era—the "new (or later) life" era—the era, comprising the Tertiary and Quaternary periods, from about 60 million years ago to the present day, in which there was the development of life as it is now.

This term is occasionally spelt **Caenozoic**; the form **Cenozoic** (see CENO-) is now becoming common.

CALC-, CALCI-, CALCAREO-
lime, stone, limestone, hence chalk (L. *calx*, *calc-* (n.); *calcarius* (adj.)).

A **calculus** is "a little stone" (L.). The term is used in medicine for a stony concretion (coalescence of solid particles) in a part of the body, e.g. in the kidney. The use of small stones on an abacus (counting board) for reckoning purposes has given us the well-known word **calcul'ate**. The term **calculus** is also used for a special branch of mathematics.

calc'ine—to roast a substance (e.g. a mineral) so as to change it into a crumbling powder (a **calx**) as in the roasting of limestone to produce quick-lime.

calc'ium—the metal element present in lime and chalk.

calc'ite—a crystalline form of calcium carbonate.

calci'ferous—producing, or containing, calcium salts (especially calcium carbonate).

calci'philous—"chalk loving"—(plant) which grows readily (or almost exclusively) on a chalky soil.

hyper'calc'aemia—an abnormally high amount of calcium salts in the blood.

calcareous, calcarious—(rocks) consisting chiefly of limestone.

calcareo-argillaceous — composed of clay with a proportion of lime.

CALOR-, CALORI-
heat (L. *calor*).

calorie—a metric unit for the measurement of a quantity of heat.

calor'ific value (of a fuel)—the quantity of heat produced by burning a certain amount (e.g. 1 lb.) of the fuel.

calori'meter—an instrument or apparatus for the measurement of a quantity of heat.

CALYC-, CALYCI-
The **calyx** of a flower (Gk. *kalyx*, *kalyk-*, cup of flower) is the ring of green sepals which encase the bud and form a cup below the petals.

calyc'ine—relating to, or having, or on, a calyx.

calyci'florate — having flowers with **calyces** (plural of *calyx*).

calyc'ule—"a little calyx"—a group of small bracts (leaflets) closely beneath a calyx.

calyc'anthemy—"calyx flowering"—an unusual condition in which the calyx becomes coloured and resembles the petals.

CALYPT-, CALYPTO-, CALYPTR-
covered, hidden (Gk. *kalyptos*).
An **apo'calypse** is a revelation ("an uncovering"), especially that made to St. John in the island of Patmos.

calyptra—a covering (lid) over the capsule of a Moss or Liverwort.

calypto'branchiate—having the gills hidden by a gill-cover.

calyptr'ate—(flies, etc.) having the balancers covered by scale-like parts.

Eu'calypt'us—a genus of plants which includes the Australian Gum-tree. (So called because the flower is "well covered" by a kind of cap before it opens.)

CAMPAN-
a bell (L. *campana*).

campan'ology—the science of making bells or of ringing them.

campan'iform — bell-shaped, dome-shaped.

Campan'ula—"a little bell"—a kind of plant with blue or white bell-shaped flowers, e.g. Canterbury Bell.

CAMPYLO-
curved, bent (Gk. *kampylos*).

campylo'tropous—"curved turning"—said of a part of a plant (e.g. an ovule) which curves over so that the normal apex is downwards.

campylo'spermous—said of the carpels (which contain the ovules and young seeds) of some flowers, e.g. some of the family Umbelliferae, in which the seed inside produces a long groove.

CAN-, (CANI-)
a dog (L. *canis*).

can'ine—(1) pertaining to a dog. (2) one of the four strong, pointed teeth between the incisors (cutting-teeth) and the molars (grinding-teeth).

Rosa canina—the Dog-Rose.

Can'idae—the family of dog-like animals including Dog, Fox, Wolf, etc.

CANCELL-
made up of cross-lines, like a lattice (L. *cancello* (v.) from *cancelli*, cross-bars, lattice).
When we **cancel** something we make cross-lines over it, literally or figuratively.

cancell'ate, cancell'ous—like a lattice

or network, made up of inter-crossing tissue, having a spongy structure.

CANCER, CANCRI-

(1) a crab (L. *cancer*).

Cancer—a crab; the constellation of stars known as The Crab.

cancri'form—crab-shaped.

cancer'ite—a fossil crab.

(2) The Latin word **cancer** was also used, as it still is in English, for a tumour or disordered growth, particularly of surface cells (e.g. of skin, bowel), which invades surrounding tissue and may spread to other parts.

cancer'ous—of the nature of, or affected with, cancer.

CANTHUS, CANTHO-

edge, corner, hence corner of the eye (L. *canthus*; Gk. *kanthos*).

canthus—the angle at the outer or inner junction of the eye-lids.

epi'canthus—a fold of skin above, and sometimes covering, the inner angle of the eye.

cantho'plasty—the operation of enlarging the opening between the eye-lids.

CAPILL-

a hair (L. *capillus* (n.), *capillaris* (adj.)).

capill'ary tube—a tube with a thin, hair-like bore.

capill'arity—the rise of a liquid up a narrow tube as a result of surface tension.

CAPIT-

a head (L. *caput*, *capit-*).

A **capital** is a head (chief) town of an area.

capit'ate — resembling a pin-head; having flowers grouped on a capitulum (as below).

de'capit'ate—to cut off the head.

capit'ulum—"a little head"—a tightly packed group of flowers as in the head of a Daisy; the head of a bone (especially of a rib).

CAPR-, CAPRI-

a goat (L. *caper*, *capri*).

A **capricious** person behaves in a fanciful, unaccountable way "like a goat". We sometimes say he 'cuts a caper'.

capri'form—resembling a goat.

Capri'corn—"the horned goat"—the constellation of stars resembling a goat.

capr'oic, capri'lic and **capr'ic acids**—fatty acids, the molecules having 6, 8 and 10 carbon atoms respectively, found (in a combined state) in butter made from goat's milk.

CARB-, CARBO-, CARBON

charcoal, coal, hence relating to, containing, or derived from, the element carbon (L. *carbo*, *carbon-*).

carbon'aceous—like coal or charcoal; containing carbon.

carbon'iferous rocks—the system of rocks which includes the coal-bearing strata.

carb'ide—a compound of carbon and another element, e.g. calcium carbide CaC_2.

carb'orundum—a compound of carbon and silicon used as an abrasive (for smoothing and polishing).

carbon'ic acid—a solution of carbon dioxide (CO_2) in water, H_2CO_3. The salts are **carbonates**.

carb'ol'ic acid—a popular name for phenol, C_6H_5OH.

carbon'yl—the group of atoms CO:, e.g. carbonyl chloride $COCl_2$.

carb'oxyl — the group of atoms -CO.OH, the characteristic group of the organic acids.

carb'inol—methyl alcohol CH_3OH. Used particularly in the naming of higher alcohols, e.g. **tri'methyl-carbinol** $C(CH_3)_3OH$ in which three methyl groups take the place of the three hydrogen atoms in carbinol.

carbo'hydrates—the group of chemical compounds which includes sugars, starch and cellulose. In composition they consist of carbon+water (hence the name), e.g. cane sugar may be represented by $C_{12}(H_2O)_{11}$, but they are not formed by such direct combination.

carb'urettor—a device for correctly

mixing the fuel and air for passing into the cylinder of an internal combustion engine.

carb'uncle — "a little fiery coal" — originally, a red gem; now, an inflamed tumour or a pimple on the nose.

CARCIN-, CARCINO-

(1) a crab (Gk. *karkinos*).

carcino'logy—the study of crustacean animals (e.g. crabs).

(2) a disordered growth, a tumour.

carcin'oma—a disorderly growth, especially of surface cells (e.g. of skin, bowel) which invades surrounding tissues and may spread to other parts, a cancer.

carcino'genesis — the production and development of cancer.

carcino'matous—like, of the nature of, cancer.

CARDI-, CARDIA-, CARDIO-

the heart (Gk. *kardia*).

cardi'ac—pertaining to the heart, e.g. cardiac muscle.

cardio'logist—a doctor who specialises in heart diseases.

cardia'graph, cardio'graph—an instrument which records the heart beats.

brady'cardia—slow heart beat.

peri'cardium—the membranous sac which surrounds the heart; **pericarditis**—inflammation of the pericardium.

cardi'oid—like a heart; a heart-shaped curve.

CARINA

the keel of a boat (L. *carina*).

carina — the boat-shaped structure formed by the two lower petals of a pea (etc.) flower; the keel-like ridge on the breast-bone of birds.

carin'ate—shaped like a keel; having a keel-like projection.

CARN-, CARNI-

flesh (L. *caro*, *carn-*).

A **carnival** was originally a festive period just before Lent, once only Shrove Tuesday, the day when it was necessary "to put away flesh (meat)" (*carnem levare*). The name of the flower **Carnation** (a kind of cultivated Clove Pink) is probably a corruption of *Coronation* (a reference to its crown-like petals), the corruption having occurred because of the fleshy colour of some of the varieties. Similarly, **carnelian** is a corruption of *cornelian* (a flesh-coloured gem-stone).

carn'ose—of a fleshy texture.

carni'vorous—flesh-eating.

Note. *Carnallite*, the mineral chloride of potassium and magnesium, was named after the Prussian Von Carnall.

CARP-, CARPO-

This element has two quite distinct meanings.

(1) the wrist (Gk. *karpos*).

carpus (L.)—the wrist.

carp'als — the small bones which make up the wrist.

meta'carpals — "behind (farther from the body than) the wrist"—the bones of the hand between the wrist and the fingers.

(2) a fruit (Gk. *karpos*).

carp'el—a part of a flower which contains the ovules; singly, or with other carpels, it forms the pistil (female part of the flower).

peri'carp—"around the fruit"—the fruit-wall.

acro'carpous—bearing fruit at the ends of the branches.

carpo'phore — "fruit bearer" —a general name for the stalk of a fruit-structure, especially in lower plants.

carpo'gonium—the female organ in the Red Algae; an early stage of the fruit body of Lichens and some Fungi.

CARYO-, KARYO-

a nut (Gk. *karyon*).

cary'opsis—a dry, nut-like fruit, containing one seed with the fruit-wall closely sticking to it, e.g. a wheat grain.

Caryo'phyllaceae—the plant family which includes the Pink, Campion and Chickweed. Named after the Clove (Gk. *Karyophyllon*, "nut-leaf") in which the dry fruit opens and forms teeth which curl back. Hence:

caryophyll'ene — a terpene (q.v.), $C_{15}H_{24}$, occurring in oil of cloves; **caryophyll'in** — a crystalline substance extracted from cloves.
The element is also used to denote the nucleus of a cell.

karyo'lymph—the liquid part of the cell nucleus.

karyo'kinesis—"nucleus movement" —the series of changes which takes place in the nucleus during the process of cell division.

CASEIN, CASEINO-
cheese (L. *caseus*).

casein — the coagulated (solidified) substance produced by the action of certain agents (e.g. rennet) on milk and made into cheese.

caseino'gen—"casein producer"—the principal protein of milk.

CATA-, CATH-, KATA-, (KATH-)
down, against (Gk. *kata, kata-*).

Note. Many words formerly spelt with *k* are now more usually spelt with *c*.

A **cataract** (waterfall) is a "dashing down" of water; a **cathedral** contains the bishop's throne, i.e. where he "sits down"; a **catapult** is an instrument for hurling stones against an enemy.

cata'dromous—(fish, e.g. Freshwater Eel) which moves down the river to spawn.

cata'plexy — "a striking down" — a sudden attack of weakness in which a person falls to the ground and remains motionless.

cata'rrh — "a down flow" — inflammation of a mucous membrane and discharge of mucus, as in a cold.

cata'lysis—"a loosening down"—the speeding up (or, occasionally, slowing down) of a chemical process by the presence of a substance which is not itself permanently changed in the process.

cath'ode—"a down way"—the electrode by which an electric current leaves an electrolytic cell, discharge tube, etc.

kata'batic — (winds) caused by the downward motion of air (as when cold air flows into a valley).

kata'bolism—the breaking down of complex substances into simpler substances in a living body.

CATEN-
a chain (L. *catena*).

caten'ate—joined with links into a chain.

caten'ary—the curve in which a loose, uniform chain hangs when supported at its two ends.

CATH- See CATA-.

CATHET-, CATHETO-
a thing sent down (Gk. *kathetos*, from *kathienai* to send down).

cathet'er—a tubular instrument for passing into the urinary bladder.

catheto'meter — an instrument for measuring small vertical (up and "down") distances.

CAUD-
a tail (L. *cauda*).

caud'al — pertaining to the tail.

caud'ad—towards the tail.

caud'ate—having a tail (or tail-like part), **nudi'caudate** — having a nude (bare) tail.

sacro-caudal—pertaining to the sacrum (q.v.) and the tail region.

CAUL-, CAULI-
a stalk, a stem (L. *caulis*; Gk. *kaulos*).

cauli'form—having the shape of a stem.

caul'escent—developing into a stem, having a clearly seen stem.

a'caul'ine, acaulose—without a stem (or nearly so).

amplexi'caul—"embracing the stem" —(leaves) wrapping round the stem at their bases.

The Latin word *caulis* also denoted a cabbage. A **cauliflower** is thus a "cabbage flower". The word has come to us through the French and has suffered a number of variations of spelling over the centuries.

-CELE
hernia, hence a swelling out (Gk. *kēlē*).

entero'cele—a swelling out of a piece of the intestine through a hole or weak place in the cavity wall which contains it, a form of hernia.

cephalo'cele—a swelling out of the brain-covering through a hole in the skull.

galacto'cele—a swelling in the breast because of a blockage of a milk duct.

tracheo'cele—an air-containing swelling in the neck due to a bulging of the wall of the trachea (wind-pipe).

CELER-
swift, hence speed (L. *celer*).

celer'ity—swiftness, speed.

ac'celer'ate — to increase speed.

accelerometer — an instrument for measuring rates of change of speed.

de'celer'ate—to decrease speed.

Note that *celery* is not derived from this root; it comes from the French, ultimately from the Greek *selinon*, parsley.

CELEST-
of the sky, the heavens (L. *caelestis*).

celest'ial—pertaining to the sky, e.g. the **celestial poles** are the two points in the sky in line with the Earth's axis.

celest'ine, celestite — the mineral strontium sulphate, possibly so called because it is sometimes pale blue in colour.

CELLUL-
Relating to cells of plants and animals (L. *cellula*, a small room).

cellul'ar—made up of cells.

uni'cellular—consisting of only one cell. **multi'cellular**—composed of many cells.

cellul'ose—a complex carbohydrate (polysaccharide) which forms the walls of plant cells.

cellul'oid—a well-known plastic made from nitro-cellulose, camphor and alcohol.

CENO-, -CENE
new, recent (Gk. *kainos*).

Used especially in forming the names of geological periods.

Ceno'zoic era—the "recent life" era—the era from about 60 million years ago to the present day in which there was development of life as it is now. (Also called Caino'zoic, Caeno'zoic.)

Palaeo'cene—"ancient new (life)"—the first (oldest) period of the Cenozoic era.

Pleisto'cene—"most new"—the latest period (from 1 to 2 million years ago) of the Cenozoic era; the Great Ice Age.

The gradual change in spelling from KAINO-, through CAINO- and CAENO-, to CENO- has resulted in a misnomer. The root CENO- really means "empty" (Gk. *kenos*)—a cenotaph is an empty tomb—but the Cenozoic era was far from empty of life.

CENO- is also a variant spelling (particularly in the U.S.A.) of COENO- (q.v.). Hence the 'modernisation' of the spelling of both CAINO- and COENO- has resulted in confusion; the origins of words incorporating CENO- are obscured and their meanings rendered less obvious.

CENT-, CENTI-
(1) one hundred (L. *centum*).

A **century** is a hundred years (or a hundred runs at cricket).

per cent.—a proportion expressed as so many parts per hundred.

Centi'grade scale—a thermometer scale in which there are one hundred degrees between the freezing point and the boiling point of water.

centi'pede—a member of the class Chilopoda — a wingless, crawling creature with many (if not exactly a hundred) feet.

(2) one hundredth.

centi'metre—one hundredth of a metre.

centi'gramme—one hundredth of a gramme.

CENTR-, CENTRI-, CENTRO-
a centre (L. *centrum*; Gk. *kentron*).

con'centr'ic—(circles) having the same centre.

ec'centr'ic—not mounted at its centre, not having its axis (etc.) through the centre.

geo'centr'ic—having the Earth at the centre.

con'centr'ate—to bring "together towards one centre"—to increase the strength of (e.g.) a solution by decreasing the volume of water (etc.) which it contains.

centri'fugal—acting or moving away from the centre.

centri'petal—acting or moving towards the centre.

centro'lecithal—having the yolk in the centre.

centrum—the middle part of a vertebra.

centro'some—"central body"—a very small body near the nucleus of an animal cell; it divides when the nucleus divides.

CEPHAL-, CEPHALO-

the head (Gk. *kephalē*).

cephal'ic, cephalous—pertaining to the head, e.g. **cephalic index**—a figure indicating the relation between the breadth and length of the head.

a'cephalous—lacking a head or head region.

brachy'cephalic—having a short head (length not more than 1¼ times the breadth).

cephalo'thorax—the combined head and thorax of Crabs, Spiders, etc.

Cephalo'poda — the "head-foot" animals—the class of marine animals, including the Octopus and the Squids, in which the mouth appears to be in the middle of the much-branched foot.

The root ENCEPHAL- ("within the head", i.e. the brain) is given in its proper alphabetical position.

Note. The Greek κ is sometimes represented in English by *k* but, particularly in modern spelling, more often by *c*. With this acceptance of the convention of English spelling it seems reasonable also to accept the conventions of English pronunciation. Thus *c* is pronounced soft (=*s*) before *e* (as in *cede*), *i* (as in *city*) and *y* (as in *cycle*). Yet some biologists and most members of the medical profession insist on pronouncing a hard *c* (=*k*) in words derived from the Greek. (Words derived from CEPHAL- and ENCEPHAL- are familiar examples.) Surely this is misguided pedantry. The biologist who says *kephalopod*, or the doctor who says *enkephalitis*, is hardly likely to say that he goes to the *Kinema on his bikykle*.

-CEPS

head (from L. *caput*). Used particularly for the head (point of attachment) of a muscle.

bi'ceps—a muscle with two heads, especially that in the upper arm by which the arm is bent.

tri'ceps—a muscle with three heads, especially that in the upper arm by which the arm is straightened.

Clavi'ceps—"club head"—the fungus which causes a disease of rye. (Dark, round-headed bodies project from the ear.)

-CEPTOR, RECEPTOR

a receiver, particularly of sensations (L. *captor, receptor*).

extero'ceptor—a sense organ which receives stimuli (e.g. light) from outside the body.

intero'ceptor—a sensory nerve centre which receives impressions (e.g. of pain, position) from within the body.

chemo-receptor—a sensory organ (e.g. a taste-bud) which responds to chemicals.

CER-

wax (L. *cera*).

cer'aceous—of, or like, wax.

cer'iferous—bearing or producing wax.

cer'umen—a waxy substance formed by certain Bees; a waxy substance secreted by the body (e.g. in the ear).

cer'otic acid—a fatty acid $C_{25}H_{51}.COOH$ which occurs free in beeswax and combined with **ceryl alcohol** $C_{25}H_{51}.CH_2OH$ in Chinese wax (from the Tallow tree).

Note:

(1) The word *cereal* is derived from *Ceres*, the goddess of corn. This name *Ceres* was also given to a small planet (asteroid) discovered in 1801 and adapted to form the name *cerium* for a chemical element discovered in 1803.

(2) The term *ceramics*, the art of pottery, is derived from the Greek *keramos*, pottery, *keramikos* (adj.).

CERA-, CERATO-, -CEROS, KERA-, KERAT-, KERATO-

horn (Gk. *keras*, *kerat-*).

A **rhinoceros** is a well-known, large animal which has one or two horn-like structures in the region of the nose.

Cerastes—the Horned Viper of N. Africa, a poisonous snake with a horny projection over each eye.

brachy'cer'ous — "short horned"—having short antennae, as have some of the Diptera (Flies).

cheli'cerae—"claw horns"—the gripping and biting organs which form the first pair of appendages of Spiders, Scorpions, etc.

cerato'trichia — "horny hairs" — unjoined horny rays supporting the fins of fish with cartilaginous skeletons.

kerat'in—the horny substance from which hair, finger-nails, etc., are made.

kerato'genous—horn-producing; keratin-producing.

hyper'kerato'sis—overgrowth of the horny layer of the skin.

In medical terms the root usually refers to the cornea of the eye.

kerat'itis—inflammation of the cornea.

kerato'plasty—"cornea moulding"—the grafting of a new cornea on to an eye.

kerato'cele—a swelling of the innermost membrane of the cornea through an ulcer.

CERC-, CERCO-

a tail (Gk. *kerkos*).

cerc'al—pertaining to the tail.

lepto'cercal—having a long, thin tail.

Cerco'pithecus — a genus of long-tailed African monkeys.

cerc'aria—the final larval stage of some parasitic Flatworms. (It possesses a tail for propulsion.)

CERVIC-

the neck (L. *cervix*, *cervic-*).

cervic'al—pertaining to the neck, e.g. the cervical vertebrae.

The *cervix uteri* is the neck of the uterus (womb). In medical terms the root usually refers to this neck.

cervic'itis—inflammation of the neck of the uterus.

cervic'ectomy—the surgical cutting away of the neck of the uterus.

CEST-

a girdle, a ribbon (L. *cestus* from Gk. *kestos*).

cest'oid—ribbon-like.

Cest'oda—the class of **cestode** worms (Tapeworms).

CET-, CETO-

a whale (L. *cetus* from Gk. *kētos*).

Cet'acea—the large order of animals which includes Whales and Dolphins.

ceto'logy—the study of whales.

sperma'cet'i—a white, fatty substance contained in the heads of sperm-whales.

cet'yl alcohol—a hexadecyl ("sixteen") alcohol, $C_{16}H_{33}OH$. The compound (ester) formed from cetyl alcohol and palmitic acid is the chief constituent of spermaceti.

CHAET-, CHAETO-

a hair, a bristle (Gk. *chaitē*).

chaet'iferous, chaeto'phorous—bristle-bearing.

chaeto'taxy—the arrangement of the bristles on an animal.

Chaeto'poda—a large group of the Annelida (ringed animals) the members of which, e.g. Earthworms, Bristle-worms, have conspicuous bristle-like feet.

spiro'chaetes—"spiral bristles"—bacteria which have a twisted, spiral form.

CHALCO-

copper, brass (Gk. *chalkos*).

chalco'graphy—the art of engraving on copper.

chalco'pyrite—a yellow ore of copper consisting of sulphides of copper and iron.

CHEIL-, CHEILO-, CHIL-, CHILO-
a lip (Gk. *cheilos*).

cheil'itis—inflammation of the lips.

cheilo'gnathus—harelip, a cleft in the upper lip (and often the palate) from birth.

cheil'ec'tropism—a turning outward of the lip.

Chilo'poda—"lipped feet" animals—the order (group) of Centipedes. (The first pair of legs is modified into a pair of poisonous claws.)

Chilo'gnatha—"lipped jaw" animals—the order (group) of Millipedes. (They have a plate-like lower lip formed by the fusion of a pair of mouth-parts.)
The older spelling CHEIL(O)- is normally retained in medical terms.

CHEIRO-, CHIRO-
a hand (Gk. *cheir*).

Cheiro'pod, Chiro'pod—"hand footed"—a general name for a mammal which possesses hands.

Cheiro'ptera, Chiro'ptera — animals with "hand wings"—the order (group) of mammals which includes Bats. So called because the wing contains bones which correspond to the hand bones of Man.

Cheiro'therium — "hand beast" — a large, extinct, four-footed animal whose footprints resemble a human hand.

chiro'pody—the treatment of corns, warts, nail defects, etc. on the hands and (especially) the feet. (The exact origin of the word ("hand foot") is uncertain.)

Note. The word **surgery** may be traced back through Old French and Latin to the Greek word *cheirourgia*, a working by hand; see -URGY.

CHEL-, CHELI-
a claw, a talon (Gk. *chēlē*).

chela—any form of gripping appendage in various arthropod animals (e.g. a claw of a Crab). Hence **cheli'ferous**—bearing such appendages.

chel'ate—having a claw for gripping.

cheli'cerae—the biting and gripping organs which form the first pair of appendages of Spiders, Scorpions, etc.

CHEMI-, CHEMICO-, CHEMO-
The word **chemistry** dates from the beginning of the seventeenth century (when it was spelt *chymistrie*); the term **chemist** is somewhat older. Chemist comes from the Latin *chemista* from the earlier *alchemista*. See *alchemy* (AL-).

chemi'luminescence—the production of light (without heat) in certain chemical reactions.

chemico-agricultural—relating to the applications of chemistry to agriculture.

chemo'receptor—a sensory organ (e.g. a taste-bud) which responds to chemical substances.

chemo'taxis—the movement of an organism towards a concentration of a chemical substance.

chemo'therapy—the treatment of a disease by means of chemicals (drugs).

CHILO- See CHEILO-.

CHIMO-, CHIMONO-
winter (Gk. *cheimōn*, winter, a wintry storm).

chimono'philous — "winter loving" — growing chiefly during the winter.

chimo'pelagic plankton — plankton (floating plants and animals) occurring in the open sea only in winter.

CHIRO- See CHEIRO-.

CHLAMYD-, CHLAMYDO-
a mantle, a cloak, a covering (Gk. *chlamys, chlamyd-*).

chlamyd'ate—having a mantle (as have some Molluscs).

chlamydo'spore—a fungal spore which has a thick outer covering during its resting state.

Chlamydo'monas—"a covered single thing" — a microscopic, free-moving plant consisting of a single cell with a cell wall.
The root is used in botanical terms to indicate the envelopes (coverings) of a

flower, i.e. the corolla (petals) and calyx (sepals).

chlamyd'eous—having one or more envelopes to the flower.

homo'chlamydeous — having the envelopes all of the same kind, i.e. corolla and calyx cannot be distinguished.

hetero'chlamydeous, di'chlamydeous—"having different (or two) cloaks"—(flower) having distinct corolla and calyx.

CHLOR-, CHLORO-

green, pale green (Gk. *chlōros*).

chloro'phyll—the green colouring matter of leaves (and other parts) of a plant.

chloro'sis—"a (disordered) state of greenness"—an unhealthy state of a plant due to a deficiency of chlorophyll shown by yellowness of the plant.

chlor'enchyma — plant tissue which contains chlorophyll.

chlor'oma—a greenish tumour in the bone.

Chloro'phyceae — the plant group which comprises the Green Algae.

chlor'ine—a non-metallic chemical element, a yellow-green gas under normal conditions with a characteristic smell.

In chemical terms the root refers to chlorine.

chlor'ide—a compound formed by the combination of chlorine with another element (e.g. sodium chloride NaCl) or with a group of atoms (e.g. methyl chloride CH_3Cl).

chlor'ate — a salt of **chloric acid** $HClO_3$, e.g. potassium chlorate $KClO_3$.

chloro'benzene—a compound, C_6H_5Cl, in which a chlorine atom replaces a hydrogen atom of benzene (C_6H_6).

chloro'form, tri'chlor'methane—a well-known anaesthetic, $CHCl_3$.

CHOL-, CHOLE-

bile (yellow-green liquid passed from the liver into the small intestine) (Gk. *cholē*).

A person who was **melan'cholic** (habitually tending to be sad and depressed) was supposed to be suffering from "black bile".

chol'agogue—a drug which stimulates the flow of bile.

chole'cyst'itis—inflammation of the gall (bile) bladder.

chol'uria—the presence of bile in the urine.

chole'sterol—a complex alcohol found in nerve tissue and in gall-stones.

CHONDR-, CHONDRI-, CHONDRIO-, CHONDRO-, -CHONDRIA

(1) grain, grit, granular (Gk. *chondros*).

chondr'ite—a kind of stony meteorite containing mineral grains.

mito'chondria—"thread grains"—small grain-like or rod-like bodies found in all living cells. Also called **chondrio'somes** ("grain bodies").

myo'chondria — grains found in irregular masses in muscle tissue.

(2) cartilage (soft bone substance).

chondr'oid—of the nature of cartilage.

chondr'in—the tough, elastic substance which is the basis of cartilage.

chondri'fication—the development of cartilage.

chondr'oma—a tumour (diseased growth) made up of cartilage cells.

chondro'cranium — the cartilaginous brain-box of lower animals.

a'chondro'plasia — "lacking cartilage growth"—a dwarf condition in which the arms and legs are small but the head and body are of normal size.

Hypo'chondria is a depressed state of mind, sometimes without cause, or due to unnecessary worrying about health. The word means "below the cartilage (of the ribs)"—the lower, softer organs of the body being supposed to be the seat of melancholy.

CHORD

a string, a **cord** (L. *chorda*, from Gk. *chordē*).

The root is used especially in science in reference to the **notochord** (see below) or the spinal cord of the higher vertebrate animals.

noto'chord—"back cord"—an elastic, stiffening rod, present in the backs of a few animals (e.g. Lancelet) throughout life and present in vertebrate animals in embryonic stages but later replaced by the bony backbone.

Chord'ata—the great group (phylum) of animals, including all the vertebrate animals and a few others, which at some stage possess or possessed a notochord.

chord'oma — a tumour (diseased growth) arising from the remains of the notochord in the skull and spinal column.

chord'otomy—the surgical cutting of nerve fibres in the spinal chord to relieve pain.

CHOR-, CHORION

skin, leather, membrane (Gk. *chorion*).

chorion—one of the membranes round the unborn child in the mother.

chor'oid coat—the middle of the three coats forming the wall of the eye.

Note. *Chorea* (St. Vitus' Dance) and *choreography* (design of ballet) are derived from the Greek *choros*, dance.

CHORO-

country, land (Gk. *chōros*).

choro'logy—the scientific study of the geographical distribution of things, e.g. of plants and animals.

choro'metry—the art of surveying a country.

CHRO-, CHROM-, CHROMAT-, CHROMATO-, CHROMO-

colour (Gk. *chrōma*, *chrōmat-*).

meta'chro'sis—the ability shown by some animals, e.g. Chameleon, to change colour.

mono'chromat'ic light—light of only one colour (wavelength).

chromat'in—a substance in the nucleus of a cell which is readily stained by dyes.

chromo'some—"coloured body"—one of the rod-like bodies, carrying hereditary qualities, formed from chromatin particles when the nucleus is about to divide.

chromato'phore, chromo'plast—a body in the protoplasm of a cell which contains a definite pigment.

chromo'phore — "colour bearer" — a group of atoms (e.g. -N:N-) which, as a part of a molecule, is responsible for the colour of a dye.

chrom'ium—a metallic element many of whose compounds are brightly coloured.

In chemical names the root often refers to the element chromium.

chrom'ite—an oxide of chromium and iron used as a source of chromium. (Also called **chrome iron ore.**)

chrom'ate—a salt corresponding to **chromic acid** H_2CrO_4, e.g. potassium chromate K_2CrO_4.

CHRON-, CHRONO-

time (Gk. *chronos*).

chrono'meter—an accurate time-measuring instrument.

chrono'logy—the science of working out dates.

chron'ic — (disease) which is deep-seated and lasts a long time. (The word is often misused to mean severe, bad.)

syn'chron'ise—to cause to act (happen, etc.) at the same time. Also see SYNCHRO-.

CHRYS-, CHRYSO-

golden (Gk. *chrysos*).

A **Chrys'anthemum** is a "golden flower". A **chrysalis** is a "gold-coloured thing"—originally "the golden sheath of a butterfly", now extended to denote the state into which the larvae of many insects pass before assuming the adult state.

chryso'beryl — a yellow-green jewel stone.

chryso'lite—"golden stone"—the mineral olivine, especially the pale yellow varieties.

chryso'phyll—a yellow plant-pigment associated with chlorophyll.

CHYL-, CHYLO-

juice (Gk. *chylos*, from the root *chy-*, to pour).

chylo'caulous—having a succulent stem.

chylo'phyllous — having succulent leaves.

The root more frequently refers to **chyle**—a white, milky liquid formed by the digestive processes.

chylo'poesis—the formation of chyle. Also called **chylification.**

chyl'uria—the presence of chyle in the urine.

Note. *Chylos* and *chymos*, originally synonymous, were differentiated by Galen.

CHYM-, CHYME, -CHYMA
juice (Gk. *chymos*, from the root *chy-*, to pour).

chyme—the partly digested material which is passed from the stomach to the intestines. (Also see *chyle* above.)

chym'ification — the formation of chyme.

ec'chym'osis—"a state of juice out"—a coloured area of the skin caused by blood which has escaped from damaged blood vessels.

par'en'chyma—"juice in at the side"—thin-walled cells which make the soft tissue of the body of a plant or animal. (For further examples, see -ENCHYMA-.)

-CIDE
the killing of, the killer of (L. *caedo*, to kill, *-cido* in compounds).
Suicide is the killing of one's self.

fungi'cide—a substance used to kill fungi.

insecti'cide—a substance used to kill insects.

vermi'cide—a substance used to kill worms.

CILI-, CILIO-
an eye-lash (L. *cilium*), hence a hair-like organ.

cilia (sing. **cilium**)—small, lash-like projections of a cell which (e.g.) cause the locomotion of lowly animals or waft a current of water towards the animal.

cili'ary—pertaining to, or resembling, cilia.

Cilio'phora—a class of the Protozoa the members of which always "bear cilia".

CINQUE- See QUINQUE.

CIRCIN-, CIRCU-, CIRCUL-
Relating to a circle, in a ring (L. *circinatus*; *circus*; *circulus*).

circin'ate — "made round" — with leaves rolled up from tip to base (as the young leaves of Ferns).

circu'it—a closed, ring-like route.

circul'ar—in the shape of a circle.

circul'ate—to go round in a circle or closed route.

CIRCUM-
round, about (L. *circum*, *circum-*).
Circumstances are "the things standing around us". (Strictly, we can be in the circumstances but not, as many people say, under them.)

circum'ference—the line which forms a circle; the length of this line.

circum'nutation — "a swaying in a circle"—the movement of a tip of a growing stem so that it traces out a spiral in space.

circum'polar stars—stars which, as seen from a given place, move round the celestial pole (marked approximately by the Pole Star) and do not set below the horizon.

CIRRI-, CIRRO-, CIRRUS
a curl, a tuft of hair (L. *cirrus*).

cirrus—detached curl-like or feathery clouds at a great height; in zoology, any curled or tufted filaments.

cirro-stratus—a thin sheet of whitish, filamentous cloud at a great height.

Cirri'pedia—a class of marine crustacea (the Barnacles) which, when adult, possess a shell from between the valves of which the legs can be protruded like a curled lock of hair.

CIS-
on this side of (L. *cis*).
This prefix is seen in **cis-Atlantic**—on this side of the Atlantic. (Contrast *trans-Atlantic*.)

cis-lunar space—the space between the Earth and the orbit of the Moon.
In the name of a chemical compound *cis-* indicates that two atom groups are

situated on the same side of the axis of the molecule. (Contrast *trans-* (q.v.).) Example. There are two forms of the acid with the composition $(CH)_2$. $(COOH)_2$. In the formula shown below,

HC—COOH
||
HC—COOH

the two acid groups (-COOH) are on the same side of the molecule. This is the *cis*-form. (The acid is maleic acid.) In the *trans*-form (fumaric acid) the acid groups are on opposite sides of the molecule.

CIS-, -CISION
cut, a cutting (L. *caedo*, *caes-*, to cut, *-cido*, *-cis-* in compounds).
Precise means cut off short; a **decision** is a "cutting off".

in'cisor—one of the front cutting-teeth.

in'cision—a cutting in by surgery.

ex'cision—a cutting out (of a part) by surgery.

CITR-, CITRI-
a lemon (L. *citrus*).

Citrus—the genus of plants which includes the **citron**, lemon, orange, etc.

citr'ine — lemon-coloured; a yellow variety of quartz.

citri'form—lemon-shaped.

citr'ic acid—the acid in lemon juice. The salts are **citrates**.

CLAD-, CLADO-, -CLADE
a small branch (of a plant) (Gk. *klados*).

clad'ode, phyllo'clade — a flattened branch which resembles, and acts as, a leaf.

brachy'cladous — having very short branches.

clado'carpous—bearing fruit on small side-branches.

Clado'phora — "branch bearer" — a green Alga with a much branched body.

Clado'cera—"branched horns"—the order (class) of animals which comprises the Water Fleas.

CLAST-, -CLASE, -CLASIS
the breaking of, broken (Gk. *klasis*; *klastos*).

clast'ic rocks—rocks composed of broken pieces of older rocks.

pyro'clastic rocks—rocks formed from broken pieces of rocks of volcanic origin.

ortho'clase — a common form of potash feldspar characterised by its splitting in two directions at right-angles.

plagio'clase—other forms of feldspar (containing soda or lime) which split in two directions at a slanting angle.

osteo'clasis—the breaking of a bone so as to reset it in a better way.

osteo'clast—a cell which breaks down (destroys) bone.

CLAUSTRO-
an enclosure, a barrier (L. *claustra* (plural)).

claustro'phobia—fear of being shut in an enclosed space.

CLAV-, CLAVI-
a club, a cudgel (L. *clava*).

clav'ate, clavi'form—club-shaped.

clavi'cle—"a little club"—the collar bone (which is roughly club-shaped).

Clavi'ceps—"club head"—a fungus which causes disease of rye. Dark, club-shaped bodies replace the grains in the ear.

-CLE See -CULE.

CLEISTO-
closed (Gk. *kleistos*).

cleisto'gamic—with "closed marrying" —(flowers) which do not open and in which self-fertilization takes place.

CLIN-, -CLINE, CLINO-, KLINO-
to lean, to slope (L. *clino*, Gk. *klinō*).
This root is seen in **incline** (to lean, to lean towards) and **decline** (to lean away, to refuse).

de'clin'ation—the angular distance of a star from the celestial equator; the angle of dip of a magnetised needle.

clino'meter—an instrument for measuring the angle of slope of land.

geo'syn'cline—a large downward fold of the surface layers of the Earth.

thermo'cline—"heat slope"—the layer in a deep lake in which there is a sharp drop in temperature from that near the surface to that in the depths.

mono'clinic crystals—crystals whose shape is based on three unequal axes, two at right-angles but the third in a slant direction to one of the others.

tri'clinic crystals — crystals whose shape is based on three unequal axes all at slanting angles with one another.

klino'stat—an instrument by which a plant, usually held horizontally, may be turned slowly so as to study the effects of gravity and other stimuli on the direction of growth.

The Greek *klinē* meant a bed (on which one leans). Hence CLIN- sometimes refers to a bed, especially a sick-bed.

clin'ical—relating to lectures, treatment, etc., given at the sick-bed.

-CLUDE, -CLUSION

to shut, to close, (a closing) (L. *claudo*, *claus-*, to shut; *-cludo*, *-clus-* in compounds).

To **include** is to "shut in"; to **exclude** is "to shut out".

in'clusion — a particle of (usually) foreign matter embedded in a larger mass of matter, e.g. a fragment of foreign rock in an igneous rock, a particle of a non-metallic compound in iron or steel, a food particle within a cell.

oc'clude—"to shut against"—to block up a hole (e.g. a leaf-pore by a growth of tissue) or a duct (so that liquid cannot flow); to hold a gas on the surface of a solid.

oc'clusion (*Meteorology*)—a sector of warm air which has been squeezed up from the ground ("shut away") by the overtaking of a warm front by a cold front.

CNIDO-

a sting (Gk. *knidē*, a nettle).

cnido'blast—a stinging cell of an animal such as a Jellyfish.

cnido'phore—a projection bearing a group of cnidoblasts.

CO-

with, together (L. *cum*).

CO- was originally the form of COM- (q.v.) used before Latin stems beginning with *h* or a vowel. It is seen in **coalesce, coagulate, cooperate** and **cohesion.** It is now a living prefix, used with or without a hyphen, before words of various kinds and origins.

co'axial—having the same axis.

co'efficient—"producing an effect together"—a numerical multiplier used to find the magnitude of a physical property, e.g. the coefficient of expansion, of friction, etc.

co-planar—situated, or acting, in the same plane (flat surface).

co-enzyme—an additional substance necessary to the action of an enzyme.

co-polymer—a plastic made by polymerising (building a large molecule) two plastic-making substances together (e.g. vinyl acetate and vinyl chloride).

COCCI-, COCCUS

a berry, a small round thing (L. *coccum*; Gk. *kokkos*).

cocci'ferous—berry-bearing.

coccidi'osis — a disease of poultry caused by small round Protozoa. (*Coccidi-*, a diminutive of *cocci-*.)

coccus—a one-seeded portion formed by the breaking up of a dry fruit; a small round bacterium. (Plural **cocci.**)

pneumo'coccus—the round bacterium which causes pneumonia.

staphylo'cocci—cocci (round bacteria) which tend to form clusters ("like a bunch of grapes"). They are responsible for various kinds of inflammation and festering.

COCCY-, COCCYG(O)-, COCCYX

The **coccyx** is the tail-bone; it is formed by the fusion of the end vertebrae of the spine. So called because in Man it resembles the bill of a cuckoo. (Gk. *kokkyx*, *kokkyg-*, cuckoo.)

coccyg'eal—pertaining to the coccyx.

coccy'dynia, coccygodynia — severe pain in the coccyx.

coccyg'ectomy—the surgical cutting away of the coccyx.

COEL-, COELI-, COELOM, -COEL(E)
a hollow, especially the belly or other large cavity of the body (Gk. *koilos*).

coelom—the body cavity formed by the splitting of the middle layer of the germ cells of triploblastic (three-layered embryo) animals, i.e. in Earthworms and higher animals. **coelom'ate**—having a coelom.

coeli'ac—pertaining to the belly or abdomen.

Note. A **coeliac disease** is a wasting disease in children in which there is failure to absorb fat from the intestines.

Coel'enterata—animals with "hollow bellies"—the phylum (large group) of animals, including Hydra, Corals and Jellyfish, which consist essentially of two layers of cells surrounding a single body cavity which has only one opening.

blasto'coel(e) — the cavity within a blastula (q.v.).

haemo'coel(e) — the body cavity of Arthropods (Insects, Crabs, etc.) filled with blood.

coelo'spermous—having hollow seeds, having seeds which are hollowed out on one side.

Coel'acanth—a kind of fish which flourished in Devonian times (about 300 million years ago); there is one living representative. So called ("hollow spines") because the fin-rays (thin bones supporting the fins) were found to be bony only on the surface thus leaving internal cavities in the fossils.

COELO-
the heavens, the sky (L. *caelum* or *coelum*).

coelo'stat—an instrument (consisting of a turning mirror) which reflects light from the same part of the sky into a telescope in spite of the turning of the Earth.

COENO-, (CENO-)
held in common, shared (Gk. *koinos*).

coeno'cyte—"a shared cell"—a protoplasmic plant-body containing a number of nuclei but not divided into separate cells.

coeno'bi'um—"a shared life"—a group of algal cells living together as one organism.

coeno'gamete—a gamete (reproductive unit) containing a number of nuclei.

COL- See COLO- and COM-.

COLEO-
a sheath (Gk. *koleos*).

Coleo'ptera — animals with "sheath wings"—the order (group) of Beetles.

coleo'pter—an experimental type of aircraft consisting of a jet engine and a 'wing' shaped like a barrel round the body.

coleo'ptile—"a sheathed feather"—the first sheath-like leaf of a grass seedling.

coleo'(r)rhiza — "sheathed root" — a protective tip on the radicle (root) of the embryo of some flowering plants (e.g. Grasses).

COLL-, COLLO-
glue (Gk. *kolla*).

coll'oid—a substance "like glue"—a substance which readily assumes the colloidal state, i.e. a state between a true solution (in water) and a suspension of coarse particles. Substances with large molecules, e.g. starch, form a glue-like jelly instead of a true solution.

collo'type — a method of printing pictures from a gelatine plate.

coll'enchyma—"glue tissue"—tissue in Sponges consisting of cells in a jelly-like material; plant tissue (e.g. in a young stem) in which the cells have been strengthened by the addition of cellulose.

glyco'coll — "sweet glue" — amino-acetic acid, CH_2NH_2COOH, formed by boiling glue with acids or alkalis. It has a sweet taste.

COLO-, COL-
the **colon** (the greater part of the large intestine) (Gk. *kolon*).

col'ic—gripping pains in the belly.

col'itis—inflammation of the colon.

colo-enter'itis—inflammation of the colon and the small intestine.

COLOR-, COLORI-
colour (L. *color*).

colori'meter — an instrument for measuring the hue and brightness of colours.

colori'fic — producing a colour or colours.

-COLOUS
living in, growing on (L. *-colus*, from *colo*, to dwell).

calci'colous—(plant) which flourishes in soils which contain lime or chalk.

cauli'colous — (fungus) which grows on the stem of another plant.

cysti'colous—living in a cyst.

COLP-, COLPO-
bosom, womb (Gk. *kolpos*). Used especially to denote the vagina (the passage from the womb to the outside).

colp'itis—inflammation of the vagina.

colpo'scope—an instrument for inspecting the inside of the vagina.

colp'eurynter — an instrument for dilating ("broadening") the vagina.

colpo'rr(h)aphy—the sewing up of a torn vagina after a difficult childbirth.

COM-, CON-, COL-, COR-
This very common prefix (L. *cum*, *com-*) gives the meanings:
(1) with, together, as in **combine, contract** ("to draw together"), **compare**;
(2) very, completely, as in **combustion** ("complete burning").
It takes the forms COM- before *b*, *p* and *m*, COL- before *l*, COR- before *r*, and CON- before other consonants. Also see CO-.

con'centric—(circles) having the same centre.

con'tagious—(disease) which is passed on from one person to another when they "touch together".

con'genital—(qualities) with which one is born.

col'linear—(points) which are in the same straight line.

com'plement—that which completes or fills up, e.g. 40° is the complement of 50° because it completes the right-angle 90°.

com'minuted—broken up into small fragments (which are "very minute").

cor'rode—to "gnaw thoroughly"—to wear away, to destroy gradually, e.g. by rust, by an acid.

CONCH-, CONCHI-, CONCHO-
a shell (L. *concha*, from Gk. *konchē*).

concho'logy—the study of shells.

conch'oid—shell-like; a mathematical curve (first studied by Nicomedes) which has roughly the shape of a shell.

conchoidal fracture—breakage of glass or a bone which results in a shell-shaped edge or surface.

conchi'ol'in — the horny substance forming the outer layer of the shells of Molluscs.

CONDYL-
a knuckle, a rounded end of a bone which forms a joint with another bone (L. *condylus*; Gk. *kondylos*).

condyle—the rounded end of a bone.

condyl'ar, condyl'oid—(bone) having a rounded end.

CONI-, CONIC, CONICO-
a cone (the mathematical solid and the seed-body of pine/fir trees) (L. *conus*; Gk. *kōnos* (n.), *kōnikos* (adj.)).

conic'al—shaped like a cone; pertaining to a cone.

conic sections, conics—the curves (circle, ellipse, parabola, hyperbola) obtained by the cutting of a cone by a plane.

conic'oid—a surface such that every section made by a plane is a conic.

conico-cylindrical—roughly cylindrical but with a tendency towards a cone shape.

Coni'ferae—the chief order of the Gymnospermae. It comprises those plants, e.g. fir, pine, spruce, etc., which bear cones.

CONIDI-, CONIDIO-
A **conidium** is a one-celled spore developed at the end of a fungal branch and never enclosed in a container. The name means a "small dust thing" (from Gk. *konis*, dust). (Also see KONI-.)

conidi'al—pertaining to conidia.

conidio'phore—a fungal branch which bears conidia.

CONTRA-

opposite, against (L. *contra*, *contra-*). This prefix is seen in **contradict** ("to say against").

contra'ceptive—an agent which prevents conception (fertilization of the ovum by a spermatozoid).

contra-rotating—rotating in opposite directions (e.g. two aeroplane propellers).

COPRO-, -COPROS

dung, faeces (waste matter passed out from the bowel) (Gk. *kopros*).

copro'lite—"dung stone"—fossilised faeces used as a fertilizer for the soil.

copro'philous—(liking) growing in or on dung.

copro'phagous — (beetle) which eats dung.

ornitho'copros—the excrement of birds.

copro'lalia—a speaking of filthy words (e.g. by the insane).

COR- See COM-.

CORI-

leather (L. *corium*).

cori'aceous—leathery, e.g. coriaceous leaves.

CORN-, CORNI-, CORNEO-

horn (L. *cornu* (n.), *corneus* (adj.)). A **corn** on the foot is a horny thickening of the skin; a **cornet** (wind instrument) is a "little horn".

corn'ate—shaped like a horn; bearing horns.

corni'form—shaped like a horn.

Capri'corn—the "horned goat"—the star-group resembling a goat and forming the 10th division of the zodiac.

cornea—L. *cornea tela*, horny tissue—the transparent, horny, front part of the eye.

CORNEO- may refer to horn or to the cornea.

corneo-calcareous—chalky with a mixture of horn.

corneo-iritis—inflammation of the eye affecting both the cornea and the iris.

CORO-, CORON-, CORONA

a crown, a garland (L. *corona*). A **coronation** is the crowning of a sovereign.

corona—this word has various applications in science, e.g. a ring of light round the Sun or Moon when viewed through a thin haze, a tufted appendage on the inner side of petals (as in Campion), the name given to various crown-like parts of the body.

Coron'atae—a class of Jellyfish which have lobed margins (like crowns).

coro'lla—"a small crown, a garland"—the general name for the whole set of petals of a flower, often coloured.

corollary — (L. *corollarium* — money given for a garland, a gift)—a proposition (or fact) which follows "as a gift" from a previous proposition which has already been demonstrated or proved. In medical terms **coronary** refers to the heart (not the head) e.g. **coronary thrombosis** (the formation of a clot in one of the arteries of the heart).

CORP-, CORPUS

a body, substance (L. *corpus*). *Corpus* (plural *corpora*) occurs in its original Latin form in the names of certain parts of the body, e.g. **corpus callosum** ("thick-skinned body")—the band of nerve fibres joining the two sides of the brain.

corpus'cle—a small body, a particle.

CORTEX, CORTIC-, CORTICO-

bark, rind, cork (L. *cortex*, *cortic-*).

cortex—outer layers, e.g. the outer layers of an organ, the soft tissue between the skin and the stronger central part of a plant stem. Adj. **cortic'al**.

cortic'ate—having bark; bark-like.

de'cortic'ated nuts—nuts from which the shells have been removed.

corti'sone—a complex drug (for relieving arthritis) derived from **corticosterones** which are produced by the outer shell of the suprarenal glands.

COSM-, COSMO-
the world, the universe (Gk. *kosmos*).

cosmo'logy—the scientific study of the universe as a whole.

cosmo'gony—the science which concerns the origin of the stars, planets, etc.

cosm'ic rays—radiation which enters the Earth's atmosphere from outer space.

The **microcosm** is the "small world", i.e. Man, as distinct from the **macrocosm** ("large world") or universe. **Microcosmic salt** — a phosphate of sodium and ammonium—was originally obtained from man's urine.

COST-
a rib, the ribs (L. *costa*).

cost'al—pertaining to the ribs.

sub'costal—below the ribs.

cost'algia—pain in the ribs.

COTYL-, COTYLEDON
Gk. *kotylē*, a hollow thing, a small cup; *kotylēdōn*, a cup-shaped hollow such as a sucker on the feelers of a polyp or a socket of a joint.

cotyl'oid—cup-shaped. (The description is applied especially to the socket of the hip joint and to the cavity of a similar joint in an insect's leg.)
In botany the term **cotyledon** has come to denote a leaf of the embryo in the seed of a flowering plant—a striking example of a word which is now used with a substantially changed meaning.

cotyledon'ous—having a cotyledon (or cotyledons).

Di'cotyledons—flowering plants whose seed-embryos have two cotyledons.

hypo'cotyl—the part of the stem of a seedling between the cotyledon(s) and the young root.

COUNTER-
against, opposite to (L. *contra*, *contra-*).

counter'act—to act against, to defeat, to neutralize.

counter-clockwise — in a direction opposite to that in which the hands of a clock turn.

counter-irritant—a substance used to produce mild irritation and so counteract more serious pain.

CRANI-, CRANIO-, CRANIUM
The **cranium** (L. *cranium*, from Gk. *kranion*) is that part of the skull which surrounds and protects the brain.

peri'cranium—the fibrous layer of tissue which surrounds the cranium.

chondro'cranium — the cartilaginous brain-case of lower vertebrate animals.

Crani'ata—the large group (subphylum) of animals which possess a backbone and a skull; the Vertebrata.

cranio'tomy—the cutting (or breaking) of the skull of an unborn child.

CREAS-, CREAT-
flesh, meat (Gk. *kreas*, *kreat-*).

creat'ine—an organic substance found in flesh (muscle).

pan'creas—an organ, lying behind the stomach, which pours digestive juices into the small intestine (duodenum). Called the sweetbread when used as food. (The word means "all flesh". An explanation was given by Banister in 1578: "(The pancreas) is all carnous or fleshy . . . it is made and contexted of glandulous flesh".)

crea'sote, creosote — "saviour of flesh"—an oily liquid obtained from coal-tar, with strong antiseptic qualities. Hence **cresols**—hydroxy'toluenes, $CH_3.C_6H_4.OH$, which are used (e.g. in Lysol) as antiseptics.

CRESC-, -CRETE, -CRETION
(1) to grow (L. *cresco*, *cret-*).

cresc'ent—the shape of the growing Moon (between New and First quarter).

ex'cresc'ence—an abnormal or diseased outgrowth on a plant or animal body.

ac'cretion—growth by the adding of new material externally.

con'cretion—"a growth together" —a collection of organic and stony matter in a body organ.

con'crete — a composition of

cement, gravel, etc., used for building. (Also other meanings.)

(2) The elements -CRETE and -CRETION also appear as derivatives of the L. verb *cerno*, *cret-*, to sift, to separate.

dis'crete — "separated apart" — separate, individually distinct.

ex'crete—to get rid of poisonous or waste substances from out of a cell, tissue or organism.

se'crete—to extract, form or collect a substance and discharge it for the service of the organism or for excretion, as by a gland.

CRET(A)-
chalk (L. *creta*).

Cret'aceous period—a period (about 135–60 million years ago) when the chalk rocks (e.g. in S.E. England) were formed.

CRIB-, CRIBRI-
a sieve (L. *cribrum*).

cribri'form—like a sieve, perforated with small holes.

crib'ellum—"a little sieve"—a perforated plate through which a spider draws out its thread.

CRIN-, CRINO-, -CRINE
(1) a lily (Gk. *krinon*).

crin'oid—like a lily, shaped like a lily.

Crin'oidea—the Sea Lilies (a class of the Echinoderms). Many fossil forms are known.

(2) to distinguish, to separate, to put apart (Gk. *krinō*).

exo'crine gland—a gland which "separates" a substance and passes it "out" into a body cavity or on to the surface of the body.

endo'crine gland—a gland which passes its secretions directly into the blood. **endocrinology**—the study of endocrine glands and of their secretions.

(3) hair, fibre (L. *crinis*).

crin'ite—hairy.

crino'line—(originally) a stiff fabric made of horsehair.

CRUCI-
a cross (L. *crux*, *cruc(i)-*).
A matter is **crucial** when a decision must be taken, i.e. when one of the "cross roads" must be chosen. A **crucifix** (L. *cruci fixus*) is one fixed to a cross.

cruci'ate, cruci'form—shaped like a cross.

Cruci'ferae "the cross-bearers"—the family of flowering plants (including Wallflower, Cabbage) whose flowers bear four petals like a cross.

A **crucible** is a familiar piece of chemical apparatus but the origin of its name is obscure. The Romans called a melting-pot a *crucibulum*. One feels that this name must be related in some way to *crux*, *crucis*, a cross, but the connection is not clear.

CRY-, CRYO-
frost, ice (Gk. *kryos*).

cryo'genic—relating to the production of ice or of low temperatures.

cryo'plankton — plant life (mainly Algae) which lives on the surface of snow and ice.

cryo'lite—"ice stone"—the mineral form of aluminium fluoride, so called from its appearance.

hyper'cry'algesia — abnormally increased sensitivity to cold.

CRYPTO-, KRYPTO-
hidden (Gk. *kryptos*).
A **crypt** is a hidden underground cell; a **cryptogram** is a hidden message (e.g. in cipher).

crypto'branchiate — having the gills hidden (concealed).

Crypto'gams—plants with "hidden marriage"—the large group of plants, including Ferns, Mosses, Fungi, Algae, which do not have flowers and do not reproduce by means of seeds.

crypto'mere—a character which is inherited but which remains "hidden", i.e. is not apparent.

crypto'crystalline—consisting of very minute ("hidden") crystals.

krypton—a chemical element, one of the rare gases in the air, which for a long time was undetected.

CRYSTALLO-

The Greek word *krystallos* basically meant ice (cf. CRYO-). It also seems to have been used to denote certain **crystalline** minerals which, it might have been thought, had been formed by freezing. In scientific terms CRYSTALLO- refers to **crystals.**

crystallo'graphy—the study of the forms, structures and properties of crystals.

crystallo'geny — the formation of crystals.

crystallo'gram—a photograph of the X-ray pattern produced by a crystal.

CTEN-, CTENO-

a comb (Gk. *kteis, kten-*).

cten'oid—shaped like a comb, bearing comb-like structures.

Cteno'phora—the "comb bearers"—a class of marine coelenterate animals (e.g. the Sea Acorn) which bear eight rows of swimming plates.

cten'idium — "a small comb" — any small comb-like structure, especially a type of gill of invertebrate animals consisting of a central axis with a row of filaments on each side.

CUB-, CUBI-

(1) The well-known words **cube, cubic, cuboid** (like a cube, brick-shaped) are derived from the Greek *kubos*, a cube or die.

(2) The Latin verb *cubo*, *cubit-*, to lie down, has given rise to **incubate** (to lie upon, or sit upon, eggs to hatch them) and **cubicle** (essentially a small sleeping-room).

-CULA, -CULUM, -CULUS, -CULE, -CLE

The Latin suffix -CULUS (-A, -UM) forms a diminutive, i.e. a word denoting a small version of the thing indicated by the main part of the word.

Auri'cula — a kind of cultivated Primula the leaves of which are supposedly like "little ears".

oper'culum—"a little cover"—a gill-cover.

fasci'culus—"a small bundle" of (e.g.) nerve fibres or muscle fibres.

In English the suffix becomes -CULE and this, in turn, often becomes -CLE.

animal'cule—a microscopic animal.

sac'cule—"a little sac (bag)".

mole'cule — "a little mass" — the smallest particle of a chemical substance which can exist independently, usually consisting of two or more atoms (like or unlike) joined together.

denti'cul'ate—"having little teeth", e.g. a leaf with a denticulate edge.

parti'cle—"a small part".

corpus'cle—"a small body"

ossi'cle—"a small bone".

ventri'cle—"a small belly"—a hollow part of an organ (especially one of the two lower chambers of the heart).

-CUMBENT

lying (L. *cumbens*).

re'cumbent—reclining, lying back.

pro'cumbent—(plant) lying forward with its stem on the ground.

CUMUL-, CUMULO-

a heap (L. *cumulus*).

This root is seen in **accumulate** (to "heap together", to amass, to store). An electric **accumulator** stores electricity.

cumul'ative—increasing by successive additions.

cumulus—a thick cloud (like "a heap") with a well-marked rounded top and (usually) a horizontal base.

cumulo'nimbus—a great mass of piled-up cloud, sometimes ragged at the top, dark and irregular at the bottom; a thunder-cloud.

CUNE-, CUNEI-

a wedge (L. *cuneus*).

cune'ate—wedge-shaped.

cunei'form, cuniform—in the form of a

wedge (or wedges). (Used especially to describe the ancient Babylonian and Assyrian writing.)

CUPR-, CUPRO-

copper (L. *cuprum*).

The chemical symbol for copper, **Cu,** is derived from the Latin name.

cupr'ic oxide—black copper oxide, CuO. **cupr'ous oxide**—red copper oxide, Cu_2O. (See -IC and -OUS.)

cupr'ite—a common ore of copper.

cupro-nickel—an alloy of copper and nickel used, e.g. for turbine blades and condenser tubes.

CURR-, CURS-

to run (L. *curro*, *curs-*).

An **excursion** is a "running out"; a **current** is that which "runs" (flows) along.

con'curr'ent—"running together", e.g. concurrent lines; existing together.

pre'curs'or—that which "runs before" and so forewarns.

curs'or—the movable glass slide of a slide-rule.

curs'orial—having limbs adapted for running, e.g. cursorial birds.

Curs'ores—a general name for birds (e.g. Ostrich) which can run.

CUT-, CUTI-

the skin (L. *cutis*).

cut'aneous—pertaining to the skin.

sub-cutaneous—just below the skin.

cuti'cle—"a little skin"—the outer layer of the skin; a thin waterproof layer over the skin of a leaf. Adj. **cuticular**.

CYAN-, CYANO-

dark blue (Gk. *kyanos*).

cyan—the blue-green colour formed when red is taken from white light. (Also called Minus Red.)

cyanines — a group of blue dyes. (**Cyanin** is the colouring matter of the Cornflower.)

haemo'cyanin—a blue pigment, containing copper, in the blood of Crustaceans and Molluscs.

cyano'sis—blueness of the skin due to lack of oxygen in the blood.

Cyano'phyceae—the group of Blue, green Algae. (Also called the Myxophyceae.)

Prussian Blue and similar pigments are derived from **cyanides**—compounds containing the characteristic group of atoms -CN. In chemical names CYAN(O)- indicates the presence of this group of atoms.

cyan'ide—a compound formed by the combination of the -CN group with an element (or with a group of atoms), e.g. potassium cyanide KCN.

hydro'cyanic acid — a solution of hydrogen cyanide (HCN) in water. Also called Prussic acid.

cyano'gen — a colourless, poisonous gas, $(CN)_2$, with a smell of almonds.

cyano'acetic acid—an acid in which one hydrogen atom of acetic acid has been replaced by the cyano-group, $CH_2(CN).COOH$.

CYBERNET-

a pilot, a director (Gk. *kybernētēs*).

cybernet'ics—the study and theory of instruments which automatically control the working of a machine (e.g. a 'robot' mechanism for controlling the flight of an aeroplane). (The study has been extended to include living mechanisms, e.g. the human brain.)

CYCL-, CYCLO-

a circle, a ring (Gk. *kyklos*).

A **bicycle** has two circular wheels; a **cycle** of events is a series which comes round over and over again. An **encyclopaedia** provides "all round education".

cyclo'meter — an instrument for measuring the turns of a wheel (e.g. that for measuring the distance travelled by a bicycle).

cycl'oid—the curve traced out by a point on the circumference of a circle which rolls along a straight line. (Also see **epi'cycloid.**)

mega'cycles—millions of cycles (e.g. of alternating current).

cyclo'coelic—having a spirally coiled intestine.

Cyclo'stomata—the group of fish with round, jawless mouths, e.g. Hagfish, Lampreys.

cycl'ic compounds—compounds whose molecules consist of a ring (or rings) of atoms, e.g. benzene. **hetero'cyclic compound**—a compound whose molecule is based on a ring of atoms which are not all alike, e.g. pyridine which contains a ring of five carbon atoms and one nitrogen atom.

In the names of chemical compounds, the prefix *cyclo-* indicates a closed ring (as distinct from a chain) of atoms.

Example. *Hexane* is a paraffin hydrocarbon in which six carbon atoms are joined in a chain: $CH_3.CH_2.CH_2.CH_2.CH_2.CH_3$. *Cyclo*-**hexane** is a hydrocarbon in which six carbon atoms are joined in a ring as shown. (Note. This is not the same as benzene which has only one hydrogen atom joined to each carbon atom.)

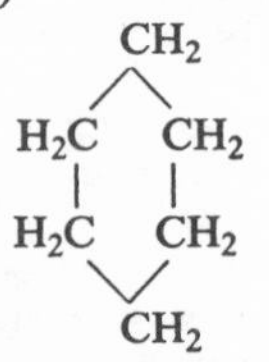

CYMA, CYMO-, KYMO-

the swell of the sea, a wave (Gk. *kyma*).

cyma—a curved moulding consisting partly of an outward, and partly of an inward, curved circle.

cymo'phane—"wavy appearance"—a mineral (a form of chrysoberyl) which gives a wavy optical effect (like that of a cat's eye).

cymo'meter—an early form of instrument for measuring wavelengths (e.g. of radio waves).

kymo'graph — an instrument which draws a wavy line to show heart-beat, breathing, etc.

CYN-, CYNO-

a dog (Gk. *kyōn*, *kyn(o)-*).

cyno'podous—having dog-like claws.

cyno'phobia—dread of dogs.

Cyn'oidea—the animals of the dog family, e.g. Dog, Wolf, Fox. (Also called the Canidae.)

Cynosure ("dog's tail") is the name sometimes given to the star-group Ursa minor—the group which includes the Pole-star. Hence, figuratively, a cynosure is something that serves to direct, a centre of attraction.

CYST, CYSTI-, CYSTO-

a bladder, a bag (Gk. *kystis*).

cyst—a bladder-like structure (e.g. containing the products of inflammation).

cysti'form—in the form of a cyst.

nemato'cyst—the stinging cell (a bag of poisonous liquid) of a simple animal such as a Jellyfish.

cysto'spore—a spore, normally motile, which is **encysted,** i.e. enclosed in a protective coat.

In medical terms the root generally refers to the urinary bladder.

cyst'itis—inflammation of the urinary bladder.

cysto'scope—an instrument for inspecting the inside of the urinary bladder.

cysto'stomy—the making of an opening in the bladder by surgery.

CYTO-, -CYTE

a hollow vessel (Gk. *kytos*). Used especially to denote a plant or animal cell. (Plant cells were originally thought to be empty.)

cyto'logy—the study of the structure and functions of cells.

cyto'genesis—the production and development of cells.

cyto'plasm—the protoplasm of a cell other than that of the nucleus.

cyto'lysis — "cell loosening" — the breaking down and destruction of cells. **haemo'cytolysis**—the breaking down of red blood cells.

erythro'cyte—a red blood cell. **leucocyte**—a white blood cell.

phago'cyte—a cell which can change its shape and engulf ("eat") foreign bodies.

D

d-
An abbreviation for dextro-rotatory (q.v.), e.g. *d*-**lactic acid.**

DACRYO-
a tear (of the eye) (Gk. *dakryon*).
dacry'oma—stoppage of the tear duct.
dacryo'cyst'itis—inflammation of the tear sac.

DACTYL-
a digit, i.e. finger or toe (Gk. *daktylos*).
dactyl'ate—having digits.
penta'dactyl—having five digits on each hand (or foot) as has Man.
poly'dactyl'ism — a state of having more than the usual number of digits.
Ptero'dactyl — "winged fingers" — an extinct winged reptile.
The **dactyl** (the metrical foot—⏑⏑) is also derived from this root.

DASY-
hairy, thick with hair, rough with hair (Gk. *dasys*).
dasy'phyllous—having hairy leaves; having small, close-set leaves giving a hairy appearance.
Dasy'ure—"rough hairy tail"—a small marsupial animal of Australia and Tasmania.
dasy'paedes—"hairy children"—birds which when hatched have a complete covering of down.

DE-
This prefix (from L. *de*, *de-*) has a range of meanings:
down (e.g. **depend**=to hang down (from)),
away (e.g. **deflect**=to bend away),
completely, thoroughly (e.g. **denude**= to make completely nude),
and in the sense of reversing a previous action, condition, etc. (e.g. **decapitate**=to take away the head of, **decontaminate**=to cause not to be contaminated).
The prefix is widely and freely used in this last sense.

(1) down, away.
de'posit—to "place" (lay) down (eggs, soil particles, etc.).
de'generate—to become less strong or less complicated.
de'ciduous — (trees) whose leaves fall in the winter.
de'clination—the angle of slope up or down, e.g. the angular distance of a star from the celestial equator, the angle of dip of a magnetised needle.

(2) to take away, to cause not to be.
de'cerebrate—to remove the cerebrum (brain) of an animal; having had the cerebrum removed.
de'corticated nuts — nuts from which the shells have been removed.
de'hydrate—to remove water from.
de'humidify—to cause not to be humid, to remove moisture (e.g. from air) artificially.
de'vitrification — the loss of the transparent, glassy quality of glass and minerals due to the formation of tiny crystals.
de'magnetise—to cause to cease to be a magnet.
de'carbonise—to remove carbon from (e.g. the cylinder of an internal combustion engine).

DECA-, DECEM-, DEKA-
ten.
Although the Latin *decem* is familiar to us in December (the tenth month of the Roman year), it is not often found in English words. The Greek *deka* (which often becomes DECA- in English) is more common.
decem'fid—cleft (divided) into ten parts or segments.
deca'hedron—a solid figure with ten faces.
Deca'poda—an order of Crustacea (Crabs, Shrimps) which have ten legs; an order of Cephalopoda (e.g. Cuttlefish, Squids) which have ten arms.
deka'metre—a length of ten metres.

DECI-, DECIM-
tenth.
The Latin *decimus* (tenth) has given rise

to **decimal** (relating to tens and tenths) and **'decimate** (to select by lot every tenth man for punishment). The shorter prefix DECI- is particularly used in forming the names of fractional units.

deci'metre—a tenth of a metre.

deci'gramme—a tenth of a gramme.

deci'normal solution—a solution which is one-tenth the strength of a 'normal' solution (a solution of standard strength as defined by chemists).

DELT-, DELTA

The capital form of delta, the fourth letter of the Greek alphabet, is written Δ. The **delta** of a river (e.g. the Nile) is a triangular area of river-borne material deposited at the mouth of the river.

delt'oid—having the approximate form of an equilateral triangle, e.g. a deltoid leaf, the deltoid muscle of the shoulder.

DEM-, DEMO-

the people (Gk. *dēmos*).

Democracy is government by the people; a **demagogue** is a leader of the people (especially a political agitator who appeals to the greed or prejudices of the masses).

epi'dem'ic — "over the people" — an outbreak of an infectious disease which spreads widely among the people in a region.

pan'dem'ic—affecting "all the people" —an epidemic which affects a wide area such as a whole country or continent.

demo'graphy — the study of the numerical information about the population, e.g. births, deaths, diseases.

DENDR-, DENDRO-, -DENDRON

a tree (Gk. *dendron*).

Rhodo'dendron — "the rose tree" — a genus of large shrubs which have (mainly) red flowers.

dendr'oid, dendr'iform—tree-like, tree-shaped.

dendr'ite—a small branch of a nerve cell by which it joins up with another nerve cell; a crystalline growth of branched formation. **dendritic markings** —tree-like markings, e.g. on rocks.

dendro'lite—"tree stone"—a fossilised tree (or part of tree).

Lepido'dendron—"scaly tree"—a giant Lycopod which attained tree-like proportions in Carboniferous (coal measure) times. (The modern Club-mosses are diminutive survivors of the plant-group.)

DENT-, DENTI-

a tooth (L. *dens*, *dent-*).

This root is well-known in **dentist** (person who deals with diseases of teeth, etc.) and **dentifrice** (powder or paste used for cleaning the teeth). The name **dandelion** comes from the French *dent de lion*, a lion's tooth.

dent'ate—having a toothed edge.

denti'form—of the form of a tooth.

dent'ition — the kind, number and arrangement of the teeth of an animal.

dent'ine—the bone-like substance of which teeth are composed.

denti'lingual—pertaining to, or formed by, the teeth and the tongue.

denti'cle—"a small tooth", any small tooth-like structure.

DERM-, DERMAT-, DERMATO-

the skin (Gk. *derma*, *dermat-*).

dermat'itis—an inflammation of the skin.

hypo'derm'ic injection—the putting of a drug into the body by means of a needle passed into the under-layer of the skin.

epi'dermis—the outside layer of skin.

ecto'derm—the outside layer of cells of a gastrula (an early stage of a developing embryo); the tissues directly formed from this layer.

pachy'derm—an animal with a thick skin, e.g. an elephant. Adj. **pachydermatous.**

taxi'dermy—the art of preparing and mounting animal skins.

dermato'phyte — a parasitic fungus which causes skin diseases.

DES-

This prefix occurs in modern chemical nomenclature with the meaning of 'without'. Thus DESOXY- means "without oxygen" or "with less oxygen than".

des'thio'biotin—a compound similar to biotin (a vitamin) but with less sulphur (*thio*) in it.

DEUTER-, DEUTERO-, DEUTO-
second (Gk. *deuteros*).
Deutero'nomy is the second book of the law in the Old Testament.

deuter'ium—the second form of hydrogen ('heavy hydrogen'). The atomic nucleus (a **deuteron**) consists of a normal hydrogen nucleus (a proton) and a neutron.

deutero'gamy—"second marriage"—any process which takes the place of normal fertilization.

deuto'cerebron—in higher Arthropods (e.g. Insects), the fused nerve cells of the second division of the head forming part of the so-called brain.

DEXTR-, DEXTER-, DEXTRO-
right hand (L. *dexter*).
As most people are able to perform tasks better with the right hand than the left, the root DEXTR(O)- is used in general words to denote manual skill. Thus **dexterity** may be defined as manual skill or neatness of handling, and to be **dext(e)rous** means to be neat-handed. A person who is **ambidextrous** is able to use both hands equally well—as if he has two right hands.

In scientific terms the original meaning of right-handedness is retained.

dextrorse—(plant stem, shell, etc.) which turns or twists to the right.

dextro'cardia — an unusual state in which the heart lies on the right-hand side of the chest.

dextro-rotatory substance—substance (e.g. ordinary glucose) whose solution turns (twists) the plane of polarisation of a beam of polarised light to the right.

dextro-lactic acid—a form of lactic acid, as found in muscle tissue, which is dextro-rotatory.

dextrose—an alternative name for glucose. **dextrans** — complicated substances built up from dextrose units.

dextrin—a gummy substance formed as an intermediate product in the breakdown of starch into maltose and dextrose.

DI-
(1) two, twice (Gk. *dis*, *di-*).

Di'ptera—the order of insects (e.g. Housefly, Gnat), which have two transparent flying wings.

di'cotyledonous plant—a plant (belonging to one of the two great groups of flowering plants) whose embryo within the seed has two cotyledons (seed-leaves).

di'crotic pulse—pulse (or heartbeat) having a double beat, as in the case of some fevers.

di'ode—a radio valve with two electrodes.

In chemical names the prefix denotes the presence of two atoms (or atom groups) of the same kind in the molecule.

carbon dioxide—the oxide of carbon which contains two oxygen atoms in the molecule, CO_2. (Contrast carbon monoxide, CO.)

di'chlor'benzene — a compound formed by replacing two of the hydrogen atoms of benzene (C_6H_6) by chlorine atoms, $C_6H_4Cl_2$.

di'methyl ether—an ether which contains two methyl (CH_3-) groups in the molecule, $CH_3.O.CH_3$.

di'chromate—A *chromate* is a salt of chromic acid which may be regarded as a compound of one molecule of chromium trioxide with water (i.e. CrO_3+H_2O). A *dichromate* is a salt corresponding to an acid composed of two molecules of chromium trioxide with water. ($2CrO_3+H_2O$, or $H_2Cr_2O_7$), e.g. potassium dichromate $K_2Cr_2O_7$.

(2) A form of DIS- (q.v.) used somewhat inconsistently before certain consonants, e.g. **divert** ("to turn away"), **digress** ("to walk away"), **dilate** ("to widen out").

(3) A form of DIA- (see below) used before a vowel.

di'actinic—transparent to actinic rays.

di'optric — "through seeing" — transmitting light by refraction.

di'electric — a medium through which electric forces can pass (but which is not a conductor).

DIA-
through, right through, across (Gk. *dia*, *dia-*).

Diameter (the measure or line across a circle) and **diagonal** (a line across a figure from one corner to another) are common examples of the use of this prefix. A **diagram** is a "drawing through" a thing, i.e. essentially a cross-section.

dia'gnosis—"knowing through"—the identification of a disease (etc.) from the symptoms and other evidence.

dia'rrhoea—"a flowing through"—an excessive flow of loose matter from the bowels.

dia'scope—an instrument for projecting pictures by sending light through a transparent slide or film.

dia'thermanous–"letting heat through" —having the property of transmitting radiant heat.

dia'phragm — "a hedge across" — a membrane across the body of a mammal separating the chest from the abdomen; a membrane which forms part of an instrument (e.g. of a microphone).

Note. The word *diamond* may be traced back to late Latin *diamas*, *diamantem*, from the earlier *adamas*, *adamantem*, itself from the Greek word meaning "untameable", i.e. unyielding, hard. (Compare the English word *adamant*.) The name, with varying spellings (e.g. *adimantem*), was applied, in turn, to various hard substances, e.g. corundum and diamond.

DICHO-
in two, apart (Gk. *dicha*).

dicho'tomy—a splitting into two, e.g. of a group of things for the purpose of finer classification, of the apical (tip) cells of a branch to give rise to two branches, etc.

dicho'gamy—"separate marrying"—the state of having the male and female parts of a flower ripen at different times so that self-fertilization is impossible.

DICTYO-
a net (Gk. *diktyon*).

dictyo'spore — a many-celled spore divided into segments by a network of walls.

dictyo'stele—a stele (pipe-like tissue in a stem or root) in the form of a network, as in certain Ferns.

Dictyota—a much-branched, ribbon-like brown seaweed.

DIDYM-, DIDYMIS, -DYMIUM
(1) twin (Gk. *didymos*).

didym'ous — occurring in pairs; divided into two parts (as are some fruits).

didym'ium—the name given to a supposed chemical element separated from lanthanum in 1841, so called because of its 'twin' association with lanthanum. In 1885 it was discovered that didymium was itself a mixture of two elements; these were named **praeso'dymium** (from the pale green colour of its oxide) and **neo'dymium** ("new dymium").

(2) the testicles (Gk. A special use of the plural of *didymos*).

peri'didymis—the fibrous coat of a testicle.

epi'didymis—a long, narrow structure attached to a testicle and consisting chiefly of the coils of the duct through which the semen passes. Hence **epididym'itis** (inflammation of) **epididym'ectomy** (surgical removal of), etc.

DIF-
A form of DIS- (q.v.) used before *f*, e.g. **different** ("bearing (tending) away"), **diffuse** ("to pour (spread) out").

DIGIT-, DIGITI-
a finger or toe (L. *digitus*).

digit'ate — divided into parts like fingers; having divided fingers or toes.

digiti'grade—walking on the toes (not the sole) of the foot.

Digitalis—the wild Foxglove, a well-known plant whose purple flowers are shaped like finger-covers. **digitalis**—a drug prepared from the leaves of the Foxglove; it contains **digitalin** and related compounds.

In mathematics a **digit** is a number below 10, i.e. 0 to 9 (if the cipher 0 is included). So called from the habit of counting on the fingers.

DIN-, DINO-

terrible (Gk. *deinos*).

Dino'ceras—"terrible horns"—a huge extinct animal which apparently had three pairs of horns.

Dino'saurs — "terrible lizards" — a group of fossil reptiles of huge size or strange form.

Din'ornis—an extinct ostrich-like bird.

DIPL-, DIPLO-

double, twofold (Gk. *diploos*).

A **diploma** was originally a folded (double) sheet of paper, e.g. a letter of recommendation. The words **diplomatist** (originally a person bearing official documents) and **diplomacy** are derived from this.

diplo'blastic — having two primary layers in the embryo.

diplo'cocci — cocci (round bacteria) which associate in pairs.

dipl'opia—seeing things double.

dipl'oid — having a double set of chromosomes (hereditary bodies) in the nucleus as in body cells. (Reproductive cells are haploid; they have a single set of chromosomes.)

DIS-

This very common prefix (from L. *dis-*) is similar in meaning and application to DE-:

apart, asunder, away (as in **dissolve, disperse**),
completely, thoroughly (as in **distort**),
not, reversal of (as in **disease, discolour**).

The prefix takes the forms DI- or DIF- before certain letters.

dis'sect—to cut apart, i.e. into many pieces.

dis'locate—to put (e.g. a bone) out of place.

dis'infectant—a substance used to kill bacteria on infected things.

dis'sociation — the breaking up of associated molecules (groups of molecules) into simpler parts.

DISC, DISCO-

a disc (Gk. *diskos*).

The familiar spelling **disc** is gradually being replaced by the spelling **disk**—a curious example of a reversion to the *k* of the Greek.

disc'oid—round and flat, shaped like a disc.

disco'dactylous—having the ends of the digits flattened and rounded to form sucking discs.

Disco'mycetes—a group of Fungi the members of which bear asci (club-shaped spore-forming organs) in a cup-shaped 'fruit'.

DODECA- See DUODEC-.

DOLICHO-

long (Gk. *dolichos*).

dolicho'cephalic—long-headed, having the breadth of the skull less than $\frac{4}{5}$ of the length.

dolicho'colon — an excessively long colon (large intestine).

-DON, -DONT See ODONT-.

DORS-, DORSI-, DORSO-, -DOS

the back (L. *dorsum*).

We **endorse** a document by signing on the back of it.

dors'al—pertaining to the back or to the back of any part.

dors'algia—pain in the back.

dorsi'grade—walking on the backs of the digits (as do Sloths).

dorsi'ventral—(part of a plant, e.g. a leaf) which has different back and front sides.

dorso-lateral—pertaining to the back and the sides.

-DOS is a variant of this root. It is seen in

intrados and **extrados**, respectively the inner and outer (upper) curves of an arch.

DROM-, -DROME

The Greek *dromos* was a course, especially a running-course. A **hippodrome** ("horse course") was basically a course for chariot racing though now it is a more general place of entertainment. A **dromedary** is an Arabian camel used for riding, e.g. by messengers.

aero'drome — "a course for aeroplanes", i.e. a place where aeroplanes land, are maintained, stored, etc.

cata'dromous — (fish) which moves down the course of a river, towards the mouth, e.g. to spawn.

pro'dromal — "running before" — (symptoms) forewarning of a coming disease.

syn'drome—"a running together"—the set of symptoms and signs which are characteristic of a disease.

palin'drome—"a running back again" —a word, number, etc., which reads the same backwards as forwards, e.g. *level*, 1221. **palindromic disease**—a disease which apparently ceases and then comes on again.

DROS-, DROSO-

dew (Gk. *drosos*).

droso'meter—an instrument for measuring the amount of dew.

Drosera—the Sundew, a plant whose leaves excrete drops which look like dew.

Droso'phila—"dew lover"—the Fruit Fly which is attracted by vinegar, cider and fruit juices. (One of the species, *D. melanogaster*, has been much used in studies of heredity.)

-DUCE, -DUCT, -DUCTION

to lead (L. *duco*, *duct-*).

This Latin verb has given rise to many well-known words, e.g. **produce, reduce, induce, conduct, reproduction,** though some of these words, e.g. reduce, have specialised meanings when used in science. A **duct** is a pipe or channel for "leading" (carrying) a liquid, e.g. water from a reservoir; a **viaduct** is a large bridge leading one way (e.g. road, railway) over a valley or over another way.

con'duct'or—that which leads (carries) heat or electricity along.

ab'ductor—a muscle which moves a part (e.g. the thumb) away from the main axis (e.g. the hand).

ovi'duct—the tube by which eggs are led from the ovary and discharged.

duct'less glands—glands (endocrine glands) which discharge their products directly into the blood.

duct'ile—(metal) which can be drawn into a wire.

trans'ducer—"a leader across"—an apparatus or device which conveys power from one system (e.g. an electrical system) and supplies it to another (e.g. an acoustic system). A loudspeaker is an example of a transducer.

DUODEC-, DODECA-

L. *duodecem*, twelve; *duodecimus*, twelfth. Gk. *dōdeka*, twelve.

These prefixes would have been more common if we had adopted the **duodecimal** system instead of the decimal system, i.e. counting by twelves instead of by tens. There are relics of such a system in our measures: 12 pence to the shilling, 12 inches to the foot, and special names for 11 and 12 (instead of one-teen and two-teen). For some purposes we count in dozens. The decimal system probably prevailed because, within it, one could easily count on the fingers.

duodecimo—a size of paper formed by folding a standard sheet into twelve.

duodenum—the first part of the small intestine (immediately after the stomach). So called from its length: *duodenum digitorum*—twelve each of fingers, i.e. a length equal to twelve finger-breadths.

dodeca'gon — a figure with twelve angles (and sides), like the English three-penny piece.

dodeca'hedron — a solid figure with twelve faces.

DUPL-, DUPLI-

twofold, double (L. *duplus*; *duplex*).

To **duplicate** means to make double, to

make an exact copy; **duplicity** is double-dealing.

dupl'et—a pair, e.g. a pair of electrons shared between two atoms.

dupli'cident — having two pairs of incisors (cutting-teeth) in the upper jaw (as have Rabbits and Hares).

-DYMA, -DYSIS
putting clothes on or off (Gk. *dyō*).

ep'en'dyma—"clothing put over"—the lining membrane of the cavities of the brain and the central canal of the spinal cord.

ec'dysis—"the putting off of clothes"—the casting off of a coat as by snakes, caterpillars, etc.

-DYMIUM See DIDYM-.

DYNAM-, DYNAMO-
power, strength, force (Gk. *dynamis*).

dynam'ics—the study of forces and of the movements they cause.

dynam'ite—a very powerful explosive.

dynamo'meter — an instrument for measuring forces.

dynamo—(Short for dynamo-machine, itself short for dynamo-electric-machine)—a small electric generator, especially one which provides direct current.

thermo'dynamics — the mathematical study of the relation between heat and mechanical energy (as, for example, in the working of a steam-engine).
The name of the scientific unit of force, the **dyne**, is derived from the first part of this root.

-DYNIA See -ODYNIA.

DYS-
bad, ill, out of order, difficult (Gk. *dys-*).

dys'pepsia—weak digestion of food, indigestion.

dys'entery—"bad (out of order) intestines"—a disease involving inflammation of the intestines, violent pains, and bloody evacuation from the bowels.

dys'phagia—difficulty in swallowing.

dys'pnoea—difficult, laboured breathing (as by a sufferer with a weak heart).

dys'uria—difficulty in passing urine.

-DYSIS See -DYMA.

E

E-
A shortened form of the Latin prefix *ex-* (out, away, free from) used before *b*, *d*, *g*, *l*, *m*, *n*, *r*, *v*, and sometimes *s*. It is well-known in such words as **eject** ("to throw out"), **emit** ("to send out") and **erase** ("to rub out").

e'bullition—thorough boiling, boiling away.

E'dentata—animals with "teeth out"—the order (class) of mammals, including Ant-eaters and Sloths, which are without teeth in the front part of the jaw.

e'gress—a way out, a going out; in astronomy, the end of an eclipse or transit.

e'jaculation—the forceful expulsion of spores, semen, etc.

e'liminate—to put "outside the threshold"—to remove, to get rid of, e.g. to remove one symbol (quantity) from two algebraic equations by combining the equations, to remove a simple molecule (e.g. of water) from two other molecules.

e'masculate—to take the "male out"—to remove the male parts of an unopened flower, to remove the testicles of an animal.

e'nucleate—to remove the nucleus of a cell.

e'rosion — "a gnawing away" — the lowering of a land surface by weathering (wind, rain, etc.).

e'scribe—to draw (e.g. a circle) outside a triangle so that it touches one side and the extensions of the other two sides.

e'viscerate—to disembowel, to remove the viscera (internal organs).

EC-
out, out of (Gk. *ek*, *ek-*). Also see ECTO-.

ec'centric—away from the centre, not mounted at its centre. (Also figuratively, descriptive of a person who is odd or unusual.)

ec'dysis—"a taking off"—the casting off of a coat or moulting (as by snakes and many Arthropods).

ec'topic — "out of place", displaced from the normal position, e.g. **ectopic**

pregnancy—the development of the fertilized egg in the Fallopian tube instead of in the uterus.

ec'lipse—"a leaving out, a failing to appear"—the interception of the light from one body (e.g. the Sun) by the passing of another body (e.g. the Moon) between it and the observer.

ec'zema—"an out-boiling"—a general term for various kinds of inflammation and eruptions of the skin.

ECHIN-, ECHINO-

a sea urchin, a hedgehog; hence prickly (Gk. *echinos*).

echin'ate—bearing evenly distributed bristles.

Echinus—the Sea Urchin.

Echino'dermata—the phylum (large group) of creatures, such as Starfish, Sea Urchins, which have "prickly skins" and are radially symmetrical.

-ECIOUS, -ECIUM See -OECIOUS, -OECIUM.

ECO-

The Greek *oikos* (a house) has given us the words **economy** (basically, the management of a house) and **ecology**—the study of a living thing in relation to its environment (i.e. to where it lives, to other things living in the area, etc.). The element -OECIOUS (q.v.) comes from the same root.

-ECTASIS

This suffix (derived from Gk. *ektasis*, an extension) is used in medical terms to denote an extension of, or enlargement of, a structure or passage of the body due to disease.

bronchi'ectasis—enlargement of the bronchi (main branches of the wind-pipe).

tel'angi'ectasis—enlargement of the arteries and capillaries.

oesophag'ectasis—enlargement of the oesophagus (food-pipe).

ECTO-

outside (Gk. *ektos*). (Contrast ENDO-.)

ecto'derm—the outside layer of cells of a gastrula (early stage of development of the embryo); the tissues directly formed from this layer.

ecto'phyte—a parasitic plant growing on the surface of its host.

ecto'plasm—the clearer, outer part of the protoplasm of a cell.

-ECTOMY

A suffix (from Gk. *ek-*, out, and *temnō*, to cut) denoting the surgical cutting out, or removal, of a part of the body. In modern surgery the suffix may be added to the appropriate name of almost any part of the body. A few examples will suffice.

appendic'ectomy, append'ectomy—the surgical removal of the appendix.

gastr'ectomy—the surgical removal of the whole, or a part, of the stomach.

irid'ectomy—the surgical removal of a part of the iris of the eye.

hyster'ectomy—the surgical removal of the womb.

-EDEMA See OEDEMA.

EF-

The form of EX- (out, away from) used before *f*.

ef'ferent—"carrying outwards"—e.g. an efferent nerve carries impulses away from the central nervous system.

ef'fervescence — "a boiling out" — a vigorous escape of bubbles of gas from a liquid.

ef'florescence—"an out-flowering"—the production of flowers; the formation of white crystals on a wall; the loss of water from a crystal (e.g. washing soda) so that a powder is formed on its surface.

ef'fluent—"an out-flowing"—a stream flowing from a lake; liquid sewage flowing from a sewage tank.

EGO-

I, one's self (L. and Gk. *egō*).

ego'tism—self-conceit, frequent talking about one's self and drawing attention to one's self.

ego'centric—being centred on one's self.

ego-involved — being intensely interested in something so that one figuratively becomes part of it.

ELAEO-, ELAIO-
oil (Gk. *elaion*, olive oil).

elaeo'meter — an instrument for measuring the purity of oils.

elaeo'plast, elaioplast—a body in a cell which forms oils and fats.

elaeo'dochon—an oil gland, e.g. the oil gland of a duck for oiling its feathers.

ELECTRO-
The word **electricity** is derived from the Greek *ēlektron*, amber. Dr. William Gilbert, at the time of Elizabeth I, was the first to describe the unusual properties possessed by amber when vigorously rubbed (properties which we now associate with static electricity); the word electricity was in use early in the seventeenth century. Modern terms, such as **electrovalency** and **electron** itself, are really derived from the word **electric** and have only a very indirect connection with amber.

electro-chemistry—the study of the relation between electricity and chemical changes.

electr'ode—a conductor by which an electric current is led into, or out of, an electric cell, gas tube, etc.

electro'lysis—"loosening by electricity"—the conduction of electricity by a solution or melted solid, accompanied by chemical changes and the formation of chemical products.

electro'statics—the science which deals with the behaviour of electric charges which are at rest.

electro'cardiogram — the written record, made by an instrument, of the electric changes during the contraction of the heart muscles.

ELEUTHERO-
free (Gk. *eleutheros*).

eleuthero'dactyl—having the hind toe free.

Eleuthero'zoa—a sub-group of the Echinoderms (q.v.) comprising forms which are free and active.

-ELLA, -ELLUM, -ELLUS
A Latin suffix forming a diminutive, i.e. a word indicating a small version of the thing denoted by the main part of the word. **Umbrella**, derived through Italian from Latin *umbra*, shade, is a familiar example.

colum'ella—"a small column", e.g. the central column of the sterile tissue in the spore-container of a Moss, the central part of a root-cap.

lam'ella—"a small lamina"—a thin plate, scale, or layer (especially of bone or tissue). **Lamelli'branchiata**—a class of Molluscs having plate-like gills; the members are **lamellibranchiate**.

cereb'ellum—"the small brain"—the smaller hind-brain of vertebrate animals.

haust'ellum — "a little sucker" — the enlarged end of the tongue of a Fly.

nuc'ellus — "a small nucleus" — the tissue occupying the middle of an ovule.

EM- See EN-.

EMBRYO, EMBRYON-
An **embryo** is an undeveloped plant (as in a seed) or an animal in an early stage of development (before it leaves the egg or mother's womb). The word is derived from the Greek *embryon* (probably from the verb *bryō*, to swell with) meaning (a) the fruit of the womb, and (b) a newly born animal such as a kid or lamb. The word **embryo** was at first used mainly in the second sense (b) but now it refers only to the 'unborn' plant or animal.

embryon'ic—pertaining to an embryo; undeveloped, in an early stage of development.

embryo'logy—the study of the formation and development of embryos.

embryo'tomy—the cutting up of an embryo in the womb to effect its removal.

-EMESIS, EMET-
a vomiting, a spitting (Gk. *emesis*).

haemat'emesis—the spitting of blood.

hyper'emesis—excessive vomiting.

emet'ic—(drug) which causes vomiting.

emet'ine—a drug, obtained from the roots of the Brazilian Ipecacuanha, used as an emetic.

-EMIA See -AEMIA.

EN-, EM-
in, into (L. *in*, *in-*; Gk. *en*, *en-*).
This prefix is familiar in such words as **enclose, encamp, encase, embedded**; only a few examples from science need be given.

em'bolism—a thing "thrown in"—the blocking of a blood vessel by a mass carried from some other part of the system.

em'pyema—a collection of pus in a cavity of the body.

en'arthro'sis—"a state of being jointed in"—a ball-and-socket joint (as at the shoulder).

en'ergy—"the work in"—the ability of a body to do work.

ENANTIO-
opposite (Gk. *enantios*).

enantio'morphous—having "opposite shapes"—(crystals) having shapes which are mirror images of each other.

enantio'tropic—existing in two contrasted forms, one stable above, and one stable below, a certain temperature.

ENCEPHAL-, ENCEPHALO-
the brain (Gk. *enkephalos*, within the head; *enkephalon*, that which is within the head, the brain).

encephalon—the brain.

mes'encephalon—the mid-brain of a vertebrate animal.

encephal'itis — inflammation of the substance of the brain.

encephalo'gram — an X-ray photograph of the skull and brain. It is produced by **encephalography**.

encephalo'malacia—a softening of the brain by disease.
See the comment under CEPHAL- concerning the pronunciation of these terms.

-ENCHYMA
an infusion, a substance poured in, a juice (Gk. *enchyma*, from *en* (in) + *chyma* (that which is poured). Used in forming the names of the various cellular tissues of animals and (especially) plants.

par'enchyma—tissue "at the side"—tissue consisting of blunt, thin-walled cells, forming the 'packing material' in a plant; soft, spongy tissue in an animal. Adj. **parenchymatous.**

scler'enchyma—"hard" tissue—plant tissue composed of woody cells used for support and protection.

coll'enchyma — "glue" tissue — tissue (e.g. in a young plant stem) the cells of which are strengthened by the addition of cellulose (not wood).

aer'enchyma — "air" tissue — loosely constructed tissue by means of which air can circulate inside a plant; tissue composed of thin-walled, rather corky cells present in the stems of some water-plants.

Note. Biologists do not seem to have agreed whether in the pronunciation of these terms the stress should be put on the second syllable or on the first and third.

ENDO-
within (Gk. *endon*). (Contrast ECTO- and EXO-.)

endo'cardiac—within the heart.

endo'crine — (glands) which secrete internally ("separate within") directly into the blood.

endo'derm—the inner layer of cells of a gastrula (an early stage of development of the embryo); the tissues formed directly from this layer.

endo'plasm—the inner, granular part of the protoplasm of a cell.

endo'scope—an instrument for inspecting the internal parts of the body.

endo'skeleton—a skeleton within a body, as in vertebrate animals (including Man).

endo'thermic reaction — a chemical reaction in which heat is absorbed (taken in).

-ENE
A suffix used in forming the names of hydrocarbons. It is properly applied to:
(a) straight-chain hydrocarbons in which there is a double join between carbon atoms, -C:C-, (contrast -ANE),
(b) 'aromatic' hydrocarbons built up from a ring or rings of carbon atoms.
The first group includes **ethylene** ($CH_2:CH_2$), **propylene** ($CH_2:CH.CH_3$), and **butylene** ($CH_2:CH.CH_2.CH_3$). These names are often abbreviated to **ethene, propene, butene** to bring them into line with the names of the corresponding paraffin hydrocarbons (-ANE).

A hydrocarbon which contains two double joins is denoted by the suffix -DIENE, e.g. **butadiene** ($CH_2:CH.CH:CH_2$), an important compound in the synthesis of rubber.
The second group includes **benzene** (C_6H_6), **toluene** ($C_6H_5.CH_3$), **naphthalene** ($C_{10}H_8$) and **terpene** (turpentine, $C_{10}H_{16}$).

-ENNIAL
yearly (cf. *annual*) (L. *annus*, a year).

bi'ennial—(plant) which lasts for two years, producing seeds in the second year.

per'ennial — (plant) which lasts through several years (until it is dug up or dies).

ENTER-, ENTERO-
the bowel, the intestine (Gk. *enteron*).

enter'ic—pertaining to the intestines, e.g. enteric fever.

enter'itis—inflammation of the small intestine.

entero'ptosis—a dropping of the intestines within the abdomen as a result of weakness of the muscles.

entero'stomy — the forming of an opening in the intestine by surgical operation.

dys'entery—"bad intestines"—a disease involving inflammation of the intestines, violent pains, and bloody evacuation from the bowels.

Coel'enter'ates—animals with "hollow bowels"—the large group (phylum) of animals, including Jellyfish and Sea Anemones, which possess a single cavity in the body.

ENTO-
within (Gk. *entos*).

ento-parasite—an internal parasite.

ento'branchiate—having internal gills.

ento'gastric—within the stomach or intestines.

ENTOMO-
an insect (Gk. *entomon*).
The Greek *entomos* means "cut into pieces", i.e. cut into segments, notched —descriptive of the general form of the body of the creatures which we now call insects. (The word *insect* itself has a similar origin—L. *animal insectum*, a cut-into animal.)

entomo'logy—the study of insects.

entomo'phagous—insect-eating.

entomo'lite—"insect stone"—a fossil insect.

EO-, EOS-
dawn, daybreak (Gk. *eos*).

eo'lith—"dawn stone"—(one of) the earliest stone implements made by Man.

Eo'cene period—"dawn of recent (life)"—the name given in 1832 to the earliest period (about 60 million years ago) of the Cenozoic era. The fossil plants include many of the modern genera and the fossil vertebrate skeletons show that mammals had become dominant. (In 1874 it was found necessary to insert an earlier period, the Palaeocene, in the Cenozoic era as a transition between the Cretaceous and Eocene periods.)

eos'in—a fiery red, fluorescent dye.

EPI-, EP-
upon, over (Gk. *epi*, *epi-*).
(The form EP- is used before *h* or a vowel.)
An **epidemic** is a disease over (affecting) all the people at once, an **epitaph** is an inscription over a tomb, an **epilogue** is a concluding speech 'over' what has gone before. The prefix is much used in scientific terms.

epi'centre—the point on the Earth's surface immediately over the point where an earthquake begins.

epi'cycloid—the curve traced out by a point on the circumference of a circle which rolls on the outside of another circle.

epi'dermis—the thin outer layer of skin of a leaf, a stem, or (less usually) of an animal.

epi'glottis—a flap at the back of the tongue which covers the glottis (beginning of the wind-pipe) during the act of swallowing.

epi'phyte — a plant (often tropical) which grows on another plant (but is not parasitic upon it).

epi'scope—an instrument for projecting an image of a solid thing (e.g. a book) by sending light on to the thing from above and then passing the reflected light through a lens.

epi'thelium—tissue forming the outer layer of a membrane, blood vessel, etc., in the body.

ep'hemeral—lasting "over one day" only, short-lived, e.g. an ephemeral disease, an ephemeral insect (such as the May Fly which has a very short adult life).

EPOXY-

This curious prefix, which appears in modern chemical nomenclature, apparently means "with oxygen over". It indicates a compound in which an oxygen atom is joined across two of the carbon atoms in a chain, e.g. *epoxy-***propane** as shown.

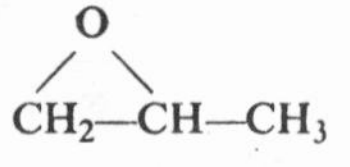

EQUAT-, EQUI-

equal (L. *aequo*, *aequat-*, to equal; *aequus*, equal).

This element is almost self-explanatory. It must not be confused, however, with derivatives of the Latin *equus*, a horse, as in *equine* and *Equisetum*.

equat'or — the great circle which divides the Earth into two equal parts and is **equidistant** from the two poles.

equat'ion—a symbolic statement affirming that two expressions connected by the = sign are equivalent to each other, e.g. $x+y=3$, $C+O_2=CO_2$. (Note that in a chemical equation the atoms on each side are equivalent to each other but the substances denoted by the symbols are not.)

equi'nox—a time of the year (about March 21 and September 22) when days and nights are equal in length.

equi'lateral—(figure, especially a triangle) having sides of equal length.

equi'librium—a steady state in which forces balance.

ERG-, ERGO-

work (Gk. *ergon*).

en'ergy—the ability to do work.

erg—the scientific unit of work.

erg'ate — a worker-ant (a sterile female).

syn'ergic, synergetic — "working together"—said of muscles which work together to produce a particular movement.

syn'erg'idae—two naked cells which help in the process of fertilization of a plant ovule.

all'ergy—"other working (reactions)" —an unusual sensitiveness to certain substances, resulting in inflammation, destruction of tissue, etc.

ergo'nomics—the science of human work. The study of the physical and mental problems of work, of the effects of the conditions of work, etc.

The chemical element **argon,** which is one of the inert (inactive) gases of the air, was named after the Greek *argos*, idle, not working, from *a-* (not) and *ergon* (work).

Also see -URGY.

ERIO-

wool (Gk. *erion*).

erio'phorous—bearing a thick cover of hairs resembling wool or cotton.

erio'carpous—having fruit with a hairy covering.

erio'meter—an instrument for measuring the diameters of fibres.

ERYTH-, ERYTHR-, ERYTHRO-
red (Gk. *erythros*).

eryth'ema — redness of the skin in patches.

erythr'ite — red, crystalline cobalt arsenate.

erythro'cyte—a red blood cell.

erythro'penia—a lack (deficiency) of red cells in the blood.

phyco'erythr'in—a red pigment present in the cell-sap of Red Algae (seaweeds).

-ESCENT, -ESCENCE
A suffix of Latin origin used in forming adjectives from verbs which denote the beginnings of actions. It thus has the meaning of "becoming, growing into".

sen'escent—growing (becoming) old.

nigr'escent—becoming black.

turg'escent—becoming turgid (swollen, distended).

efferv'escent—"coming (to a state of) boiling out"—(liquid) giving off bubbles of gas.

The suffix is also used in words describing a play of colour, e.g. **irid'escent**—flashing with the rainbow colours.

-ESCENCE forms the corresponding nouns, e.g. **senescence**—the process or state of growing old.

-ESIS See -SIS.

ETH-, ETHER, ETHYL
The Latin *aether*, Greek *aither*, was "the clean, upper air". The liquid **ether,** made from alcohol (spirit of wine) and sulphuric acid, is a volatile, clean-smelling substance and so it was named after the *aether*. (Later when scientists felt it necessary to invent a medium to carry light (and other) waves through space the same word was adopted.)
Ordinary ether has the formula $C_2H_5.O.C_2H_5$. Other **ethers** have other hydrocarbon groups joined by the oxygen atom. **Thio-ethers** are of similar structure but have a sulphur atom in place of the oxygen atom.
Ordinary alcohol from which ether is made is called **ethyl alcohol,** C_2H_5OH. (*Ethyl*=ether-substance; see -YL.) The characteristic **ethyl group** of atoms C_2H_5- is found in a number of compounds.

ethyl chloride—C_2H_5Cl. (Also called chloro-ethane.)

ethyl'amine — an amine (q.v.), $C_2H_5NH_2$, in which the ethyl group takes the place of one of the hydrogen atoms of ammonia.

lead tetra'ethyl—$Pb(C_2H_5)_4$, used to improve the quality of motor spirit.

The corresponding paraffin hydrocarbon, C_2H_6, is called **ethane** and the corresponding hydrocarbon with a double join between the carbon atoms, CH_2: CH_2, is called **ethylene.**

Note. *Ethology* (the science of character formation) is derived from the Gk. *ethos*, character, nature.

ETHM(O)-
a strainer, a sieve (Gk. *ēthmos*).

ethm'oid — like a sieve, e.g. the **ethmoid bone**—a square-shaped bone at the root of the nose through the many holes in which the nerves of smell pass.

ETHNO-
a race, a tribe (Gk. *ethnos*).

ethno'logy—the study of races of mankind, their characteristics, inter-relations, etc.

ETHYL See ETH-.

EU-
well, good (Gk. *eu*, *eu-*).
A **eulogy** is a speech or writing in praise of a person (usually after he is dead); a **euphemism** is the substitution of a mild word or expression for a harsh one.

eu'genics—the study of the ways of producing a better race of Man.

eu'phoria — a feeling of well-being (though not necessarily indicating good health).

eu'thanasia—easy or painless death.

eu'tectic mixture—"well melting"—a mixture (e.g. of two metals) of such proportions that it solidifies (and melts) at one temperature like a pure substance.

Eu'mycetes—the large group of higher (more complicated) Fungi. (It includes the Ascomycetes and the Basidiomycetes.)

eu'dio'meter—a graduated tube for measuring the composition of gases by observing the change in volume which takes place when the gases interact under the influence of an electric spark. (The name was given by Priestley to the apparatus which he used for determining the proportion of oxygen ('good air') in the air. It is derived from *eu-* (well, good), *Dios* (god of the sky), *meter* (measurer).)

EURY-
broad, wide (Gk. *eurys*).

eury'bathic—(animal, plant) able to live in a wide range of sea-depths.

eury'thermous—(animal, plant) able to live within a wide range of temperatures.

an'eurysm—"a widening up"—an unhealthy or diseased enlargement of an artery.

EX-
out, out of, free from (L. *ex*, *ex-*; Gk. *ek*, *ek-*, *ex*, *ex-*).
A very common prefix, well-known in such words as **exit, exclude, except, extend**. It is not necessary to give many examples from science.

ex'anthema—"an out-flowering"—an eruption on the surface of the body.

ex'crescence—an abnormal or diseased outgrowth of tissue.

ex'creta—"that which is sifted or separated out"—the waste matter which is got rid of by a cell, tissue or organism.

ex'tensor—a muscle which stretches ("pulls out") a limb or other part of the body.

ex'ponential function — a function (relationship) between two quantities, e.g. x and y, in which one of them is "placed out", e.g. $y=2^x$ (i.e. 2 to the power x).

ex'tensometer — an instrument for measuring the increase in length when a bar is stretched.

ex'foliation — "the coming out of leaves"—the coming off in thin layers of, e.g. the bark of a tree, flakes of rock in icy weather.

EXO-
outside (Gk. *exo*, *exo-*). (Contrast ENDO-.)

exo'skeleton—a hard, bony or horny covering on the outside of the body, e.g. finger nails in Man, the shell of a Crab.

exo'genous—"producing (growing on) the outside", e.g. increasing the thickness of a plant stem by the addition of new tissue on the outside.

exo'thermic reaction—a chemical reaction in which heat is given out.

exo'crine — (glands) which secrete into a cavity or on to the surface of the body (i.e. "outside", not directly into, the blood).

EXTRA-
outside a thing, not within its scope (L. *extra*).
Anything which is **extraordinary** is outside the usual order, beyond that which is ordinary; **extravagance** is a "wandering beyond" that which is modest or reasonable.

extra'cranial—outside the skull.

extra-galactic—outside the Galactic system (the group of stars to which the Sun belongs).

extra'vasation—an abnormal escape of liquid (e.g. blood, lymph) from its containing vessel.

extra'polation—the making of calculations, from known terms, of other terms which lie outside the range of the known terms, e.g. estimating the temperature at 9 p.m. from observations made of the temperature at 6, 7 and 8 p.m.

F

-FACIENT
making (L. *faciens*).

rube'facient—producing redness (e.g. of the skin).

calori'facient—heat-producing.

tubi'facient — tube-building (as are certain Bristle Worms).

FACIO-
the face (L. *facies*).
facio-lingual—pertaining to the face and the tongue.

FACT-, -FACTION, -FACTURE
to make, making (L. *facio*, *fact-*).
fact'or—a component which goes to make up a whole.
manu'facture — "hand making" — originally, the making of articles by physical labour, though now also by machinery.
arte'fact, arti'fact—a man-made stone implement; a body or structure seen in a specimen (e.g. under the microscope) which is not natural but is caused by faulty preparation of the specimen.
putre'faction—the process of becoming rotten.
lique'faction—the process of becoming liquid (e.g. by condensation from a vapour).
petri'faction — the process of being made (turned) into stone.
The suffix -FACTION (which is relatively uncommon) should, strictly, only be used for nouns corresponding to Latin verbs ending in *-facio* (e.g. *liquefacio*, to make liquid). *Petrifaction*, though correct English, is etymologically incorrect. Most English verbs ending in *-fy* (e.g. *magnify*) correspond to Latin verbs ending in *-fico*, *-ficare*, and nouns are formed by the use of the suffix *-fication*. (See -FY.)

FALCI-
a sickle (L. *falx*, *falc-*).
falci'form—shaped like a sickle (e.g. the beak of an eagle or vulture).

FARIN-, FARINA
meal, flour (L. *farina*). (In chemical sense, starch.)
farina—flour made from corn, nuts or starchy roots.
farin'aceous foods—foods made from flour.

FAUN-, FAUNA
Fauna is a collective term for the animals occurring in a particular region or period, e.g. the fauna of S. America. (The term is also used for a book describing such animals.) The word is taken from the name *Fauna*, the sister of the demigod *Faunus* (worshipped by shepherds and farmers); compare *Flora*.
faun'ist—one who studies the fauna of a district.
faun'ology—the study of the geographical distribution of animals.
avi'fauna—the bird population of a district or area.

FAV-
a honeycomb (L. *favus*).
faveolate, fav'ose — shaped like a honeycomb.
Favos'ites—a genus of fossil corals, abundant in Silurian and Devonian rocks, having polygonal cells with perforated walls.
favus — a contagious skin disease (especially of the scalp) due to a fungal infection, so called because of the resemblance to a honeycomb.

FEBR-, FEBRI-
fever (L. *febris*).
febr'ile—pertaining to fever.
febri'fuge—"putting fever to flight"—a drug for reducing fever.
Anti'febrin—a substance (acetanilide) used against fever.

FER-, -FERA, -FEROUS
to carry, to bear; bearing (L. *fero*).
This root is seen in **transfer** ("to carry across"), **refer** ("to carry back") and **differ** ("to bear apart").
circum'ference—the line which "bears round" a circle; the length of this line.
ef'ferent—"carrying away or out of", e.g. an efferent nerve carries impulses away from the central nervous system.
auri'ferous—(rock) which bears gold.
pili'ferous — (surface, membrane) which bears tiny hairs.
coni'ferous—cone-bearing (as are, e.g. fir trees).
luci'fer'in — a protein-like substance which occurs in the light-producing organs of certain animals (e.g. Glow

Worm). It produces light when oxidised under the influence of the enzyme **luciferase.**

Cruci'ferae — "the cross-bearing" plants—the family of plants, including Wallflower and Cabbage, whose flowers bear four petals in the form of a cross.

Roti'fera—very small pond creatures which bear a ring of cilia (hair-like organs) round the mouth. Also called the Wheel animalcules.

Note. *Feral* (wild, untamed) is derived from L. *fera*, a wild beast.

FERR-, FERRO-
iron (L. *ferrum*).
The chemical symbol for iron, **Fe,** is derived from the Latin name.

ferr'ous oxide—black iron oxide, FeO. (See -OUS.)

ferr'ic oxide—red-brown iron oxide, Fe_2O_3. (See -IC.)

ferr'ite—a form of almost pure iron occurring in the substance of cast-iron and steels. (Also a compound of iron and other metals, having important magnetic qualities, used for the cores of a computer.)

non-ferrous metals—metals other than iron and steel, e.g. copper, zinc.

potassium ferro'cyanide—a complex cyanide of iron and potassium, $K_4Fe(CN)_6$.

ferro-concrete—concrete strengthened with iron bars.

ferro'type — a method of printing pictures using a thin iron plate.

FERRUG-, FERRUGIN-
iron rust (L. *ferrugo, ferrugin-*).

ferrugin'ous—red-brown, like rusty iron. **ferruginous clay**—an impure form of clay rock coloured with iron compounds.

FIBR-, FIBRO-, FIBRINO-
fibre (L. *fibra*).

fibr'oid—having a fibrous structure or resembling such a structure.

fibr'il—a small fibre, e.g. of a root.

fibr'oma—a tumour composed of fibrous tissue.

fibro'my'oma—a tumour composed of fibrous tissue and unstriped muscle tissue.

fibro'sis—a formation of fibrous tissue as a result of injury or inflammation. **fibros'itis**—rheumatic inflammation of fibrous tissue.

fibr'in—an insoluble protein substance which forms a network of fibres when blood clots. It is formed from **fibrino'gen.**

-FIC, -FICATION See -FY.

-FID
cleft, split into parts (L. *-fidus*).

bi'fid—divided by a deep cleft into two parts.

quadri'fid—cleft into four parts. So also **sexfid** (six), **decemfid** (ten), etc.

pinnati'fid—"cleft like a feather"—said of a leaf-blade which is divided, by cuts about half way towards the middle, into two rows of leaflets.

FIL-, FILA-, FILI-, FILARI-
a thread (L. *filum*).

fila'ment—a slender, thread-like body, e.g. the thin wire inside an electric lamp, the thread-like body of some pond-weeds.

bi'filar—supported by two vertical strings, e.g. a bifiiar pendulum.

fili'branchiate — having thread-like gills.

Filaria—a genus of thread-like parasitic worms. **filari'asis**—a general name for diseases (e.g. elephantiasis) caused by such worms.

FILIC-
a fern (L. *filix, filic-*).

filic'al—pertaining to Ferns.

Filic'ales, Filic'inae, Filices—the plant group comprising the Ferns.

FISS-, FISSI-
split (L. *findo, fiss-*).
Fissure is perhaps the best known example of a word derived from this root though, with the advances in atomic physics, **atomic fission** and **fission products** are becoming household terms.

fiss'ile—tending to split, able to be split.

fissi'ped — with "split feet" — having the digits (e.g. the toes) separate.

fissi'lingual—having a forked tongue.

FLABELL-, FLABELLI-
a fan (L. *flabellum*).

flabell'ate, flabelli'form—shaped like a fan.

FLAGELL-
a whip (L. *flagellum*).

flagellum—a thread-like extension of a cell, especially that of very simple organisms used for propulsion.

flagell'ate—having a flagellum or other thread-like projection.

FLAV-
yellow (L. *flavus*).

flav'escent — turning yellow; having yellow spots on a surface (e.g. a leaf) which is normally green.

flav'one—a yellow pigment found in plants.

FLECT-, FLEX-
to bend (L. *flecto, flex-*).
To **deflect** is "to bend aside"; **flexible** means "able to bend"; **flex** is flexible, insulated electric wire.

re'flection, reflexion* — a "bending back again", e.g. the bending back again of light by a mirror.

flex'or—a muscle which bends a limb or other part of the body.

re'flex action—an action, such as blinking, which is an automatic (involuntary) response to a stimulus.

* The principal parts of the Latin verb are *flecto* (I bend), *flectere* (to bend), *flexi* (I bent) and *flexum*. Nouns, especially in Latin, are normally derived from the supine (*flexum*). Hence, etymologically, **reflexion** is the correct spelling. But the influence of a large number of correctly formed nouns ending in *-ection* (e.g. *election, affection*) has been such that **reflection** is now the more common spelling. This spelling has the advantage that it does not obscure the connection with the verb **reflect**. (*Connection*, itself, is another example of the same point.)

FLOCC-, FLOCCUL-
wool (L. *floccus*).

flocc'ose — (plant) bearing a dense wool-like covering of hairs.

floccul'ent—like tufts of wool, having tufts similar to those of wool.

floccul'ation—the joining together of small particles (especially when floating in a liquid) to form larger, woolly masses.

FLOR-, FLORI-, -FLOROUS
a flower (L. *flos, flor-*. *Flora* was the goddess of flowers).
Floral and **florist** are examples of well-known words derived from this root. **Florid** means "adorned with flowers", or, more commonly, ruddy and flushed.

flora—the plant population of a given area; a book describing such plants.

flori'ferous — bearing, or producing, flowers.

flori'gen—a substance in plants believed to cause plants to produce flowers.

albi'florous—bearing white flowers.

in'flor'escence—the whole group of flowers on a stem.

ef'flor'escence—"a flowering out"—the formation of a white powder on walls; the loss of water from a crystal (e.g. washing soda) so that a powder is formed on its surface.

FLU-, FLUX
to flow (L. *fluo, flux-*).
Words flow freely from a **fluent** speaker; when we **influence** a person some of our power (etc.) flows into him. The term **influenza** is derived, via the Italian, from the same root.

flu'id—a form of matter which can flow, i.e. a liquid or a gas.

flux—a substance added to a solid to make it melt (and so flow) more easily. (The term is also used in special senses to denote a flow of light, magnetic power, etc., through a given area.)

con'flu'ence—a flowing together, e.g. of rivers.

ef'fluent — flowing out; that which flows out, e.g. a stream flowing from a lake, the outflow from a sewage tank.

FLUO-, FLUOR-, FLUORO-

The name **fluor** (or **fluorspar**) was originally given to a class of gem-like minerals which melted easily and so flowed (L. *fluo*, to flow). The name is now especially applied to a mineral compound of calcium and fluorine—calcium fluoride. Hence the element FLUOR- usually pertains to the chemical element fluorine.

fluor'ine — a non-metallic chemical element, having similar properties to those of chlorine, named from its occurrence in fluor.

fluor'ide — a compound of a metal (usually) and fluorine, e.g. calcium fluoride CaF_2.

fluo'boric acid — a complex acid formed from hydrogen fluoride and boron trifluoride.

fluoro'hydrocarbons — compounds similar in structure to hydrocarbons but with fluorine in place of the hydrogen.

poly'tetra'fluoro'ethylene — Ethylene has the formula $CH_2:CH_2$; in tetrafluoroethylene the four hydrogen atoms are replaced by four fluorine atoms, $CF_2:CF_2$. When this is polymerised (made into a giant molecule) a useful plastic is formed which may be called by the above long name or by the shorter trade name Teflon.

Fluorspar and similar minerals have the property of absorbing light of one colour (especially near the violet end of the spectrum) and re-emitting, and so shining with, light of another colour. This process is called **fluorescence**. Occasionally the element FLUOR- refers to this process.

fluoresce'in — a compound which forms a red solution in alcohol but which solution shines green when a beam of light is passed through it.

FLUX See FLU-.

FOLI-, -FOIL

a leaf (L. *folium*).

Silver **foil** is a very thin sheet of silver (or similar metal).

foli'age—the leaves on a plant.

foli'ate—leaf-like; bearing leaves.

foliated—made up of thin leaves (e.g. mica).

acuti'foliate — having sharp, pointed leaves.

tre'foil—a plant (similar to clover) with a leaf made up of three parts.

ex'foliation—a coming off in thin layers, e.g. bark from a tree, thin pieces of rock in icy weather.

fol'ic acid—a substance first detected in the 'foliage' of spinach, later found in grass, mushrooms, yeast and liver. (It is the essential substance of certain vitamins. Chemically it is called pteroylglutamic acid and is related to the colouring matter in butterflies' wings. See- PTERIN.)

FORAMEN, FORAMIN(I)-

an opening, an orifice (L. *foramen, foramin-*).

foramen—an opening or perforation, especially in horny or bony structures.

foramen magnum—the main opening in the base of the skull through which the spinal cord passes.

Foramini'fera—"the pore bearers"—a group (order) of marine Protozoa consisting of one-celled animals each with a chalky shell pierced by a number of pores through which fine pseudopodia (projections of protoplasm) are thrust.

FORCIP-

forceps, pincers (L. *forceps, forcip-*).

forcip'ate—having the form of a pair of pincers.

FORM-, FORMIC-

an ant (L. *formica*).

formic'arium—a container in which ants are kept for observation and study.

formic'ation—a tingling sensation such as would be produced by ants crawling on the skin.

The essential acid of ant stings is **formic acid**, H.COOH. Hence the root FORM-, especially in chemical names, usually indicates a relation with formic acid.

form'ates—salts of formic acid, e.g. sodium formate H.COONa.

form'aldehyde — a colourless gas, H.CHO, whose solution (commonly called **formalin**) is a disinfectant. When oxidised it produces formic acid.

chloro'form — a well-known anaesthetic, $CHCl_3$. (*Chloro'form=formyl chloride*; *formyl* is the obsolete name for the group HC⁝ (cf. formic acid, HCOOH).)

-FORM, (-IFORM)

(1) having the form of (L. *forma*).

cruci'form—having the form of a cross.

falci'form—shaped like a sickle.

This suffix is always preceded by *-i-*, which letter arises naturally from the Latin stem to which the suffix is added. When the suffix is used with an English (or other non-Latin) word, an *i* is inserted as a connecting vowel, e.g. **cubiform** (cube-shaped).

(2) referring to the number of forms.

uni'form—of one form, character, etc.

multi'form—having many forms, of many kinds.

-FORMES, (-IFORMES)

A Latin plural suffix (*-formes*, from *forma*) used in forming the names of orders (groups) of birds, e.g. **Aves anseriformes**=the duck-form birds; hence **Anseriformes** is used as the name of the order.

Galli'formes—the order of poultry birds, e.g. Turkey, Pheasant, Grouse. (L. *gallus*, a cock.)

Passeri'formes—the large order of perching birds, including the Sparrow. (L. *passer*, a sparrow.)

FORMIC- See FORM-.

FRACT-, FRAG-, FRANG-, FRING-

to break (L. *frango, fract-*).

This Latin verb gives rise to many English derivatives. Among the more common words are **fragile** (easily broken), **fragment** (a piece broken off), **fraction** (a piece "broken" from a whole) and **fracture** (a breaking, e.g. of a bone).

When a prefix is used, e.g. *infringo* (to break in, to violate) the Latin present tense takes the form *-fring-* instead of *-frang-* but this rule is not always followed in English.

frang'ible—breakable, fragile.

re'fract'ion—the "breaking back", or deviation, of light as it passes from one transparent substance into another. (Verb: to **refract**.)

re'frang'ible—able to be refracted.

bi're'fringence—the formation of two refracted beams when light passes into certain minerals (e.g. calcite).

Refractory means "breaking back" in the sense of 'not yielding, stubborn'. **Refractory substances** can withstand high temperatures, etc., and are used for lining furnaces.

FRONT-, FRONTO-

the forehead (L. *frons, front-*).

pre'front'al leucotomy—the cutting of the white nerve fibres in the front area of the brain.

fronto-nasal—pertaining to the forehead and the nose.

FRUCT-, FRUCTI-, FRUGI-

fruit (L. *fructus*; *frux, frug-*).

fruct'ose—fruit sugar. (Also called laevulose.)

fructi'fication—a general term for the structure which develops (e.g. after fertilization) and contains seeds or spores. Used especially for the spore-containing structures of lower plants.

frugi'vorous—fruit-eating.

fructi'colous—(fungi) living on fruit.

FUGIT-, -FUGAL, -FUGE

to flee, to run away (L. *fugio*); to flee hastily, to shun (L. *fugito*); to put to flight (L. *fugo*).

A **fugitive** is one who is running away. A **refugee** runs away from an unfavourable political climate and seeks **refuge** in another country.

fugit'ive dyes—dyes which "run away", i.e. **fade** or wash away easily.

centri'fugal—acting or moving away from the centre.

luci'fugous — (plant, animal) which shuns light.

febri'fuge—a substance which "puts fever to flight", i.e. acts against and reduces fever.

FUM-, FUMIG-, FUMAR-
smoke (L. *fumus* (n.), *fumigo* (v.), *fumarium*, a smoke-chamber).

fumig'ate—to use chemical fumes to destroy insects, moulds, bacteria, etc. The substances used are **fumigants.**

fumar'oles—"small smoke-chambers" —small holes in the side of a volcano through which smoke comes.

fumar'ic acid—an acid, HOOC.CH: CH.COOH, which occurs in the **Common Fumitory** (*Fumaria officinalis*). The smoke of the plant was believed by the ancients to have the power of expelling evil spirits.

FUNG-, FUNGI-
a fungus (L. *fungus*).

fung'al—relating to fungi.

fung'oid—consisting of, or resembling, fungi.

fungi'cide — a substance used for destroying fungi.

FUNI-
cord, rope (L. *funis*).

funi'form—rope-like.

funi'cle—"a small rope"—a small stalk which joins an ovule to the wall of the ovary.

funi'cular railway—a railway in which the carriage is pulled up (e.g. a cliff) by a rope.

FURC-
a fork (L. *furca*).

furc'ate—forked.

bi'furc'ate—forked into two.

FURFUR-
bran (outer covering of grain) (L. *furfur*).

furfur'aceous—covered with bran-like scales; scurfy.

furfur—dandruff.

furfur'al, furfur'aldehyde—a colourless liquid, $C_4H_3O.CHO$, made from farm waste (bran, straw, etc.).

FUS-, -FUSE, -FUSION
to pour, to cast metal, hence to melt (L. *fundo, fus-*).

trans'fusion — "a pouring across" — the transference of liquid from one vessel to another, e.g. of blood from one person to another.

dif'fusion—"pouring apart"—spreading out, e.g. the spreading out of light when reflected from a rough surface, the spreading of one gas into another.

fuse—a deliberately weak part of an electric circuit which melts (and breaks the circuit) when an excessive current flows.

fus'ible—easily melted.

fusion — originally, the process of melting. The meaning has been extended to include the melting of two substances together so that they become one, and hence the blending of two things to become one. Thus we speak of the fusion of petals, of the nuclei of cells, and of atomic nuclei.

FUS-, FUSEL-
a spindle (L. *fusus*).

fus'iform — spindle-shaped, cigar-shaped.

fusel'age—the framework of an aeroplane (made in spindle form).

fuse—a tube, case or cord filled (or saturated) with combustible matter for firing an explosive, etc. (The word has come through the French and means "a spindleful".)

Note. *Fusel oil* is a mixture of several higher alcohols produced in small amounts in alcoholic fermentation. The word is derived from the German for "bad spirit".

-FY, -FIC, -FICATION
to make (L. *facio, fact-*; hence *-fico, -ficatum*).
When compounds were formed from this Latin verb its spelling was sometimes left unchanged, e.g. *liquefacio* (to make liquid) but more usually it underwent a slight change in spelling, e.g. *magnifico* (to raise in value, to make bigger). English verbs corresponding to both these types end in -FY.

lique'fy—to make into a liquid, to turn into a liquid.

magni'fy—to make bigger, to make to appear bigger.

English nouns corresponding to Latin verbs ending in *-fico* normally have the suffix -FICATION.

magni'fication—the process of magnifying, the result of magnifying.

nidi'fication—the building of nests.

fructi'fication—the bearing of fruit; the structure containing the spores of lower plants.

-FIC is an adjectival suffix meaning "making".

calori'fic—making heat, e.g. **calorific value of a fuel**—the quantity of heat which a fixed amount (e.g. 1 lb.) of a fuel can produce.

sopori'fic—producing, or tending to produce, sleep.

The suffixes -FY and -FICATION are now freely used with stems of various origins, an *i* being inserted when necessary as a connecting vowel.

acidify—to make acidic.

denitrify—to set free nitrogen from nitrogen compounds (especially by bacteria in the soil).

solidification—the act of becoming a solid.

esterification—the formation of esters from alcohols and organic acids.

gasification—the manufacture of a gas, the conversion of a solid into a gas.

G

γ-, GAMMA-

γ, the third letter of the Greek alphabet, is sometimes used to denote the third member of a series or group of things. (Other members are denoted by α and β.)

γ-rays, gamma-rays—one of the three kinds of radiation given off by radioactive substances, consisting of rays of very short wavelength similar to X-rays.

γ-brass, gamma-brass—a hard, brittle form of brass containing 60–68 per cent. zinc.

Gammexane—a form of benzene hexachloride, $C_6H_6Cl_6$, used for killing insects.

GALACT-, GALACTO-, GALAXY

milk (Gk. *gala, galakt-*).

galact'ose—a simple sugar derived from milk-sugar.

galacto'rrhoea—an excessive flow of milk from the breast.

galacto'cele—a swelling in the breast due to a blocked milk gland.

Galaxy—the Milky Way, the faint band of many stars which stretches across the sky.

extra-galactic—outside the **Galactic system** (the system of stars, including the Sun, bounded by the Milky Way).

GALL

The term **gall** has several meanings. It denotes the liquid (bile) secreted by the liver, a painful sore or swelling, and an abnormal outgrowth in plants caused by plant or animal parasites. The word in this third sense has come through French from Latin *galla*.

gallic acid—tri'hydroxy'benzoic acid, $C_6H_2(OH)_3COOH$, found in nut galls, tea and other plants.

pyro'gallic acid, pyrogallol — tri'hydroxy'benzene, $C_6H_3(OH)_3$, formed by heating gallic acid; used as a developer in photography.

GALLI-

cock, hence poultry bird (L. *gallus*).

Galli'formes — the order (class) of birds which includes Turkey, Pheasant, Grouse.

galli'um—a soft bluish-white metal element. (Named after the discoverer Lecoq de Boisbaudran from Latin *gallus*.)

GALVAN-, GALVANO-

The name of Luigi Galvani (1737–98), whose observations on the twitching of the leg muscle of a dead frog led to the invention of the electric cell, was commemorated in several scientific terms relating to electricity but most of these terms are now out of date.

galvan'ize—to coat metal with zinc as a protection against rust. (Originally done electrically.)

galvano'meter—an instrument for detecting or measuring very small electric currents.

GAMETE, GAMET-, GAMETO-, GAMO-, -GAM, -GAMY

to marry (Gk. *gameō*; *gametē*, a husband; *gametēs*, a wife).

Bigamy is the taking in marriage of two wives (or husbands); **polygamy** is the taking of many wives.

gamete—"a husband or wife"—a reproductive cell. Such cells, often of two kinds designated as male and female, unite in pairs.

gamet'angium—the organ in which gametes are formed.

gameto'genesis—the formation of gametes.

gameto'phyte—the form of a plant (e.g. of a Fern) which produces gametes and reproduces sexually.

gamo'genesis—reproduction by means of the union of gametes.

gamo'tropism—the tendency of gametes to move towards each other.

Crypto'gams—plants with "hidden marriage"—the large group of plants, including Ferns, Mosses, Fungi, Algae, which do not reproduce by means of seeds and which do not have flowers.

cleisto'gamy—"closed marriage"—the production of flowers which do not open and in which self-fertilization occurs.

The element GAMO- is sometimes used with the meaning of 'joined, united, fused'.

gamo'petalous—(flower) whose petals are joined together (e.g. Primrose but not Buttercup). So also **gamo'sepalous**—having joined sepals.

GAMMA- See γ-.

GAST(E)R-, GAST(E)RO-

the stomach (Gk. *gastēr*, *gast(e)ro-*).

Note. The *e* is usually omitted in modern spelling.

gastr'ic—pertaining to the stomach, e.g. gastric juices.

gastr'itis—inflammation of the stomach.

gastr'ectomy—the surgical removal of the whole or part of the stomach.

gastro'ptosis—the dropping of the stomach in the abdominal cavity.

Gastro'poda—the "stomach-foot" animals—a class of Molluscs the members of which have a shell in one piece and move about on a muscular outgrowth on the underside of the body. (Snails, Slugs, etc.).

gastr'ula—"a little stomach"—the double-walled stage of a developing embryo.

GEL, GELAT-

to freeze, to stiffen (L. *gelo*, *gelat-*), hence to form a jelly.

gelat'ion—solidification by freezing.

gelat'in(e)—a substance, usually of animal origin, which dissolves in water to form a jelly. (Italian *gelatina*.)

gelatin'ous—jelly-like.

gel—a jelly-like substance formed from a colloidal solution, e.g. starch jelly.

gel'ignite—"jelly fire"—a nitro-glycerine explosive.

GEMIN-

twin (L. *geminus*).

gemin'ate—occurring, or combined, in pairs.

Gemini—the star-group also known as The Twins, the third sign of the zodiac.

GEMMA, GEMMI-

a bud (L. *gemma*).

gemma—a small cellular body, formed vegetatively (not sexually), which can separate from the parent plant and give rise to a new individual.

gemmation—the formation of gemmae.

gemmaceous, gemmi'form—like a small bud.

gemmi'ferous, gemmi'parous—producing gemmae; producing buds.

The Latin word also denoted a jewel, whence the English word **gem**.

-GEN, -GENIC, -GENOUS

L. *gigno*, *genit-*, to beget, to bring forth. Gk. *gen-*, root of *gignomai*, to give birth to, to be born.

The suffix -GEN, as used in modern scientific terms (which have come mainly through the French), has two distinct though related meanings.

(1) a producer of, that which gives rise to

hydro'gen — "water producer" — the chemical element (usually gaseous) which combines with oxygen to form water.

nitro'gen — the chemical element (usually gaseous) which is a constituent of nitre (saltpetre).

oxy'gen — "acid producer" See OXY-.

glyco'gen — "sugar producer" — a form of starch found in the liver. On decomposition it produces glucose.

chromo'gen—"colour producer"— a coloured chemical compound which contains a chromophore (colour-producing group of atoms).

fibrino'gen — a substance in the blood which forms a fibrous network when blood clots.

andro'gens—"male producers"— sex hormones (stimulating substances) which are responsible for the development of the male sex organs and the masculine characteristics.

anti'gen—any body (e.g. a bacterium) which, when introduced into the blood, results in the production of an antibody (opposing substance).

gene—a factor or element (located in a chromosome of a cell) which carries a hereditary quality.

(2) to produce, to grow, hence growth.

exo'gens—plants which increase in thickness by the production (growth) of new layers on the outside. (So also **endo'gens**—on the inside.)

-GENIC, -GENOUS are adjectival.

nitro'genous—containing, or yielding, nitrogen.

photo'genic — producing light, e.g. photogenic bacteria. (Also, having the qualities for producing a good photograph.)

amphi'genous—growing all round, as when a fungus grows on both sides of a leaf.

Note. The term *gena* (a general term for the side of the head) comes from the Latin *gena*, a cheek.

GENER-

(1) to beget (L. *genero*, *generat-*).

This root is seen in **generate**— to bring into existence, to produce (gas light, electricity, mathematical shapes, etc.). A **generator** is a machine or apparatus which generates.

de'generate—"begotten down"— having lost qualities proper to the race, having reverted to a lower type.

(2) race, kind (L. *genus*, *gener-*).

genus—a group of closely related species of plants or animals, e.g. the genus *Rana* (Frog) includes the species *R. temporaria* (common grass frog), *R. esculenta* (edible frog), *R. catesbiana* (the American bullfrog), etc. Plural **genera**.

gener'ic — (features, qualities) which are characteristic of a genus or class.

GENESIS, GENET-, -GENY

coming into being, origin, formation (Gk. *genesis*).

Genesis, the first book of the Old Testament, gives an account of the origin and creation of the world.

cyto'genesis — the formation and development of cells.

oo'genesis — the origin and development of ova (eggs).

partheno'genesis—reproduction by ova which have not been fertilized. (*Parthenos* =virgin.)

gameto'genesis — the formation of gametes (reproductive cells).

patho'genesis — the production and development of a disease.

-GENETIC is usually adjectival, e.g. **bio'genetic** (relating to the production of life).

genetics—the study of inheritance and variation and (loosely) the art of breeding.

-GENY forms nouns with the sense of "the mode (manner) of formation of" but it also occurs as an alternative to -GENESIS.

anthropo'geny — the study of the origins of man, of the history of human evolution.

oro'geny — the formation of mountains.

GENICUL-

a (little) knee (L. *geniculus*).

genicul'ar—pertaining to the knee.

genicul'ate—bent like a knee; having joints like a knee.

GENIO-

the chin (Gk. *geneion*).

genio'plasty—plastic surgery of the chin.

genio-glossal—pertaining to the chin and the tongue.

GENIT-, GENITO-

Relating to reproduction or to the reproductive organs (L. *genitalis*).

genitals—the external organs of reproduction.

genito-urinary — pertaining to the genital and urinary organs.

GEO-, GE-

the Earth (Gk. *gē*).

Geography is essentially the descriptive study of the Earth's surface; **geometry** ("Earth measurement") developed from the practical measurement of areas of the Earth's surface.

geo'centric—with the Earth as the centre.

geo'logy—the study of the composition and formation of the Earth's crust.

geo'tropism—the turning and growth of a part of a plant downwards towards the Earth (as by a root) or upwards from the Earth (as by a shoot).

geo'phagy—the eating of earth (as by Earthworms).

ge'oid—a solid shaped like the Earth, i.e. a slightly flattened ball.

epi'ge'al — (plant) which brings its cotyledons (seed-leaves) "above the ground", e.g. the Castor Oil plant but not the Bean.

apo'gee—"away from the Earth"—the point on the Moon's orbit which is farthest from the Earth.

GERAT(O)-, GER(O)-, GERONT(O)-

old age (Gk. *gēras*, *gērat-*, old age; *gerōn*, *gerōnt-*, an old man).

gerato'logy—the science of the phenomena of decadence, especially those characteristic of a species or group of animals which is approaching extinction.

ger'iatrics — the medical care and treatment of old people.

geront'ic—pertaining to the old-age period of an individual.

geronto'logy—the scientific study of old age and of diseases peculiar to old age.

GERM, GERMIN-

a bud, an offshoot, a germ (L. *germen*, *germin-*).

germ — a part (fragment) of an organism capable of developing into a new organism. (Also an obsolete and loose term for a disease bacterium.)

germ cells—reproductive cells, gametes.

germin'ate—to begin growth, as by a spore or seed.

germi'cide — a substance used for killing disease germs.

GERONT- See GERAT-.

-GEROUS

bearing, producing (L. *-ger*, from *gero* (v.)).

This suffix is always preceded by *i*. The suffix -FEROUS, with a similar meaning, is more common.

denti'gerous—bearing teeth.

seti'gerous — bearing a bristle or bristles.

calci'gerous—producing or containing calcium salts.

Globi'ger'ina—"the globe bearers"—a genus of small sea creatures (Foraminifera) which consist of a spiral mass of spherical chambers.

GEST-
to carry (L. *gesto*, *gestat-*).

gest'ation—the carrying of the unborn animal within the mother.

in'gest—"to carry in"—to take in a material, e.g. food into the stomach.

di'gest—to reduce facts, etc., into a systematic order, to arrange within the mind; to reduce food in the stomach and intestines to a form ready for absorption. (The word means "to carry apart", i.e. to separate and sort out the facts, elements of food, etc.)

con'gest'ion—a "carrying together"—a crowding, an accumulation, e.g. an accumulation of blood in a part of the body.

GINGIV-
the gums (in which the teeth are set) (L. *gingiva*).

gingiv'al—pertaining to the gums.

gingiv'itis—inflammation of the gums.

GLACI-
ice (L. *glacies*; *glacio*, *glaciat-*, to turn into ice).

glaci'al—icy; produced by ice (e.g. glacial erosion of rocks); like ice (e.g. glacial acetic acid, pure concentrated acetic acid which at room temperature is solid and looks like ice).

glacier — a slowly moving mass ('river') of ice down a mountain side.

glaci'ation—the covering of a land area with ice (as in the Ice Age).

GLAUC-, GLAUCO-
greyish-blue, greyish-green, like a grape, like a greyish-blue eye (L. *glaucus*; Gk. *glaukos*).

glauc'ous—of a dull greyish-green or greyish-blue colour; covered with a bloom (fine powdery deposit) like a grape.

glauc'onite—a green silicate mineral found on submerged banks in the sea.

glauco'ma—a disease of the eye in which internal pressure causes hardening of the eyeball and partial or total loss of sight.

GLOB-, GLOBI-, GLOBUL-
a ball, a sphere (L. *globus*; *globulus*).
A **globe** is a body which has a spherical form, e.g. the Earth or a model of the Earth.

glob'ate—having the form of a globe.

Globi'ger'ina — "globe bearers" — a genus of small sea creatures (Foraminifera) which consist of a spiral mass of spherical chambers. The remains of the animals form **globigerina ooze** on the ocean floor.

A **globule** is a "little ball", e.g. a small round drop of a liquid.

globul'ins—a class of proteins, insoluble in water but soluble in dilute salt solution, including globulin (in the crystal lens of the eye), fibrinogen (in blood), and legumin (in seeds of various plants of the family Leguminosae). Probably so called because it was first thought that a globulin was the chief protein of 'globules' of blood, i.e. of red cells.

haemo'globin — haemato-globulin — the pigment in red blood cells.

globulin'uria—the presence of globulins in the urine.

-GLOEA, -GLIA
glue (Gk. *gloia*, *glia*).

zoo'gloea—a sticky mass of bacteria.

neuro'glia — "nerve glue" — the connecting tissue of the brain and spinal cord.

GLOMER-
a ball, massed into a ball (L. *glomus*, *glomer-* (n.); *glomero*, *glomerat-* (v.)).
A **conglomeration** of people is a large number of people massed together into a crowd.

con'glomer'ate—a rock formed from rounded fragments (like pebbles) cemented into a mass.

ag'glomer'ate—a rock formed from volcanic fragments and ash cemented into a mass.

glomerule, glomerulus—"a little massed ball"—a crowded head of small flowers; a rounded cluster of spores.

GLOSS-, GLOSSO-, GLOTTIS

the tongue (Gk. *glōssa* or *glōtta*).
A **glossary** is a word-list with comments and explanations. A **polyglot** is a person who speaks with "many tongues", i.e. speaks a number of languages.

gloss'al—pertaining to the tongue.

gloss'itis—inflammation of the tongue.

gloss'ectomy—the surgical cutting out of the tongue.

hypo'glossal—underneath the tongue.

glosso'phagine — securing food by means of the tongue.

glottis—the opening at the upper part of the wind-pipe. **epi'glottis**—a flap (lid) which protects the glottis.

GLUCOS-, (GLYCOS-)

glucose, hence sugar.
Glucose (Grape sugar) is a simple sugar $C_6H_{12}O_6$; it is one of the units into which more complicated sugars and carbohydrates can be broken down. The name was based, somewhat freely, on the Gk. *gleukos*, sweet wine.

It is normal to retain the *u* in the spelling of terms directly derived from *glucose* though in modern variants its place is sometimes taken by *y*. Also see GLYC- below.

glucos'ides, glycosides—complex substances (e.g. salicin, digitalin) found in plants which can be decomposed into a sugar (often glucose) and a non-sugar part.

glucos'amine—an amino-sugar which represents a link between carbohydrates and proteins.

glucos'uria, glycosuria—the presence of sugar in the urine.
The name **glucina** was first proposed for an alkaline 'earth' discovered in beryl in 1797—because some of its salts have a sweet taste—and hence the name **glucinium** for the metallic element. The terms *beryllia* and *beryllium*, however, have replaced them.

GLUT-, GLUTEN, GLUTIN-

glue (L. *gluten, glutin-*).

gluten—a sticky substance; the sticky, nitrogenous part of flour after starch has been washed away.

glutin'ous—sticky.

ag'glutin'ation—the sticking together of small particles, bacteria, etc., into a larger mass.

glut'elins—a class of simple proteins, e.g. oryzenin in rice.

glut'amic acid—an amino-acid (q.v.) obtained from albuminous substances.

GLYC-, GLYCER-, GLYCO-

sweet, sugary (Gk. *glykeros* and *glykys*).

glycer'ine, glycer'ol—a thick, colourless, sweet-tasting liquid, $CH_2(OH).CHOH.CH_2(OH)$, often used medicinally. Hence **glycer'ides**—compounds (oils and fats) of glycerol and organic acids.

glyc'ol—a sweetish, colourless liquid $CH_2(OH).CH_2(OH)$; the molecule contains two -OH groups.

glyco'gen—a form of starch found in the liver. On decomposition it produces a sugar (glucose).

hyper'glyc'aemia—the presence of too much sugar in the blood.
Also see GLUCOS-.

GLYPH-, GLYPT-, GLYPTO-

to carve; carved (Gk. *glyptō*: *glyptos*).
Hieroglyphics are "sacred carvings"—a form of writing as used in ancient Egypt in which carved figures of objects stand for words, syllables or sounds.

ana'glyph—an embossed ornament in low relief.

tri'glyph—a tablet (block of stone) with three upright grooves cut in it.

Glypt'odon—an extinct S. American animal, like an Armadillo, with fluted teeth.

glypto'graphy—the art of engraving upon gems.

GNATH-, GNATHO-

the jaw, having to do with biting (Gk. *gnathos*).

pro'gnath'ous—having the jaw standing forward.

epi'gnathous—having the upper jaw longer than the lower jaw.

gnatho'stomatous—having a mouth provided with a jaw.

gnatho'pod—"a biting foot"—a foot-like appendage of Arthropods (e.g. Crabs) which is used to help with eating.

-GNOMY, -GNOSIS, -GNOSTIC
to know (Gk. *gignōskō*).
An **agnostic** is one who holds that nothing is known or can be known about the existence of God or of a First Cause.

dia'gnosis—"knowledge through"—identification of a disease from the signs, symptoms, and other information.

pro'gnosis — "knowledge before" — foretelling the course of a disease (or a person's capabilities, etc.) from the present signs and symptoms.

physio'gnomy—the art of judging the character of a person from the features of the face and the form of the body; hence the features themselves. Hence, the characteristic features of a plant community by which it may be readily recognised.

patho'gnomy—"disease knowing"—specially indicating a particular disease.

-GOG(UE-) See -AGOGUE.

GONIO-, -GON
a corner, an angle (Gk. *gōnia*).

dia'gon'al—"across the corners"—a line joining two (not adjacent) corners of a figure, e.g. of a square.

penta'gon—a plane figure with five angles (and sides). So also **hexagon** (six), **octagon** (eight), etc. A **poly'gon** is a plane figure with many angles (and sides).

gonio'meter — an instrument for measuring angles, e.g. those between two faces of a crystal.

GONO-, GON-, -GONY
seed, birth, generation, reproduction (Gk. *gonē*; *gonos*).

gon'ad—a sex gland (in which male or female sex cells are produced).

gono'cocci — bacteria which cause **gono'rrhoea** (an infectious disease of the sex organs).

oo'gon'ium — "egg generator" — the female sex organ of lower plants (e.g. of Algae, Fungi).

cosmo'gony—the science which concerns the origin of the stars and other members of the universe.

GRAD-, GRADI-, -GRADE
a step (L. *gradus*).
Gradual means "step by step"; **degradation** is "downwards stepping". The root has the sense of a level or degree in such words as **grade, to graduate** an instrument (put the scale marks on it) and **gradient** (the degree of slope).

gradi'ometer—an instrument used in surveying for setting out long, uniform gradients.

Centi'grade scale — a temperature scale in which there are "one hundred steps (degrees)" between the freezing point and the boiling point of water.

planti'grade—walking on the sole of the foot.

salti'grade—having feet adapted for jumping (as has a Grasshopper); progressing by jumping.

-GRAM
that which is drawn or written (Gk. *gramma*). Used especially for the written record of an instrument.
This root is seen in **telegram** ("a writing from afar"). The word **gramophone** is probably an inversion of **phonogram** (the sound record used by a phonograph). A **diagram** is essentially a drawing of a thing as seen "through" or "across".

Note that, strictly, -GRAM denotes the record which is drawn or written and -GRAPH (q.v.) usually denotes the instrument which produces it.

spectro'gram — a photograph of a spectrum.

cardio'gram—a record made by an instrument to show the beats of the heart.

mari'gram — a record made by an instrument to show the varying height of the tide.

radio'gram—an X-ray photograph.

GRAMIN-, GRAMINI-
grass (L. *gramen, gramin-*).

gramin'aceous—of, or like, grass.

gramini'vorous—grass-eating.

Gramineae—the family of Grasses.

GRAMME, GRAM
The unit of mass in the metric system (about 1/28th of an ounce). The name (from Latin) may be traced back to the Greek *gramma*, a small weight.

kilo'gramme, kilogram—one thousand grammes (about 2¼ lbs.).

milli'gramme—one-thousandth of a gramme.

GRAN-, GRANI-, GRANO-, GRANUL-
a grain (L. *granum*; *granulum*).
A **granule** is "a small grain"; substances (such as some kinds of sugar) which have been formed into little grains are said to be **granulated.**

gran'ite — a well-known, coarse-grained igneous rock.

granit'oid—(rock) having shapeless, interlocking grains as has granite.

grano'lithic — (stone) made from cement and small chippings of granite.

granul'oma—a growth of granular tissue occurring, e.g. in cases of tuberculosis.

grani'ferous — producing grain-like seed.

GRAPH, -GRAPH, -GRAPHY
a drawing, a writing (Gk. *graphē*).
A **graphic** account is a vivid and lifelike "drawing". A **graph** is a symbolic diagram to show the variation in mathematical or scientific quantities.

graph'ite—a fairly soft, crystalline form of carbon (also called plumbago) used for making pencils and 'black lead'.
-GRAPH originally denoted "a written or drawn thing", e.g. an **autograph** ("thing written by one's self"), a **photograph** ("a thing drawn by light"). In modern terms, -GRAM (q.v.) is more commonly used for the thing which is written or drawn and -GRAPH for the instrument which does the writing or drawing.

baro'graph — a form of barometer which produces a drawn record of the varying atmospheric pressure.

seismo'graph — an instrument by means of which earthquake shocks are recorded.

cardia'graph, cardiograph—an instrument which records heart-beats.

-GRAPHY is used in two senses.

(a) The process or the manner of making the writing or drawing, e.g. **photography.**

litho'graphy—the process of printing from a flat plate on which the image has been formed.

radio'graphy — the process of making X-ray photographs.

steno'graphy—"narrow writing"—shorthand.

(b) A descriptive science, e.g. **geography.**

crystallo'graphy—the study of the forms, structures and properties of crystals.

seleno'graphy — the descriptive study of the Moon's surface.

GRAVI-, GRAVID-
heavy (L. *gravis, gravid-*).
It is the force of **gravity** which pulls things towards the Earth and so gives them weight.

gravi'metric analysis—the measurement of the weights of components which make up a certain mixture or compound.

primi'gravida — a woman who is "heavy" with, i.e. bearing, her first child. So also **multi'gravida**—a woman who is pregnant for a second or subsequent time.

GYMNO-
naked, uncovered (Gk. *gymnos*).
The Greeks called the place in which the young men took exercise, usually naked, a *gymnasion* **(gymnasium).**

gymno'cyte—a cell without a cell wall.

gymno'rhinal — (birds) with "naked noses", i.e. having no feathers in the region of the nostrils.

Gymno'spermae—plants with "naked seeds"—one of the two great classes of the seed-bearing plants. The Gymnosperms (which include the Conifers such as Fir and Pine) bear their ovules and seeds on leaf-like parts in a cone and not enclosed in an ovary.

GYN-, GYNAEC-, GYNAECO-, GYNO-, -GYNOUS

a woman (Gk. *gynē*, *gynaik(o)-*). Hence, relating to women, to female characteristics, to female parts of an organism. A **misogynist** is one who professes to hate women.

gynaeceum—originally, the woman's apartment in a house; in botany, the female structure in a flower. (Often misspelt *gynoecium*; see -OECIUM.)

gynaeco'logy—the science which deals with the diseases and disorders peculiar to women.

gyn'androus—"female-male"—having both male and female organs in one individual; (flower) bearing both stamens and pistil on one column (as in Orchids).

gyno'basis — an extension of the flower-stem forming a short stalk to the ovary.

poly'gynous — (flower) having many female parts (pistils, styles or stigmas).

hypo'gynous — (flower) in which the petals, etc., join the stem below the ovary.

Although the Greek *γ* (=*g*) is pronounced hard (as the *g* in *get*), in English a *g* followed by a *y* is usually pronounced soft (as in *gem*, *orgy*). Occasionally one hears words derived from the roots GYMNO- and GYN- pronounced with a hard *g* and, it seems, the pronunciation of those derived from GYN- has not become standardised.

GYPS-, GYPSO-

chalk: later, gypsum (L. *gypsum*; Gk. *gypsos*).

Gypso'phila—"chalk lover"—a garden plant, with thread-like stalks, some species of which flourish in chalky soils.

gypsum—hydrated calcium sulphate ($CaSO_4.2H_2O$), used in the making of plaster of Paris.

gypso'ferous—yielding, or containing, gypsum.

GYR-, GYRO-

to rotate (L. *gyro*, *gyrat-*), a ring (Gk. *gyros*).

gyr'ate—to move in a circle or spiral, to whirl round.

gyro'plane—an aircraft which obtains its lift (upward support) from overhead revolving vanes. In an **autogiro (autogyro)** the vanes turn freely of their own accord.

gyro'scope, gyrostat—an instrument consisting of a heavy, rotating wheel, used as a compass or control in a ship or aeroplane. (A small model is used as a toy and for illustrating the mechanics of rotating bodies.)

H

HAEM-, HAEMAT-, HAEMATO-, HAEMO-, (HEM-, etc.)

blood (Gk. *haima*, *haimat(o)-*).
(The spellings HEM-, etc., are less common in this country than in the United States.

When the root forms the second part of a word the *h* is omitted; see -AEMIA.)

haemo'rrhage—"blood breaking"—an escape of blood from blood vessels, i.e. bleeding.

haemo'rrhoids—a disease, commonly called piles, in which there are swellings of veins in the lower part of the rectum (waste-outlet of the body). The word means "a flowing of blood", i.e. bleeding (from veins).

haemat'emesis — the vomiting (spitting) of blood or of blood-stained material from the stomach.

haemat'oma—a swelling composed of blood which has escaped into connecting tissue.

haemato'phagous—feeding on blood.

haemo'poiesis, haemopoiesia—the formation of blood.

haemo'philia—"a liking for blood"—a hereditary tendency to bleed very easily (even from slight wounds) and to continue bleeding.

haemo'cyto'lysis—"loosening of blood cells" — the breaking down and dissolving of red blood cells. (Also shortened to **haemolysis.**)

haemat'in—a complicated compound containing iron which, when combined with globin (a protein), forms **haemoglobin**—the red pigment of the blood.

haemat'ite — a red-brown (blood-coloured) iron ore consisting essentially of ferric oxide.

-HALE, HALIT-

to breathe, breath (L. *halo* (v.), *halitus* (n.)).

in'hale — to breathe in. **ex'hale** — to breathe out.

halit'osis — unusually bad-smelling breath.

HALO-, HAL-

salt (Gk. *hals, hal(o)-*).

halo'biotic — living in salt water (especially the sea).

halo'phyte—a plant which can live in soil containing salt.

halo'gens — "salt producers" — the chemical elements (fluorine, chlorine, bromine, iodine) which form salts similar to Common Salt (sodium chloride) and also possess other similar chemical properties.

hal'ide—a general term for a salt formed by the combination of a halogen with a metal.

iso'hal'ine—a line drawn on a map to pass through places in the sea which have equal saltness.

Note. The word *halo* is not derived from this root.

HAPL-, HAPLO-

one, single (Gk. *haploos*).

hapl'odont—having molars (grinding teeth) with single crowns.

haplo'caulescent — (plant) having a single stem or axis.

haplo'sis—the halving of the number of chromosomes (hereditary bodies) in the production of sex cells. Each sex cell has a **haploid** set of chromosomes.

HAPT-

to grasp, to take hold of (Gk. *haptō*).

hapt'ic perception — the perceiving (sensing) of the shape and nature of a thing by taking hold of it and feeling it.

hapt'eron—a tentacle; a part of a plant (especially a lower plant) which holds on to a support.

HAUST-

to draw, to drink, to suck (L. *haurio, haust-*).

This root is seen in **exhaust** ("to suck out").

haust'orium—a growth on a parasitic plant by which it takes food from the host on which it lives.

haust'ellum — "a little sucker" — the enlarged end of the tongue of a fly.

HEBE-, HEBET-,

blunt, dull (L. *hebes, hebet-*).

hebet'ate—having a blunt or soft point.

hebe'tude—general dullness and slowness, stupidity.

hebe'phrenia — a mental disorder in which there are periods of excitement and periods of weeping and depression.

HECTO-

hundred (Gk. *hekaton*).

hecto'gramme—one hundred grammes.

hecto'graph—an apparatus for producing many (e.g. 100) copies of writing or a drawing.

HEDON-

pleasure (Gk. *hēdonē*).

hedonism—the idea that pleasure is the main aim of life.

hedon'ic glands—glands in the skin of some reptiles which produce a pleasant-smelling substance during the breeding season.

-HEDRON, -HEDRA, -HEDRAL

a base (Gk. *hedra*, chair).

A **cathedral** is a principal church containing the bishop's throne (where he "sits down").

di'hedral—"with two bases"—(*noun* and *adj.*) the slight upward tilt of each of the two wings of an aeroplane so that they are not in the same straight line.
The root is used especially in forming the names of solid figures which have flat faces.

tetra'hedron—a solid figure with four flat faces, i.e. a triangular pyramid.

octa'hedron—a solid figure with eight flat faces.

poly'hedra—a general name for solids which have many flat faces.

Note that the plural of *-hedron* is *-hedra.*

HELIC-, HELICO-
a helix (Gk. *helix*, *helik-*).
Loosely, a **helix** is a spiral; strictly, it is a screw-curve (similar to a cork-screw).

helic'al—like a spiral or a helix, e.g. a helical spring.

helico'pter—"screw wings"—an aircraft which is supported by horizontally-turning blades.

HELI-, HELIO-
the sun (Gk. *hēlios*).
The root appears in the names of two well-known flowering plants: **Helianthus,** the Sunflower, and **Heliotrope** (whose flowers are said to turn towards the Sun).

helio'centric—with the Sun as the centre, e.g. the Solar System is now known to be heliocentric.

helio'graph—an instrument for sending signals by flashing sunlight.

helio'phyte—a plant which is able to live in bright sunlight.

helio'tropism—the turning of a part of a plant towards (or away from) the Sun.

helio'therapy—the treatment of diseases by exposing the body to sunlight.
In 1868 it was inferred, from observations of the Sun's spectrum, that a chemical element, then unknown on the Earth, existed in the Sun. It was named **helium.** In 1895 the element was found to exist in small quantities in the Earth's atmosphere.

HELMINTH-
a worm (Gk. *helmins*, *helminth-*).

Helminthes—an old-fashioned name for the large group of worm-like animals (other than Earthworms); now divided into Flatworms, Nematodes, and some smaller groups.

helminth'oid—shaped like a worm.

helminth'iasis—a disease due to parasitic worms in the body (e.g. in the bowel). **Ant'helminthic drugs** are used as a treatment against them.

Platy'helminthes—the Flatworms.

HEM- See HAEM-.

HEMI-
half (Gk. *hēmi*).
This prefix is well-known in **hemisphere.**

hemi'crystalline—"half crystalline"—(rocks) which contain some crystalline material and some glassy material.

hemi'plegia—paralysis (loss of power) on one side of the body.

hemi'an'algesia—"half without pain"—loss of the sense of pain on one side of the body.

hemi'an'opia—"half without sight"—loss of half of the field of vision.

hemi'auto'phyte—"half self plant"—a parasitic plant which, however, contains some chlorophyll (green matter) and can make some of the carbohydrate (sugar) material which it needs.

hemi'thyroid'ectomy—the cutting away, by surgery, of half of the thyroid gland.

HEPAT-, HEPATO-
the liver (Gk. *hēpar*, *hēpat-*).

hepat'ic—pertaining to the liver, e.g. hepatic artery; liver-coloured (dull purplish-red).

hepat'itis—inflammation of the liver.

hepato'ptosis—a dropping of the liver in the abdomen.

hepato'megaly—enlargement of the liver.

hepato'rrhexis—a bursting of the liver (e.g. as a result of being run over by a car).

Hepaticae—the Liverworts, a group of small plants related to the Mosses but, usually, with a thallus (plant body) which is not divided into leaf-like parts.

HEPT-, HEPTA-

seven (Gk. *hepta*).

hept'ane—a paraffin hydrocarbon (C_7H_{16}) containing seven carbon atoms.

hept'ode—a radio valve with seven electrodes.

hepta'gon—a plane figure with seven angles and seven sides.

HERB-

A **herb** (L. *herba*, grass, herb) is a plant whose stem is not woody and (usually) whose leaves, etc., are used for preparing food, medicines, scents, etc.

herb'aceous—pertaining to herbs; like a herb, i.e. consisting of soft green tissue.

herb'arium—a collection of dried plants (e.g. in a book, in a case); the place where such a collection is kept.

herb'ivorous—herb-eating.

HERPES, HERPETO-

a creeping thing, a reptile (Gk. *herpēs*; *herpeton*).

herpeto'logy—the study of reptiles.

herpes—shingles, a painful eruption of patches of small blisters which tend to 'creep' over the skin (along the course of a nerve). Simple herpes is an eruption of small blisters round the mouth.

HETER-, HETERO-

other, different (Gk. *heteros*).

This important prefix occurs in a large number of scientific terms. It is seen in the ordinary English word **hetero-geneous**—consisting of different parts, not all of one kind. HETERO- and HOMO- (q.v.) form contrasting pairs of words.

hetero'phyllous—having leaves of different forms on the same plant.

hetero'dactylous—"having different digits"—(birds) having the first and second toes directed backwards and the third and fourth forwards.

heter'odont—having different kinds of teeth for different purposes.

heter'oecious—"having different houses"—said of a fungus (especially a Rust Fungus) which lives for part of its life-cycle on one host and part on a quite different host.

hetero'zygous—"having unlike yok-ing"—possessing both of a pair of contrasting factors, e.g. blue-eye and brown-eye factors, having inherited one from each parent. Often one factor is dominant (shows itself) and the other is recessive (remains hidden).

hetero'plasty—"different moulding"—the grafting on one person of tissue (e.g. skin) taken from another person.

hetero'dyne reception—"different forces"—the changing of a high frequency current (set up by the incoming radio wave) to a current corresponding to an audible frequency by combining it with another high frequency current (produced by the receiver) of nearly the same frequency. In **superheterodyne reception** the combined current is above the frequency of audible sounds and the sound waves are obtained from it in the rest of the radio set in the usual way.

hetero'cyclic compound—a compound in which the molecule is based on a ring of atoms not all of which are carbon atoms, e.g. pyridine which consists of a ring of five CH groups and one nitrogen atom.

HEX-, HEXA-

six (Gk. *hex*).

hexa'gon—a plane figure with six angles and six sides.

hex'ane—a paraffin hydrocarbon (C_6H_{14}) with six carbon atoms.

hex'ode—a radio valve with six electrodes.

hex'ose—any simple sugar which contains six oxygen atoms.

hexa'merous—(flower) having its parts arranged in sixes.

hexa'pod—having six legs (as have insects).

benzene hexa'chloride—a compound whose molecule is formed by the addition of six chlorine atoms to a benzene molecule, $C_6H_6Cl_6$.

HIDROSIS, -IDROSIS
sweat, perspiration (Gk. *hidrōs*).

hidrosis—the formation and excretion of sweat.

hypo'hidrosis — an unusual lessening of sweating.

hyper'hidrosis, hyper'idrosis—excessive sweating.

brom'idrosis—a disorder of the sweat glands in which the sweat has an offensive smell.

HIPP-, HIPPO-
a horse (Gk. *hippos*).
A **hippodrome** was originally a "course for horses", i.e. a place for chariot racing. A **hippo'potamus** is "the horse of the river". The Christian name **Phil'ip** means "lover of horses".

hippo'phagy—the practice of eating horse-flesh.

hippo'phobia—fear of horses.

hipp'uric acid—an acid found in the urine of horses and of many other animals.

HIST-, HISTO-
a woven thing, a web, a tissue (especially a tissue of the body). (Gk. *histos*.)

histo'logy—the study of the detailed structure of tissues and organs.

histo'cyte—a tissue cell (as opposed to a germ cell).

histo'genesis—the formation of new tissue.

histo'lysis—the breaking down and destruction of tissue.

hist'oma, histioma—a tumour formed from fully developed tissue (e.g. muscles, blood vessels).

hist'amine—a substance (related to the amino-acid **histidine**) which is set free when tissues are damaged (e.g. by a cut or blow) or irritated (e.g. by foreign proteins). It expands the small blood vessels and allows blood to escape thus causing, e.g., swelling and red marks on the skin.

HODO-, -HODE, ODO-
a way, a path, a road (Gk. *hodos*).
This root is also met in the form -ODE (q.v.).

hodo'meter, odometer—an instrument for measuring the distance travelled by a wheeled vehicle; a wheeled instrument used in surveying for measuring distance.

hodo'graph—a curve used to find the acceleration of a body which is moving in a known way along a curved path.
A **method** is "a way after, a following after", i.e. a scientific enquiry.

HOLO-
whole, complete (Gk. *holos*).
A **holocaust** is a whole burnt-offering, a wholesale sacrifice.

holo'gnathous—having the jaw all in one piece.

holo'crystalline rocks—igneous rocks in which all the parts are crystalline (and none glassy).

holo'benthic—passing the whole of the life-cycle in the depths of the sea.

Holo'cene — "completely new or recent" — a name sometimes given to the later part of the Quaternary period (which began about one million years ago).

Holo'metabola—the large group of insects which undergo complete metamorphosis, i.e. pass through the distinctive stages larva (caterpillar)—pupa (chrysalis)—imago (adult).

HOMEO- See HOMOEO-.

HOMI-, HOMIN-, HOMO
Man (L. *homo, homin-*).
Scientists distinguish the present race of Man from earlier (prehistoric) types by calling him *Homo sapiens*—"the wise man", because he is regarded as wise in comparison with his predecessors.

homi'cide—the killing of a human being.

Homin'idae—the family of animals which includes the genus *Homo* and several genera of fossil ape-men.

HOMO-
same (Gk. *homos*).
HOMO- (same) and HETERO- (different) often form pairs of words with contrasting meanings, e.g. **homophyllous**—having the leaves on a plant all of the

same kind, **heterophyllous**—having leaves of different kinds on the one plant.

homo'pterous—(insect) having both pairs of wings the same (or very similar).

homo'genetic — having the same descent or origin.

homo'sexual—having sexual feelings towards a person of the same sex.

homo'zygous — having the "same yoking"—having inherited the same factor (of a pair of contrasting factors) from each parent.

homo'logous—having the same essential nature, relative position, etc. **homo'logous series** (of chemical compounds)—a series of compounds in which each member differs from the next by having an extra CH_2- group in the molecule, e.g. the paraffin hydrocarbons: ethane $CH_3.CH_3$, propane $CH_3.CH_2.CH_3$, butane $CH_3.CH_2.CH_2.CH_3$, etc.

homo'cyclic compound—a compound in which the molecule is based on a ring of atoms which are all the same, e.g. benzene which is based on a ring of six carbon atoms.

HOMOEO-, HOMEO-

like, similar (Gk. *homoios*).
(The modern tendency, especially in the United States, is to use the second spelling above; see -OE-.)

homoeo'pathy—"similar suffering"—the system of medicine based on the idea that a disease may be cured by giving drugs (in very small doses) which would produce, in a healthy person, very similar signs, symptoms, etc., to those of the disease. The idea was put forward by Dr. Samuel Hahnemann (1755–1843).

homoeo'merism—state of having "like parts"—in animals consisting of repeated parts (e.g. an Earthworm), the state of having all the parts alike (or very nearly so).

homeo'thermy—the keeping of the temperature of an animal steady in spite of changes in the temperature of its surroundings. Birds and mammals do this; they are often described as warm-blooded. (There appears to be doubt and confusion over the spelling of this word. The given spelling is the more common in this country but one well-known English dictionary gives the word **homothermous** and a well-known American encyclopaedia gives the spelling **homoiothermy**.)

HORO-

time, hour, season (Gk. *hōra*).

horo'logy—the art of measuring time or of making clocks.

horo'scope — a map of the skies, planets, etc., at a given time (e.g. at the time of a person's birth) especially as used by astrologers for predicting the influence of stars on a person's life.

HUMID-

moist, damp (L. *humidus*).

humid'ity — dampness, e.g. **relative humidity** (of the air)—a comparison of the amount of water vapour in a mass of air with the maximum amount of vapour that the air could hold at that temperature.

de'humid'ification — "unmaking of dampness" — the removal of water vapour from air (especially that drawn from a large hall for cleaning and recirculating in the hall) usually by cooling it so that some of the vapour condenses out.

HY- See HYO-.

HYAL-, HYALO-

glass-like, clear (Gk. *hyalos*, glass).

hyaline—glassy, vitreous.

hyal'ite — a colourless, transparent kind of opal.

hyal'oid—clear, transparent, e.g. **hyaloid membrane** (of the eye)—a transparent membrane which covers the solid (jelly-like) part of the eye.

hyalo'plasm—clear protoplasm without any grains.

hyalo'pterous — having transparent wings.

HYDR-, HYDRO-, HYDAT-
water (Gk. *hydōr*, *hydat-*).
This root is found in a large number of scientific terms. In some it has its basic meaning of water. Sometimes it denotes a liquid in general. In chemical terms it often denotes the element hydrogen.

(1) Water, a watery liquid.

hydr'ant—a pipe (especially in the street) by which water may be drawn from the mains.

hydro-electric power—electric power generated by a machine driven by a water turbine.

hydro'sphere—the layer of water on the Earth, i.e. the seas and oceans.

hydro'cephalous—with "water in the head"—having an abnormal collection of cerebro-spinal fluid (liquid which bathes the surfaces of the brain and spinal cord) in the cavities of the brain.

hydro'phyte—a plant which lives on or below the surface of water.

hydro'tropism—the turning of a part (e.g. a root) of a plant towards water.

hydr'aemia—a watery state of the blood; **an'hydraemia**—loss of water from the blood.

hydr'ate—a crystalline salt which contains water, e.g. crystalline copper sulphate $CuSO_4.5H_2O$.

an'hydrous—"without water"—e.g. anhydrous copper sulphate (a white powder) is obtained by driving off the water from crystalline copper sulphate.

an'hydr'ide—a substance "without water"—a chemical compound which either combines with water to form an acid or which may be obtained from an acid by getting rid of water. **Sulphuric anhydride**—sulphur trioxide, SO_3, which combines with water to form sulphuric acid. **Acetic anhydride**—a compound formed by the elimination of water from two molecules of acetic acid.

$$CH_3CO.(OH+H)O.CO.CH_3 \rightarrow (CH_3CO)_2O+H_2O.$$

The term anhydride is also loosely applied to any compound which is formed by, or whose structure corresponds to, the elimination of water from two simpler molecules.

Note. *Anhydride* is not the opposite of *hydride* (p. 82).

hydro'gen—"water producer"—the lightest chemical element (usually gaseous). It combines with oxygen, e.g. when it burns in air, to produce water.

Hydra—"a water snake"—in Greek mythology, a snake whose many heads grew again when cut off; in modern biology, a small fresh-water animal (a Coelenterate) with a number of tentacles round its mouth.

dropsy—(this word has lost its initial *hy-*)—a disease in which a watery liquid collects in cavities or tissues of the body. Adjective, **hydropic.**

HYDAT- means "watery". **hydat'id**—a small sac or cavity containing a watery liquid (especially one formed by, and containing, a Tapeworm larva).

(2) Liquids in general.

hydr'aulics—the study of the flow of liquids through pipes (*aulos* = pipe), especially when used as a means of transferring power.

hydro'meter—an instrument for measuring the densities of liquids.

hydro'statics—the study of liquids which are at rest (e.g. of the pressures within them).

hydro'dynamics—the study of liquids (and gases) which are in motion.

(3) Relating to hydrogen.

hydro'chloric acid—the acid formed when hydrogen chloride (HCl) dissolves in water.

hydro'carbons—compounds of hydrogen and carbon only, e.g. methane CH_4, benzene C_6H_6, turpentine $C_{10}H_{16}$.

hydr'ide—a compound of hydrogen and another element, e.g. calcium hydride CaH_2.

hydr'ion—a hydrogen ion.

hydr'azine—a compound of hydrogen and nitrogen N_2H_4.

hydro'quinone — an alternative name for quinol $C_6H_4(OH)_2$. It is formed by reducing (adding hydrogen to) quinone.

hydr'oxide—"hydrated oxide"—a compound consisting of (usually) a metal and a hydroxyl (hydrogen-oxygen) group -OH, e.g. calcium oxide CaO, calcium hydroxide $Ca(OH)_2$.

Note. -HYDRIC denotes an -OH group in an alcohol, e.g. ordinary alcohol C_2H_5OH is a **mono'hydric** alcohol, glycol $HOCH_2.CH_2OH$ is a **di'hydric** alcohol.

HYDROXY-, HYDROXYL-

Containing the **hydroxyl** group of atoms -OH.

hydroxy-acetic acid — acetic acid (CH_3COOH) in which one of the hydrogen atoms has been replaced by a hydroxyl group, $CH_2(OH)COOH$. (Also called glycollic acid.)

hydroxyl'amine — a compound in which one hydrogen atom of ammonia (NH_3) has been replaced by a hydroxyl group, NH_2OH.

HYET-, HYETO-

rain (Gk. *hyetos*).

hyeto'meter—a rain gauge.

hyeto'graph — an instrument which collects, measures and records a fall of rain.

iso'hyet—a line drawn on a map to pass through places having equal amounts of rainfall.

HYGRO-

wet, damp (Gk. *hygros*).

hygro'meter — an instrument for measuring the dampness of the air.

hygro'scopic—(substance) which tends to absorb moisture from the air.

hygro'petrical fauna — "wet-stone animals"—animals which live in the film of water which surrounds stones.

HYLO-

wood (Gk. *hylē*).

hylo'phagous—wood-eating.

hylo'tomous — (animal) which cuts wood.

HYMEN-, HYMENO-

a membrane (Gk. *hymēn*).

hymeno'geny—the production of a membrane by the meeting of two liquids.

Hymeno'ptera—"membrane wings"—the large group of Insects, including Bees and Wasps, which have two pairs of transparent wings.

The **hymen** is the membrane which, in mammals, initially closes, or partly closes, the entrance of the female reproductive organ.

hymeno'tomy—the surgical cutting of the hymen.

The idea of a membrane associated with the reproductive organs is seen in the following.

hymen'ium—the layer of spore-bearing (and associated) cells covering the 'gills' of mushrooms and similar fungi.

Hymeno'mycetes—the group of Fungi which includes the Mushrooms and Toadstools.

(**Hymen** was the God of Marriages; **hymeneal** means pertaining to marriage.)

HYO-, HY-

Hy'oid means "U-shaped" (Gk. *hyoeidēs*); the term is most commonly used with reference to the **hyoid bone** at the base of the tongue.

hyo'glossal—pertaining to the hyoid and the tongue.

hyo'thyroid—pertaining to the hyoid and the thyroid cartilage, e.g. the hyothyroid muscle.

Note. **Hyoscyamine (hyoscine)**, an alkaloid contained in Henbane and used as a sleep-producing drug, received its name from the Greek for Henbane, *hyoskyamos* ("pig bean").

HYPER-
over, above, exceeding, more than usual (Gk. *hyper*, *hyper-*).

hyper'acidity — excessive acidity, especially in the juices of the stomach.

hyper'aemia—a state of having too much blood in a part of the body.

hyper'algesia—unusual sensitivity to pain.

hyper'metropia—"over-measure sight" —long-sightedness. If the eye-ball is too short, light from distant things can be focused (but not by the lens in its resting condition) but light from near things cannot be focused.

hyper'tension—a blood pressure higher than usual. (It may be relieved by a hypotensive drug.)

hyper'trophy—"over feeding"—an abnormal enlargement of a cell, organ, or other part of the plant or animal body.

hyper'on—an atomic particle of greater weight than a proton (but less than twice the weight).

HYPNO-, HYPNOT-
sleep, unconsciousness (Gk. *hypnos*).

hypno'sis—a sleep-like state caused by suggestion and influence of another person (the **hypnotist**).

hypnot'ic—pertaining to hypnosis; (a drug) which induces sleep.

hypno'spore—a thick-walled spore able to live for some time in a sleep-like condition.

HYPO-, (HYP-)
under, below, less than usual (Gk. *hypo*, *hypo-*). (The form HYP- is used before a vowel.)

hypo'phyllous—attached to, or growing from, the under side of a leaf.

hypo'pharyngeal—below the pharynx.

hypo'dermic—under the skin, e.g. a **hypodermic injection**—the putting of a drug into the body by means of a needle passed into the under-layer of the skin.

hypo'tenuse — the side opposite ("stretched under") the right-angle in a right-angled triangle.

hypo'thesis — "a placing under, a foundation"—a reasonable explanation of observed happenings, used as a starting point for reasoning and for designing tests which will establish its truth (or otherwise).

hypo'cycloid—the curve traced out by a point on the circumference of a circle which rolls inside another circle.

hypo'plasia—under-development, deficient growth.

hypo'tension—low blood pressure. It is relieved by hypertensive drugs.
In chemistry the prefix indicates that the compound has fewer atoms (of a particular kind) than the normal number.

hypo'chlorites—salts of hypochlorous acid (HClO) which has less oxygen than chlorous acid ($HClO_2$).

hypo'phosphites — salts of **hypophosphorous acid** (H_3PO_2) which has less oxygen than phosphorous acid (H_3PO_3).
'Hypo', as used by a photographer, was formerly (and incorrectly) called sodium hyposulphite; its proper name is sodium thiosulphate.

HYPSO-
height (Gk. *hypsos*).

hypso'meter — an instrument for measuring height (e.g. of a mountain) by calculating it from the boiling-point of water.

hyps'odont — (animal) having teeth with very high crowns.

HYSTER-, HYSTERO-
Two probably quite different Greek roots give rise to these identical combining-elements.

(1) the womb (Gk. *hystera*).
The nervous disorder known as **hysteria** was formerly thought to be caused by disturbances of the womb.

hyster'ectomy — the surgical removal of the womb.

hystero'pexy—the fixing in place, by surgery, of a displaced womb.

(2) lagging behind, later (Gk. *hysteros*).

hystero'genic—being produced, or developing, later.

hyster'anthous—"after the flowers" —said of leaves which develop after the plant has flowered.

hyster'esis — the lagging of the magnetisation produced (e.g. in an iron bar) behind a changing magnetising force.

I

-I-

-i- occurs as a joining vowel in compound words of Latin origin, sometimes forming a proper part of the Latin stem (e.g. **brevi'pennate**) and sometimes being purely connective. It tends to be permanently associated with certain Latin suffixes as if it were part of them, e.g. with *-form* (**cub'iform**), *-ferous* (**carbon'iferous**), and *-fication* (**ester'ification**).

-I

-i occurs as the suffix forming the plurals of many Latin nouns ending in *-us*, e.g. **focus/foci, radius/radii, nucleus/nuclei, Gladiolus/Gladioli.**

Note. A few Latin nouns which occur in science, e.g. *corpus*, *genus*, belong to a different declension although they end in *-us*. The plurals are *corpora* and *genera*.

-IASIS See -ASIS.

-IATRICS, -IATRY, IATRO-

medical treatment (Gk. *iatreia*).

paed'iatrics—the branch of medical science which deals with the diseases of childhood.

psych'iatrics, psych'iatry—the treatment of mental disorders.

iatro'chemistry—the study of chemistry in order to obtain results of medical value.

-IC

(1) A suffix (occasionally from Latin, more often from Greek) used in forming adjectives, e.g. **acidic, cubic, static, toxic.**

(2) As in Greek, adjectives of this kind are sometimes taken as nouns, e.g. a **critic**, a **mechanic**, a **soporific**, an **emetic**, a **hypnotic.**

Before the fifteenth century such nouns were used as the names of certain arts and sciences, e.g. **music, logic, arithmetic**. Later the plural form -ICS (q.v.) became more usual, e.g. **mechanics, optics, dynamics.** Adjectives are often formed from such nouns by the use of *-al*, e.g. *mechanical*, *optical*.

(3) In chemical nomenclature -IC may be merely adjectival, e.g. **acetic acid, maleic anhydride, alcoholic potash,** but it also has special meanings as below.

(a) In the naming of acids, it denotes an acid which has a greater proportion of oxygen than another, similar acid, e.g. **nitric acid** HNO_3, nitrous acid HNO_2; **sulphuric acid** H_2SO_4, sulphurous acid H_2SO_3.

(b) In the naming of salts, etc., it indicates that the metal has a higher valency (joining-power) than in another, similar compound, e.g. **ferric (iron) chloride** $FeCl_3$ (in which iron has a valency of 3), ferrous chloride $FeCl_2$ (in which iron has a valency of 2). So also **stannic oxide** SnO_2, stannous oxide SnO.

ICHNO-

a footstep, a track, a trace (Gk. *ichnos*).

ichno'lite—a fossil footprint.

ichno'logy—the study of fossil footprints.

ichno'graph—a drawing of the groundplan of (e.g.) a house.

ICHTHY-, ICHTHYO-

a fish (Gk. *ichthys*).

ichthy'oid—fish-like.

ichthyo'logy—the natural history of fishes.

ichthyo'pterygia—the paired fins of fish.

Ichthyo'saurus — "fish lizard" — an extinct sea animal with a large head, tapering body, four limbs, and a long tail.

ichthy'osis—a disease in which dry scales form on the skin.

ICONO-
an image (Gk. *eikōn*).

icono'graphy—illustration of a subject by drawings and figures.

icono'scope — a form of television camera in which an optical image is formed directly on to the sensitive plate where it is scanned (repeatedly examined line by line) and converted into electrical impulses.

icono'meter — a direct-vision view-finder for photography; an optical instrument for finding the size or distance of an object in surveying.

ICOSA-, ICOSI-
twenty (Gk. *eikosi*).

icosa'hedron — a solid figure with twenty flat faces.

icosi'tetra'hedron—a solid figure with twenty-four flat faces.

-ICS
Whereas the names of many of the branches of science end in -LOGY, those of the mathematical and physical sciences more usually end in -ICS. Possibly the tradition was started with **mathematics** (1581); it has been continued with **physics, mechanics, statics, dynamics, conics, optics, acoustics, statistics, electronics, cybernetics,** and others.

The parent of the suffix is the Greek adjectival suffix *-ikos*. Adjectives were (and still are) often used as nouns. Thus *oikonomika*, actually a neuter plural adjective, meant "things pertaining to the management of a house", i.e. **economics.** Before the fifteenth century English words of this type were taken in the singular (e.g. *music*, *arithmetic*) but later plural forms were preferred. (Among such words outside the field of science are **athletics** and **politics.**)

Although most of the sciences whose names end in -ICS are mathematical or physical, there have been some such names in other fields, e.g. **obstetrics, genetics, hydroponics.** At the present time -ICS is much in fashion. Thus **psychometrics, sociometrics, paediatrics** and **ergonomics** are preferred to similar names ending in *y*.

-ID-, -ID
(1) -ID sometimes corresponds to a Latin suffix used in forming adjectives (and hence nouns) from verbs.

acid—L. *acidus*, from *acere*, to be sour.

fluid—L. *fluidus*, from *fluo*, to flow.

(2) Some Greek nouns ending in *-is*, e.g. *iris*, *rhipis* (a fan) give rise to stems in *-id-*. This is reflected in the combining-forms of the roots, e.g. in the derivatives **irid'ectomy, rhipido'ptera.**

A few English nouns end in *-id*, either instead of *-is* (e.g. **pyramid,** from *pyramis*) or as an alternative to *-is* (e.g. **Orchis/Orchid, chrysalis/chrysalid**).

(3) -ID is used to denote a member of a class or family whose name ends in -IDA, -IDAE, or -IDACEAE.

orchid—a member of the family Orchidaceae.

arachnid—a member of the class Arachnida.

annelid—a member of the class Annelida.

(4) It is also used to denote a group or shower of meteors which radiate from a particular star constellation, e.g. the **Leonid** meteors appear in the constellation Leo in the middle of August.

(5) -ID sometimes arises as a contraction of -IDE. In **plastid** it represents the diminutive suffix *-idium* (L.), *-idion* (Gk.).

-IDA, -IDAE
Suffixes, based ultimately on the Gk. *-idēs* (son of), used in forming the names of some families and classes.

Fel'idae—the Cat family.

Homin'idae—the family of animals which includes the genus *Homo* (Man) and several genera of fossil ape-men.

Annel'ida — the class (phylum) of "ringed" (segmented) animals (including Earthworms and others).

Arachn'ida—the class of arthropod animals which includes Spiders, Scorpions, Mites, and others.

-IDE

A common suffix used in the names of chemical compounds.

At the simplest level it denotes a compound formed by the combination of a metallic element and a non-metallic element. (**Oxide** appears to have been the first name of this kind.)

copper oxide—a compound of copper and oxygen, CuO.

sodium chloride — a compound of sodium and chlorine, NaCl.

zinc sulphide—a compound of zinc and sulphur, ZnS.

calcium carbide — a compound of calcium and carbon, CaC_2.

lithium hydride — a compound of lithium and hydrogen, LiH.

(Note that **anhydride** means "without water"; see HYDR-.)

By simple extension of the idea, the metallic element may be replaced by a group of atoms, e.g. **ammonium chloride** NH_4Cl, **methyl bromide** CH_3Br. Similarly, the non-metallic element may be replaced by a group of atoms, e.g. **sodium hydroxide** NaOH, **potassium cyanide** KCN.

An **amide** is essentially a compound in which one hydrogen atom of ammonia (NH_3) has been replaced by another atom or group, e.g. **sod'amide** $NaNH_2$, **acet'amide** $CH_3CO.NH_2$. (In modern nomenclature the term amide is restricted to a compound, such as acetamide, in which the hydrogen atom is replaced by the 'stem' of an organic acid.) Similarly, an **anilide** is a compound in which a hydrogen atom of aniline ($C_6H_5NH_2$) is replaced by the 'stem' of an acid, e.g. acet'anilide $C_6H_5NH(CH_3CO)$. Note that amides and anilides are not formed by the direct combination of the acid stem with ammonia and aniline respectively; a hydrogen atom is 'lost' (replaced).

A further extension of the nomenclature is seen in **glyceride** (a compound of glycerol and an acid) and **glucoside** (a compound of glucose and another substance, e.g. an alcohol). In these compounds a molecule of water has been eliminated.

IDIO-

one's own, personal, peculiar to one's self (or to one thing) (Gk. *idios*).

An **idiot** (Gk. *idiōtēs*) was basically a private person, a layman (with no professional knowledge); today an idiot is not regarded so highly. **Idio'syn'crasies** ("own together mixture") are the views, expressions, manners, etc., peculiar to an individual.

idio'glossia—a childish way of talking understood only by the child himself.

idio'blast—a plant cell of a different kind from those of the surrounding tissue, e.g. a strengthening cell among cells containing chlorophyll (green matter).

idio'pathy—a diseased condition arising within one's self (not from an outside cause).

idio'thermous — (animal) having "its own heat", i.e. warm-blooded.

-IDIUM

A Latin diminutive (Gk. *-idion*) indicating a small version of the thing named in the main part of the word.

nephr'idium — "a little kidney" — an excretory organ in lower animals.

bas'idium—"a little base"—a club-shaped cell which bears the spores in some kinds of Fungi. (See BASIDIO-.)

rhip'idium — "a little fan" — a fan-shaped head of flowers.

-IFEROUS, -IFORM, -IFY. See -FEROUS, -FORM, -FY respectively.

-IDROSIS See **HIDROSIS.**

IGN-

fire (L. *ignis*).

This root is well-known in **ignite**—to set fire to.

ign'eous rocks—rocks formed from melted material which has turned to a solid within the Earth's crust or on its surface.

gel'ign'ite—a nitro-glycerine explosive.

IL- See IN-.

ILE-, ILEO-

The **ileum** is the lower part of the small intestine; it opens into the colon (large intestine). (L. *ileum.*)

ileo'colic—pertaining to the ileum and the colon.

ile'itis—inflammation of the ileum.

ileo'stomy—the making of an opening in the ileum by surgery.

ILIO-

The **ilia** (sing. **ilium**) are the two bones which form the upper, back part of the pelvis; they can be felt as the hips. (L. *ilium.*)

ilio'femoral—pertaining to the ilium and the femur (thigh bone).

IM- See IN-.

-IMIDE, -IMINE See AMIDE, AMINE.

IN-, IL-, IM-, IR-

This prefix, which has two quite distinct meanings, takes the forms IL- before *l*, IM- before *b*, *m*, *p* and IR- before *r*.

(1) in, on, into, towards (L. *in*, *in-*).

This prefix is met in many common words and needs little elaboration here. **illuminate** (to send light in), **impregnate** (to force liquid, male cells, etc., into), **incisor** (a tooth for cutting into food), **innate** (inborn, natural), **intrusive rocks** (rocks formed from melted material which has been forced into earlier rocks), **irradiate** (to send radiations into).

(2) not, a negation (L. *in-*).

This, again, is met in many common words and needs little elaboration. **imperforate** (not perforated), **inanimate** (without life), **insomnia** (not being able to sleep), **invertebrate** (without a backbone), **infinite** (without an end, without limit), **irregular** (not regular).

-IN

A common suffix for the name of a complex organic compound, especially a **protein**. The suffix is not used systematically but the tendency is to reserve it for neutral substances as distinct from alkaloids and other basic substances whose names properly end in -INE. Only a few of the many possible examples are given.

album'ins—simple proteins which are soluble in water and whose solutions coagulate (become solid) with heat, e.g. the white of an egg.

chromat'in—the part of the material of a cell nucleus which is stained deeply with dyes.

myos'in—the protein of muscles.

kerat'in—the protein of which hair, finger-nails, etc., are made.

fibr'in—an insoluble protein substance which is formed as a network when blood solidifies.

alizar'in—the parent substance of important dyes including madder-red.

sacchar'in—a white crystalline powder which is very sweet.

digital'in—a compound (a glucoside) present in digitalis (medicine prepared from the Foxglove).

tann'in—a general name for naturally occurring substances (derived from hydroxy-benzoic acids) found (e.g.) in oak-galls and used for tanning leather.

aspir'in—a substance derived from salicylic acid used to ease pain and fever.

The suffix is also used for the names of a number of digestive enzymes (ferments), e.g. **ptyalin** (in saliva), **pepsin** (in the gastric juice), **trypsin** (in the pancreatic juice).

Note. Some words (e.g. **gelatin, saccharin, adrenalin**) now normally spelt with *-in* may sometimes be seen spelt with *-ine.*

-INA, -INAE, (-INEA)

These suffixes, which are not used systematically, may denote a group of families (sub-class or sub-order) of animals or plants or a group related to a particular genus (i.e. a sub-family).

Acar'ina—the sub-class of animals which includes the Mites and Ticks.

Filic'inae—the class of plants which comprises the various families of Ferns. Similarly, the **Lycopod'inae** comprises the Lycopods, Selaginellas and related plants.

INCUS, INCUDI-

an anvil (L. *incus, incud-*).

incus—an anvil-shaped bone in the ear.

incudi'form—shaped like an anvil.

IND-, INDIC-

Related to indigo.

Indigo, which has been known for thousands of years as a valuable blue dye, is obtained from the Indigo plant (*Indigofera tinctoria*) or from Woad (*Isatis tinctoria*). The name is derived from the Greek *indikon,* i.e. the Indian (dye).

indic'an—the substance (a glucoside) in the Indigo plant. It can be hydrolysed (made to react with water) to form glucose and **ind'oxyl.** The latter readily oxidises in air to form indigo.

ind'ole—a substance which is produced by the breakdown or fermentation of albuminous matter (e.g. in the body) and whose molecule is the basis of the indigo molecule.

-INE

(1) A suffix used in forming adjectives (L. *-inus*; Gk. *-inos*).
Most of these adjectives are based on Latin, e.g.
feminine, marine, saline (of salt), **uterine** (of the womb), **equine** (of horses), **feline** (of cats).
Some may be traced back to Greek roots, e.g.
crystalline, adamantine (very hard, e.g. like a diamond), **hyaline** (glassy), **glosso'phagine** (securing food by means of the tongue).

(2) A suffix forming nouns (based on Latin roots, often through French), e.g,, **medicine, urine, dentine** (the hard substance of which teeth are mainly composed). In **resin** the final *e* has been dropped.
The suffix is sometimes met in the names of minerals, e.g. **calamine** (zinc carbonate), **celestine** (strontium sulphate), but -ITE is the more usual suffix.

(3) The suffix is widely used in the naming of chemical substances.

(a) It appears in the names of the four halogen (salt-forming) elements: **fluorine, chlorine, bromine, iodine.** (Although most chemists pronounce these names to rhyme with 'seen', perhaps slightly clipped towards 'in' in conversation, a standard English dictionary indicates otherwise. It gives a short *i* (as in 'in') in *fluorine* and *chlorine,* allows as alternatives a short *i* or a long *i* (as in 'mine') in *bromine,* and gives a long *i* in *iodine.* It is difficult to understand this inconsistency; perhaps it is based on a subtle appreciation of the gradation of properties of these elements!)

(b) The suffix also appears in the names of certain hydrides (compounds of an element with hydrogen), e.g. **phosphine** (phosphorus hydride PH_3), **arsine** (arsenic hydride AsH_3), **stibine** (antimony hydride SbH_3).

(c) The suffix has not been used systematically in the naming of organic compounds. In Hofmann's naming of the hydrocarbons it denotes a hydrocarbon in which there is a treble join between carbon atoms, e.g. **ethine** (more commonly known as acetylene) $HC{\equiv}CH$ (compare ethene $H_2C{:}CH_2$ and ethane $H_3C.CH_3$).

The modern tendency is to reserve the suffix for the names of amines, amino-acids, alkaloids and other basic ('alkaline') compounds.

amine—a compound formed by the replacement of one of the hydrogen atoms of ammonia (NH_3) by a hydrocarbon group, e.g. **methylamine** $CH_3.NH_2$. **Phenylamine,** $C_6H_5NH_2$, is better known as **aniline**. (It is interesting again to note that a standard English dictionary gives a long *i* in both *amine* and *aniline* but most chemists do not.)

amino-acid—an organic acid in which one of the hydrogen atoms has been replaced by the amino-group (-NH_2), e.g. **amino-acetic acid** $CH_2(NH_2)COOH$, also known as **glycine**. So also **alanine, leucine** and others.

azines—complex organic bases in which the characteristic feature of the molecule is a ring of four carbon atoms and two nitrogen atoms, the latter being opposite to each other (*para*-position). They form the basis of a number of dye-stuffs, e.g. **indulines, safronines.**

Alkaloids are complex organic bases, containing nitrogen, found (usually in combination) in plants. A few examples of this large group of substances must suffice: **coniine** (in Hemlock), **nicotine** (in tobacco), **quinine** (in Cinchona bark), **morphine** (in opium), **strychnine** (in Nux Vomica), **atropine** (in Deadly Nightshade).

Note. Some names formerly spelt with -INE, e.g. **gelatine, glycerine, adrenaline,** are now frequently spelt without the final *e*.

INFRA-

below (L. *infra*).

infra'clavicular—situated below the clavicle (collar-bone).

infra'renal—beneath the kidneys.

infra-red rays—rays (e.g. in sunlight) which, in a spectrum, come below (in front of) the red rays.

INTER-

This common prefix, which is seen in **interchange, intermix, international,** has the meanings between, among, affecting both. (L. *inter, inter-*.)

inter'costal—between the ribs.

inter'digital—between the digits (fingers or toes).

inter'glacial — between the glacial (periods).

inter'node—the length of a plant stem between two nodes (the places, often swollen, where leaves are attached).

inter'cellular—between the cells of an animal or plant.

inter'action—the action of A on B and of B on A.

In **intero'ceptor** the prefix has the meaning of "within";—a nerve which brings sensations from within the body.

INTRA-

on the inside, within (L. *intra*).

intra'cellular—within a cell.

intra'cerebral—within the substance of the brain.

intra'cranial — situated within the skull.

intra'venous — within, or introduced into, a vein.

intra'molecular change—a re-arrangement of the atoms within a molecule so producing a different substance.

INTRO-

to the inside (L. *intro, intro-*).

To **introduce** means "to lead into", to place in, to bring into use.

intro'vert—a person whose mind is "turned to the inside", whose interests are mainly about his own thoughts and attitudes and how he is regarded by others. He is given to **intro'spection**—looking "within" himself, examining his own thoughts and feelings.

intr'orse—(stamen) which opens towards the centre of the flower.

IO-, IOD-, IODO-, ION-

violet; hence (particularly) pertaining to iodine (Gk. *ion*, violet; *iōdēs*, violet-like).

io'lite — "violet stone"—a mineral silicate of iron, aluminium and magnesium of a blue or violet colour. (Also called cordierite.)

ion'ones—compounds which give the smell to violets and iris roots.

iod'ine — a non-metallic chemical element whose vapour is violet in colour.

iod'ides—compounds of iodine and a

metal (e.g. potassium iodide KI) or an organic group (e.g. methyl iodide CH_3I).

iodo'form—a yellow crystalline compound, CHI_3. (Compared with chloroform, $CHCl_3$, iodine takes the place of chlorine.)

iod'ism — a condition of the body caused by too much, or over-sensitivity to, iodine.

-IOLE See -OLE.

ION, ION-, IONO-

An **ion** is an electrically charged particle (atom, molecule or radical) in a solution or gas which moves when an electric force is applied. The movement of ions is responsible for the passage of electricity through a solution and, in part, through a gas. (Gk. *iōn*, going; from *eimi*, to go.)

ion'ise — to convert into ions; to produce ions in.

cat'ion—an ion in a solution which is positively charged and moves towards the cathode (negative electrode). So also **an'ion** (towards the anode).

therm'ion'ics — (strictly) the science which deals with the sending out of electrons by hot bodies; (more generally) with the behaviour and control of such electrons (e.g. in a **thermionic** (radio) valve).

iono'phoresis — the movement of charged particles in a liquid under the effect of an electric force.

iono'sphere—a region (or layer) of the atmosphere which is ionised.

Also see IO-, violet.

IR- See IN-.

IRID-, IRIDO-, IRIS

the rainbow (Gk. *iris*, *irid-*).

(1) with rainbow colours, with many colours.

irid'escence — the production of rainbow colours as in mother-of-pearl.

irido'cyte—a reflecting cell in the covering layer of fish and some other animals which gives an iridescent appearance.

irid'ium — a brittle, steel-grey metallic element, so called because of the varying tint of its salts. An alloy of iridium and osmium (**irid-osmium**) is used for the tips of gold pen-nibs.

Iris — a genus of well-known plants, so called because of the beautiful colouring of the flowers. It is a member of the family **Iridaceae**.

(2) Pertaining to the iris of the eye.

iris—the coloured part of the eye lying in front of the lens and having the circular pupil (opening) in the centre. (Hence, the iris diaphragm of a camera which, like the iris of the eye, varies the size of the opening in the centre.)

irido'tomy—the surgical cutting of the iris. **irid'ectomy**—the surgical removal of part of the iris.

irido'choroid'itis—inflammation of the iris and of the choroid coat of the eye.

ISCH-

to hold, to restrain (Gk. *ischō*).

isch'aemia—the stopping of the blood supply to a part of the body, e.g. by the blocking of a blood vessel.

isch'uria—the retention of urine.

ISCHI(O)-, ISCHIUM

The **ischia** (sing. **ischium**) form the back part of the pelvis, i.e. they are bones on which we sit. The term comes from the Greek *ischion* which at first denoted the hip-joint and later the particular bones as now defined.

ischi'atic, ischiadic—pertaining to the hip, or (strictly) to the ischia.

isch'algia—sciatica—pain in the back of the hips due to inflammation of the sciatic nerve.

ISO-

equal (Gk. *isos*, *iso-*).

iso'dactylous—having the fingers (or toes) all the same size.

iso'podous—having the "feet" (=legs) all alike. **Isopoda**—an order of Crustaceans, including the Woodlice, which have seven pairs of equal and similarly placed thoracic legs.

is'odont—having all the teeth similar in size and form.

iso'gamete—one of a pair of uniting gametes (reproductive cells) which are almost alike in size and form.

iso'sceles triangle — a triangle with "equal legs", i.e. with two sides equal in length.

iso'chronous—occupying equal times, e.g. the swings of a pendulum are isochronous.

iso'thermal change—a change, such as the slow expansion of a gas, which takes place under such conditions that the temperature remains the same.

iso'tonic—having the same osmotic pressure (q.v.).

iso'bar — a line drawn on a map passing through places which have the same barometric pressure at a given time. So also **isotherm** (temperature), **isohyet** (rain), **isoneph** (cloud), etc.

iso'topes—forms of an element which have identical chemical properties but different atomic weights. So called because they are put in the "same place" in the Periodic Table (classification of elements).

iso'mers—chemical compounds with "equal parts"—compounds whose molecules are built up from the same number and kinds of atoms but, because of the different arrangements of the atoms within the molecules, are different chemical substances. The phenomenon is called **isomerism**. (Metamerism and stereo-isomerism are special kinds of isomerism.)

The prefix *iso-* is sometimes used in the naming of isomeric compounds.

methyl cyanide — CH_3CN;
methyl *iso*-cyanide CH_3NC.

If the molecules of isomeric compounds include a chain of carbon atoms, the prefix *iso-* denotes an isomer in which the chain is branched. (The isomer with an unbranched chain is designated as normal (*n*-).)

n-butane $CH_3.CH_2.CH_2.CH_3$;

iso-butane $\begin{matrix} CH_3 \\ CH_3 \end{matrix}\!\!>\!CH.CH_3$.

n-propyl alcohol $CH_3.CH_2.CH_2OH$;

iso-propyl alcohol $\begin{matrix} CH_3 \\ CH_3 \end{matrix}\!\!>\!CHOH$.

n-valeric acid
$CH_3.CH_2.CH_2.CH_2.COOH$;

iso-valeric acid $\begin{matrix} CH_3 \\ CH_3 \end{matrix}\!\!>\!CH.CH_2.COOH$.

-ISTOR See -SISTOR.

-ITE

(1) A suffix (L. *-ita*, Gk. *-itēs*) forming adjectives, and hence nouns, originally with the meaning of "belonging to, connected with". The following are the more important uses of the suffix in scientific terms.

(a) Names of minerals.

calc'ite — a crystalline form of calcium carbonate.

haemat'ite—a mineral form of iron oxide Fe_2O_3 (a "blood-like" stone).

argent'ite—an important silver ore (silver sulphide).

baux'ite—a clayey ore containing varying amounts of aluminium hydroxide. (Named from *Baux*, in France.)

(b) Names of fossils.

ammon'ite—a spiral-shaped fossil (a cephalopod) found particularly in the Lias beds of the Mesozoic era. (From *cornu Ammonis*—horn of Ammon (Jupiter).)

belemn'ite—a sharp-pointed fossil bone of an extinct cuttlefish. (From Gk. *belemnon*, a dart.)

(c) Names of chemical salts.

sodium nitrite—$NaNO_2$, a salt of nitr*ous* acid. (Contrast sodium nitrate $NaNO_3$.)

sodium sulphite—Na_2SO_3, a salt of sulphur*ous* acid. (Contrast sodium sulphate Na_2SO_4.)

(d) Names of parts of a body or organism.

som'ite—one of the divisions of a segmented animal (e.g. of an Earthworm).

dendr'ite—a branch-like part of a nerve cell.

(e) Miscellaneous names, e.g. of explosives (**dynamite, cordite**), of commercial products (**ebonite, vulcanite**).

(2) A suffix forming adjectives from the Latin past participle (ending in *-itus*), e.g. **composite** (from *compositus*, from *compono*, to put together). (Whence **Compositae**—the family of plants with compound flowers, e.g. Daisy, Dandelion.)

Similarly, in forming verbs based on the Latin past participle, e.g. **ignite** (from *ignio*, to set fire).

-ITIS

This well-known suffix, seen in **appendicitis, bronchitis, gastritis, neuritis,** has come to mean "inflammation of" but it is not derived from any root indicating inflammation.

It is the feminine form of the adjectival suffix *-itēs* (see -ITE (1) above), hence *gastritis nosos* means "the stomach-kind of disease". Such adjectives soon lost the accompanying noun *nosos* and themselves became the names of the diseases. Gradually the nature of the disease has been restricted to an inflammatory condition.

The suffix can be added to the appropriate name of almost every part of the body which can become inflamed; there are many examples in this book.

-IUM

The great Swedish chemist Berzelius, recognising the value of logical naming, proposed in 1811 a new system for the naming of the chemical elements. Using Latin as his basis, he proposed that the names of metal elements should end with -UM, e.g. iron should be **ferrum** and zinc **zincum**. A number of names, e.g. **sodium, tellurium, barium** already fitted in with this scheme.

The general idea was adopted in the naming of new metallic elements, e.g. **platinum** (1812), **molybdenum** (1816). Following the pattern of **sodium,** the ending -IUM has been preferred (e.g. **lithium, cerium, beryllium, radium**), though some later names (e.g. **lanthanum, tantalum**) have the ending -UM. It is interesting to note that the original name **aluminum** (from the mineral alumina) was changed to **aluminium** in this country in order to fit in with the general type but was not changed in the U.S.A.

The name **helium** (given in 1878) is really an error. The element was discovered, by means of the spectroscope, in the Sun's atmosphere and wrongly assumed to be a metal. When, some years later, the element was discovered in our own atmosphere it was found to be an inactive, non-metallic gas.

(The suffix -IUM also occurs, of course, as a termination of some Latin words and of words formed on the Latin pattern, e.g. **cilium, sporangium** (with plurals in *-ia*).)

J

JACTIT-, JACUL-

These elements (from L. *jactito*, to toss, and *jaculum*, a dart) are ultimately derived from L. *jacio* and *jacto*, to throw, to hurl.

jactit'ation—the restless tossing of a sick person; the twitching of a muscle or limb.

e'jacul'ation — the forceful shooting out of, e.g. spores, semen, etc.

Note. *Ad'jac'ent* is derived from a different root (L. *jaceo*, to lie) and means "lying near to".

-JECT

An element derived from the past participle of L. *jacio* (and its compounds), to throw.

This element is seen in **reject** ("to throw back"), **eject** ("to throw out") and **project** ("to throw forth").

in'ject—to force a fluid into a cavity, e.g. to spray fuel into the cylinder of a Diesel engine, to force a drug, etc., into a blood vessel or the tissues.

pro'ject'ile — an object which is

"thrown forward" with force, e.g. a bullet from a gun.

tra'ject'ory—"a throwing across"—the path of a projectile which moves under given forces.

JEJUN-, JEJUNO-
the jejunum — the ill-defined middle section of the small intestine. (The (*intestinum*) *jejunum*, the "fasting (empty) intestine", was so called because it was supposed to be empty after death.)

jejun'itis—inflammation of the jejunum.

jejun'ectomy—the cutting out of the jejunum by surgery.

jejuno'stomy—the making of an opening ("mouth") into the jejunum by surgery.

JUG-
to join, to yoke, to marry (L. *jugo*).

con'jug'ate—"yoked together"—e.g. **conjugate foci**—two points such that if light starts from either of them it is brought to a focus, by a lens or curved mirror, at the other of them.

con'jug'ation (*Botany*) — the joining together of two cells of a Green Alga and the fusion of their contents as a step in reproduction. The class of Green Algae which reproduce in this way is called the **Conjugatae**.
The Latin *jugulum* ("a little yoke") firstly denoted the collar bone and hence, secondly, the throat or neck. The **jugular veins** are the great veins of the neck which carry blood from the head.

JUNCT-
join (L. *jungo*, *junct-*, to join).
This root is seen in **junction** (of roads, railways, etc.). In English grammar a **conjunction** is a word (such as *and*) used to join sentences, clauses, etc. together.

con'junct'iva — "joining" tissue" — a membrane connecting the inner eye-lid and the eye-ball; **conjunctiv'itis**—inflammation of this membrane.

con'junction (*Astronomy*) — Two heavenly bodies (e.g. the Sun and a planet) are in conjunction when they have the same longitude in the sky.

K

In English words derived from Greek roots it is natural to represent the Greek *κ* by the English *k*. This *k*, however, is frequently replaced by *c*. Thus the Greek *kuklos* (circle) gives rise to the English word *cycle*; the older spelling *kinema* (based on the Greek *kineō*, to move) has given way to *cinema*.

Some roots, e.g. CYCLO-, are always spelt with a *c*. Some others, e.g. the prefix CATA- (down), are normally spelt with a *c* but may be found in certain words, or in old-fashioned spellings, with a *k*. Some others, e.g. KERA- (horn) appear with a *c* in some derivatives and with a *k* in others. A few roots, e.g. KON- (dust), almost invariably appear with a *k*.

For convenience of tabulation, and in order to make it clear that alternative spellings are merely variants of the same root, all roots which are commonly (but not necessarily invariably) spelt with a *c* are listed under C.

Attention is drawn to the comment under CEPHAL- regarding the pronunciation of derivatives in which *k* is represented by *c*.

KAIN-, KAINO- See CAINO-.

KARYO- See CARYO-.

KERA-, KERAT(O)- See CERA-.

KET-, KETO-, KETON-
A **ketone** is an organic compound whose molecule consists of the carbonyl group of atoms (-C:O-) joined to two hydrocarbon groups, e.g. acetone $CH_3.CO.CH_3$. (The name is taken from the German *keton* which is a modification of *acetone*.)

keton'aemia—the presence of ketones in the blood.

keton'uria—the presence of ketones in the urine.

keto'sis — excessive formation of ketones in the body (due to incomplete oxidation of fats) resulting in ketonaemia and ketonuria.

ket'oses — those simple sugars (e.g. fructose) which, in some reactions, behave as ketones.

keto'genic — capable of producing ketones, e.g. a ketogenic diet.

KILO-
one thousand (Gk. *chilioi*).

kilo'metre — 1,000 metres (about ⅝ mile).

kilo'gramme, kilogram — 1,000 grammes (about 2¼ lbs.).

kilo'cycles—thousands of cycles (of alternating current, waves, etc.).

kilo'ton—an explosive force (e.g. of a nuclear weapon) equal to that of 1,000 tons of T.N.T.

-KINASE
An element built up from KIN- (to move) and -ASE (an enzyme). A **kinase** is a substance which changes an enzyme-producing substance into a true enzyme, i.e. it causes the enzyme to be formed and so to act.

entero'kinase—a kinase which changes trypsinogen (in the juice from the pancreas) into trypsin (an enzyme which helps to digest proteins in the intestine).

thrombo'kinase—a kinase in the blood which changes prothrombin into thrombin; the latter helps to make the blood clot, e.g. in a cut.

KIN-, KINEMAT(O)-, -KINESIA, -KINESIS, KINET(O)-
movement (Gk. *kineō* (v.); *kinēma(t-)*, *kinēsis*, (nns.); *kinētos* (adj.)).
A **cinema** is one of the few words derived from this root in which the *k* is now normally replaced by a c. The word is an abbreviation of **cinematograph**—an apparatus for producing moving pictures.

kin'aesthetic sensations — "movement feelings" — sensations (e.g. from the muscles) which tell of the movement or position of a part of the body.

kinet'ic energy—the energy possessed by an object because it is moving.

kinemat'ics — the study of motion (e.g. distance travelled, velocity, acceleration) without reference to the forces causing the motion.

brady'kinesia—abnormal slowness in the movements of the body.

hyper'kinesia—excessive movement by a person or a muscle.

karyo'kinesis—"nucleus movement" —the series of changes which takes place in a nucleus during the process of cell division.

KLEPTO-
a thief (Gk. *kleptēs*).

klepto'mania—a state in which a person cannot resist a desire to steal (even though the articles may be unwanted or of little value).

KLINO- See CLIN(O)-.

KON-, KONI-
dust (Gk. *konis*).

pneumono'koni'osis—a disease of the lungs caused by dust (e.g. by coal-dust in a mine.

kon'ometer — an instrument for measuring the amount of dust in the air of a coal mine.
Also see CONIDI-.

KRYPT(O)- See CRYPTO-.

KURT-, KURTO-, (CYRTO-)
curved (Gk. *kyrtos*).
This root is seen mostly in terms relating to the distribution of a set of measures (e.g. heights of people, examination marks) which vary about an average.

The heights of a large number of people are 'normally' distributed about the average. The 'normal curve', which shows the number of people having various heights, is bell-shaped—many people have heights near the average and only a few have heights far from the average. The curve can be mathematically defined. Some other distributions, though evenly balanced about the average, do not agree with a 'normal' distribution.

The **kurtosis** ("state of curving") of a distribution curve is a measure (defined in a precise mathematical way) of the nature of its curving.

lepto'kurtic distribution—the measures are more closely crowded towards the average than in a 'normal' distribution so that the curve is "narrow" (but relatively taller).

platy'kurtic distribution—the measures are more spread out than in a 'normal' distribution so that the curve is "wide" (but relatively less tall).

The prefix CYRTO- (curved) is of rare occurrence.

KYMO- See CYM(O)-.

KYPHO-
bending or stooping forwards (as in old age or because of deformity) (Gk. *kyphos*). Contrast LORDO-.

kypho'sis—a deformity of the spine in which it is sharply curved in a backward direction (as in a humpback) so that the person leans forwards.

kypho'scoliosis—a deformity of the spine in which there is both kyphosis and scoliosis (sideways bending).

L

l-
An abbreviation for laevo-rotatory (q.v.), e.g. *l*-lactic acid.

LABI-, LABIO-
a lip (L. *labium*).

labi'al—pertaining to the lips; like a lip.

labi'ate — (flower) with the petals shaped and arranged in resemblance of lips, e.g. a White Deadnettle. **Labi'atae** — the family of plants with labiate flowers.

labio'dental—pertaining to the lips and teeth.

labio'glosso'pharyng'eal—pertaining to the lips, the tongue and the pharynx. (Given as an example of a horrible-looking word which, however, readily breaks down into its component parts and becomes intelligible.)

LABOR-
work, labour (L. *labor*).
This root is seen in **elaborate** (to produce by labour, to work out in detail) and **collaborate** (to work with another on a common task). The scientist's **laboratory** is his workshop.

LACHRYM-
a tear (of the eye) (L. *lacrima*).

lachrym'al—pertaining to tears.

lachrymation—the shedding of tears.

lachrymatory—causing a flow of tears, e.g. a lachrymatory gas.

This root is occasionally spelt LACRIM-.

LACT-, LACTI-, LACTO-
milk (L. *lac, lact-*). (Compare the Gk. GALACT-.)

lact'ation—the production of milk (by the mammary glands).

lacti'ferous—producing, or carrying, milk.

lact'ic acid—the acid (hydroxy'propionic acid) formed in sour milk.

lact'ose—the sugar in milk. (It is less sweet than cane-sugar.)

lacto'proteins—the proteins of milk.

lact'eals—the lymph vessels which absorb the products of fat digestion (a milky liquid) from the small intestine.

LAEV-, LAEVO-, (LEV-)
(turning, or turned) to the left. (L. *laevus*.)
Chiefly used with reference to the property of certain solutions of turning (twisting) the plane of polarisation of polarised light to the left. (Contrast DEXTRO-.)

laevo'tropism—a tending, or a turning, to the left.

laevo-rotatory—(solution) which twists the plane of polarisation of light to the left, e.g. **laevo'lactic acid** (a form of lactic acid made by fermentation of sugar by certain strains of bacteria).

laev'ulose—fruit-sugar (fructose). It is chemically similar to glucose but is laevo-rotatory.

LAMELL-, LAMELLI-, LAMIN-
Lamina—a thin plate, a layer; **lamella**—a small thin plate, a scale. (L. *lamina*; *lamella*.)

lamin'ated—made up of thin layers or plates.

Lamin'aria—a brown seaweed with large, leathery, blade-like 'leaves'.

lamin'itis—inflammation of the horn-producing membrane of the feet of horses and cattle.

lamell'ated—made up of thin plates.

lamelli'rostral—(birds) having plate-like ridges on the inner edge of the beak.

lamelli'branchiate—(shellfish) having gills which consist of thin plates, e.g. Oysters, Cockles.

LAN-, LANI-
wool (L. *lana*).

lani'ferous—wool-bearing.

lan'ol'in—a waxy substance obtained from sheep's wool used for making ointments.

LAPAR-, LAPARO-
flank, loin (Gk. *lapara*).

laparo'tomy—the cutting by surgery of the walls of the abdomen.

lapar'ectomy—the cutting out of a part of the intestines at the side.

LAPID-, LAPIDI-
a stone (L. *lapis, lapid-*).
Delapidated means in a state of disrepair or decay, i.e. like a building in which stone is being taken from stone. **Lapis lazuli** ("stone of azure") is the sapphire of the Bible, a blue stone of somewhat doubtful composition.

lapid'ary—having to do with precious stones; a worker in precious stones.

lapidi'colous—living under stones.

lapidi'fy—to make into stone.

-LAPSE
to slip (L. *labor, laps-*).

col'lapse—to fall in, to shrink, to break down.

re'lapse—to fall (slip) back, e.g. **relapsing fever**—one of several related diseases, caused by spirochaetes, in which there are recurring attacks of fever.

pro'lapse—(*noun* and *verb*)—a sinking or slipping out of place of an organ or part of the body.

LARYNG-, LARYNGO-
Pertaining to the **larynx** (Gk. *larynx, laryng-*)—the top part of the wind-pipe containing the vocal cords.

laryng'itis — inflammation of the larynx.

laryng'ectomy—the surgical cutting out of the larynx.

laryngo'stomy—the making of an artificial opening in the larynx.

LAT-, LATI-
wide (L. *latus*).

di'late—"to widen apart"—to expand, to enlarge.

di'latation—the process of dilating; expansion.

(Note that the spelling *dilation*, though not infrequently met, is irregular.)

dilat'ometer — an instrument for measuring the expansion of a substance (especially a liquid) when heated. (This word is an interesting example of the addition of the Greek element— -METER, with an o, to a Latin root.)

lati'tude—width, breadthwise measurement, e.g. the angular measurement of a point of the Earth's surface north or south of the equator.

LATER-, LATERI-, LATERO-
(1) side, sideways (L. *latus, later-*).

later'al—on, or towards, the side.

lateral inversion—the defect of a picture or image (e.g. as seen in a plane mirror) in which the right-hand side appears on the left and vice versa.

equi'lateral — (figure) having its sides equal.

bi'lateral—of, or on, or with, two sides. **multilateral**—having many sides.

lateri'grade—moving sideways (as by some Crabs).

lateri'florous — having lateral flowers.

latero'cranium—the side of the head of insects.

(2) a brick (L. *later*).

later'ite—a kind of red clay formed by the weathering of rocks in the tropics.

later'itious—brick-red in colour.

LECITH-, LECITHO-
egg yolk (Gk. *lekithos*).

lecith'in—a complex substance found in egg yolk, nerves and the brain.

a'lecith'al—having little or no yolk.

centro'lecithal—having the yolk in the centre.

lecitho'blast—a cell of the dividing egg which contains yolk.

LEGUM-, LEGUMIN-
A **legume** is a fruit formed from a single carpel, splitting along its two edges, and containing a row of seeds. Peas and beans are familiar examples. (L. *legumen*, *legumin-*, bean, pulse, from *lego*, to gather.)

legumin'ous—pertaining to, or of the nature of, legumes or legume-bearing plants. **Leguminosae** — the family of plants which bear legumes.

LEIO, (LIO-)
smooth (Gk. *leios*).

leio'phyllous—having smooth leaves.

leio'sporous—having smooth spores.

leio'my'oma—a tumour composed of unstriped ("smooth") muscle fibres.

LEMMA
(1) a skin, a peel (Gk. *lemma*, from *lepō*, to peel off).

lemma—an outer skin-like layer; a membranous scale shielding a flower of a Grass.

neuri'lemma, neuro'lemma—a thin sheath surrounding medullated (covered with a fatty substance) nerve fibre.

sarco'lemma—"flesh-sheath"—the sheath of a muscle fibre.

(2) anything taken or assumed (Gk. *lēmma*, from *lambanō*, to take).
When one is faced with a **dilemma** one has to choose between two (usually unpleasant) alternatives.

lemma (*Mathematics*)—a proposition assumed (or demonstrated) as a preliminary to the demonstration of a main proposition.

LENTI-
a lens, lens-shaped.
The Latin word *lens*, *lent-* means a **lentil** (a pea-like seed grown for food). The optical lens takes its name from its rough resemblance to the shape of a lentil.

lenti'cular—shaped like a small lens.

lenti'cle—a lens-shaped mass of one kind of rock embedded in another rock.

lenti'cel—a small lens-shaped spot, corresponding to a pore, in the corky (bark) tissue of a plant stem.

lenti'conus—an abnormal curving of the lens of the eye such that the surface is shaped more like part of a cone than a sphere.

LEPID-, LEPIDO-
a scale (as of a plant or fish) (Gk. *lepis*, *lepid-*).

lepid'ote—covered with small scales or scale-like hairs (as, e.g. the surface of the wings of Butterflies).

Lepido'ptera—animals with "scaly wings"—the order (group) of insects, including Butterflies and Moths, which have two pairs of large wings closely covered with scales.

Lepido'dendron — "scaly tree" — a genus of fossil plants (Club Mosses), having leaf-scales on the trunk and branches. They attained tree-like proportions in Carboniferous (coal-measure) times.

lepido'crocite—a form of iron oxide occurring as scaly, blood-red crystals.

-LEPSY, -LEPTIC
taking hold of, seizing (Gk. *lēpsis*).

epi'lepsy — a general name for a nervous disease in which a person falls to the ground unconscious.

cata'lepsy—a seizure in which there is loss of sensation and will-power and a stiffening of the muscles.

ana'leptic — "taking up again" — (medicine) which restores health and strength.

LEPTO-
thin, slender, fine (Gk. *leptos*).

lepto'cephalic—having a narrow skull.

lepto'dactylous—having slender fingers or toes (as a bird with long, thin toes).

lepto'dermatous—having a thin skin.

lepto'kurtic—said of a distribution of measures which are more closely crowded towards the average than in a normal distribution so that the distribution curve is narrow. (For further explanation, see KURT.)

LEUC-, LEUCO-, LEUK(O)-
white (Gk. *leukos*).

leuco'cyte — a "white" (colourless) blood cell.

leuco'cyto'penia—"white cells lack of" —an unusual lack of white cells in the blood.

leuco'cyt'haemia, leuc'aemia- leuk'aemia, leukemia — a fatal condition in which there is over-production of white cells in the blood.

leuco'tomy—the surgical cutting of white nerve fibres (especially of the front of the brain).

leuco-base, leuco-compound—a colourless compound formed by reducing (de-oxidising) a dye and which may be changed back into the dye by oxidation.

leuc'ite—a silicate of potassium and aluminium occurring in igneous rocks (especially lavas).

LEV-
(1) a variant of LAEV- (q.v.).
(2) to lift, to make lighter (L. *levo, levat-* (v.); *levis* (adj.)).

This root is seen in **levity** (lightness of weight; figuratively, want of serious thought), **elevate** and **lever** (which has come through French).

LEVIG-
smooth, to make smooth (L. *levigo, levigat-*).

levig'ate (*adjective*)—having a smooth, polished surface. (Often wrongly spelt laevigate.)

levig'ate (*verb*)—to reduce a substance to a smooth, fine powder; to make into a smooth paste.

LIEN-, LIENO-
the spleen (L. *lien*). (For explanation, see SPLEN-.)

lien'al—pertaining to the spleen.

lieno'gastric—pertaining to, or leading to, the spleen and the stomach.

LIG(A)-, LIGAT-
to tie, to bind (L. *ligo, ligat-*).

liga'ment—a band of tough tissue which binds two (or more) bones together.

ligat'ure—anything used for binding or tying; a band or cord used to tie up a blood vessel to stop bleeding.

For *ligule*, see LINGU-.

LIGN-, LIGNI-, LIGNO-
wood (L. *lignum*).

Lignum vitae ("the wood of life") is a kind of West Indian shrub whose wood is used in making medicines.

lign'eous tissue—the woody tissue of plants.

lign'in—a complicated mixture of substances forming the woody parts of plants.

lign'ite—dark brown fossilised wood (also called brown coal) representing a stage in the formation of true coal.

ligni'colous—living in or on wood (as some kinds of Termites).

pyro'lign'eous acid—the acid vapour (containing acetic acid, methyl alcohol, and other products) obtained when wood is heated with insufficient air for it to burn.

ligno-cellulose—a compound of lignin and cellulose found in wood.

LIMEN, LIMIN-
a threshold (L. *limen, limin-*).

e'limin'ate—to put "outside the threshold"—to get rid of, to expel.

limen—the limit below which a given stimulus ceases to be perceived.

limin'al—pertaining to a limen, to the threshold of perception.

sub'liminal—below the threshold of sensation, perception or consciousness.

LIMN-, LIMNO-
a lake, fresh-water (Gk. *limnē*).

limno'logy—the study of lakes and ponds and of the life in them.

limno'biotic—living in fresh-water.

limno'plankton — plankton (floating plants and animals) of the fresh-water of lakes, ponds and rivers.

epi'limnion—the top layer of a lake in which the temperature is near to that of the air.

hypo'limnion—the deep layers of a lake where the temperature remains low.

LIN-

(1) a line, in a line, lengthwise (L. *linea*, a linen thread).

This root needs little elaboration. To **delineate** means to mark out with lines; **lineaments** are distinctive or characteristic features, especially of the face.

lin'ear—pertaining to a straight line, lengthwise, e.g. **linear equation** — an algebraic equation (e.g. $y=3x+2$) whose graph is a straight line; **linear expansion**—expansion (by heat) in length (as distinct from expansion in area or volume).

col'linear—in the same straight line.

recti'linear—in, or composed of, straight lines, e.g. the rectilinear propagation of light.

(2) flax (L. *linum*).

This root is seen in **lin'oleum** ("flax+oil").

Note. The word *liniment* (substance used for rubbing on the body) is derived from L. *lino*, to smear.

LINGU-, LINGUO-, (LINGUL-)

a tongue, a language (L. *lingua*).

Linguistics is the study of languages; to be **bilingual** is to be able to speak with two tongues, i.e. in two languages.

lingua—any tongue-like structure.

lingui'form—shaped like a tongue.

linguo-dental — pertaining to the tongue and the teeth.

lingul'ate—shaped like a small tongue. **Ligulate** is a variant of lingulate. A **ligule** or **ligula** is any small tongue-like structure, e.g. a thin flap at the base of a blade of a leaf (especially in Grasses).

LIO- See LEIO-.

LIP-, LIPO-

fat (Gk. *lipos*).

lipo'genous—producing fat.

lip'ectomy—the cutting out by surgery of fatty tissue.

lip'oma—a tumour composed of fatty cells.

lipo'chromes — "fat colours" — the colouring substances of butter fat.

lip'ase—an enzyme (e.g. in the small intestine) which breaks down fats.

-LIPSE

This root (from Gk. *leipō*, to leave) appears in two important scientific terms.

ec'lipse—"a leaving out, a failing to appear"—the interception of the light from one body (e.g. the Sun) by the passing of another body (e.g. the Moon) between it and the observer.

el'lipse—the Greek verb, interpreted in the sense "to leave deserted, to be lacking in" gave rise to the noun *elleipsis*, a short-coming. The manner in which an ellipse (a mathematical curve) falls short of a parabola is explained under -BOLA.

-LITE

A suffix (derived from Gk. *lithos*, a stone) used in forming the names of minerals. It is usually preceded by the letter *o*.

cryo'lite — "ice stone" — an ice-like mineral (a double fluoride of aluminium and sodium) found in Greenland.

oo'lite—"egg stone"—a limestone consisting of round egg-like grains.

copro'lite—"dung stone"—the fossilised excreta (waste matter) of animals used as a fertilizer.

rhyo'lite — "flow stone" — a general name for glassy rocks formed from flows of lava.

LITH-, LITHO-

a stone (Gk. *lithos*).

A **monolith** is a single block of stone, especially one shaped into a pillar or monument.

litho'sphere—the crust (outer rocky layer) of the Earth.

litho'graphy—a method of printing from a flat plate (which was originally a slab of stone).

litho'genous — rock-building (as by some Corals).

litho'phagous — stone-eating, as by some birds which take small stones into the gizzard to help with mastication.

Palaeo'lithic—(age) when the earliest stone tools were made by Man.

lith'ium—the lightest metal element, similar to sodium. (It was discovered by Arfvedson in 1817 but not isolated until 1855. The name **lithia** was given to the oxide (after soda) in the belief that lithium compounds occurred only in the mineral kingdom.)

Litharge ("stone silver") is lead monoxide PbO. The name probably refers to the process of cupellation (which has been carried on for over 2,000 years). A lead-silver alloy is heated, in a stream of air, in a furnace with a hearth of bone-ash or cement. The lead is oxidised to litharge which is partly blown from the surface and partly absorbed by the bone-ash. Purified silver remains. A more romantic explanation of the name is based on the fact that litharge is reduced to lead when heated with charcoal. The ancients, on seeing the molten lead, may have mistaken it for silver.

In medical terms the root LITH(o)- usually refers to stones in the bladder or kidney.

lith'iasis—the formation of stones in the body.

litho'tomy—cutting into the bladder for the removal of stones.

LITRE

The unit of volume in the metric system. It is equal to the volume of a kilogramme of pure water at 4°C. and is almost identical with a cubic decimetre (approximately 1¾ pints). The name was coined in 1793, probably from Low Latin and ultimately from the Greek *litra* (a pound). The only derivative in common use in science is the **milli'litre**—one-thousandth of a litre (almost the same as a cubic centimetre).

LOC-, LOCO-

a place (L. *locus*),

This root is well-known in such words as **local** (at one place), **locality** (the place where a thing is) and **locate** (to find the place where a thing is). **Locomotion** is the moving from one place to another; a **locum tenens** (commonly called a **locum**) is a doctor who temporarily takes the place of another.

dis'location—a putting "out of place", e.g. the separation of two bones at a joint (especially a ball-and-socket joint), a moving out of place of atoms in a metal crystal.

radio'location—Radar—a method of finding the position (place) of a thing, e.g. an aeroplane, by means of the reflection of radio pulses which are sent to it.

locus (*Mathematics*)—a line traced out by a point which moves according to given conditions (i.e. it marks out the places where the point can be), e.g. the locus of a point which moves in a plane at a fixed distance from a fixed point is a circle.

loc'ulus—"a little place"—any small, separate cavity in (e.g.) a spore, an organ, etc.

uni'locular—containing, or consisting of, a single compartment.

pluri'locular—(spore container, ovary, etc.) which is divided into several compartments.

-LOGY, (-OLOGY)

As scientific knowledge developed, the old Natural Philosophy split into various distinctive branches and these, in turn, were broken down into specialised sub-branches. In forming the names of these branches and sub-branches great use has been made of the suffix -LOGY.

The suffix occurred originally in words adapted from Greek words ending in *-logia*. (Some of the earliest passed through a French form ending in

-logie.) In some of the Greek words *-logia* was derived from *logos*, a word, a discourse. More commonly it was derived from a form of the verb *legō*, to speak, and denoted the character, action or department of knowledge of a person who speaks. Thus *eulogia* meant "well-speaking" (whence **eulogy**, praise); *astrologia* (whence **astrology**) was the department of knowledge about which an *astrologos* ("star speaker", astrologer) spoke. Following the pattern of such words as **astrology, genealogy, theology**, -LOGY is now widely used in forming the names of departments of knowledge (e.g. various sciences) and may be freely translated as "study of".

The first branch of science to be given a name ending in -LOGY appears to have been **anthropology** ("the study of Man"). Further names were introduced during the seventeenth and eighteenth centuries and so the fashion was set for the great mass of names which were devised during the nineteenth and twentieth centuries. Only a small selection of these names of branches of science is given here in illustration.

> **biology** (the study of living things), **zoology** (of animals), **entomology** (of insects), **psychology** (of the mind), **neurology** (of the nervous system), **pathology** (of suffering and hence of diseases), **gynaecology** (of diseases and disorders of women), **geology** (of the rocks of the Earth), **spelaeology** (of caves), **limnology** (of fresh-water lakes), **conchology** (of shells), **meteorology** (of the atmosphere and hence of the weather), **chronology** (of time and dates), **cosmology** (of the universe).

It is interesting to note that the name **phytology** (the study of plants), corresponding to zoology, is rarely used; the name *Botany* is generally preferred.

In the natural formation of such names from Greek roots, the suffix -LOGY is normally preceded by the letter *o*. (*Genealogy* is an exception.) Hence the idea has grown that the suffix is -OLOGY and, indeed, the branches of learning are sometimes collectively called 'the ologies'.

As is usual when a suffix is found to be convenient and is well-known, -(O)LOGY has been added to stems of various natures and origins. Thus in **sociology** (the study of the nature and laws of human society), **radiology** (the medical use of X-rays) and **pestology** (the study of pests), it has been added to Latin stems. In **parasitology** it has been added to the effectively English word *parasite*. Some of the modern inventions (e.g. **weatherology, sexology**) are horrible words and it is to be hoped that they will not be permanently accepted into the language.

LOIM-, LOIMO-

plague, pestilence, hence infectious disease (Gk. *loimos*).

loim'ic—pertaining to infectious diseases.

loimo'logy—the study of infectious diseases.

LONGI-

long, lengthwise (L. *longus*).

longi'caudate—having a long tail.

longi'pennate—having long wings or feathers.

longi'tude—the angular 'distance' of a place east or west of the Greenwich meridian.

longi'tudinal—lengthwise, e.g. **longitudinal waves** (e.g. sound waves) in which the particles vibrate backwards and forwards along the line in which the waves are travelling.

LOPHO-

a crest, comb, tuft (Gk. *lophos*).

loph'odont—(animal) having cross-ridges on the grinding surfaces of the teeth.

lopho'branchiate—having gills in lobes or tufts.

lopho'trichous—"with tufted hairs"—(organism) having flagella (hair-like outgrowths) in a tuft or group on the surface of the cell.

LORDO-
A slight backward bending of the body (as a result of a mis-shapen spine). (Contrast KYPHO-.) (Gk. *lordos*.)

lordo'sis—a backward bending of the body caused by the excessive forward curving of the lower part of the spine.

lordo'scoliosis—a deformed state of the spine in which there is both lordosis and scoliosis (sideways curving).

LUC-, LUCI-
light (L. *lux*, *luc-*).
Lucifer is the "bearer of light". The name has been used both for Venus as the morning star and for a match. A **lucid** description or explanation is bright and clear.

trans'luc'ent — (material) through which light can pass but through which one cannot clearly see things, e.g. frosted glass.

luci'philous — (plant) which seeks ("likes") light.

luci'fugous — (plant, animal) which shuns light.

luci'ferin—a substance which occurs in the light-producing organs of some creatures (e.g. Glow Worm). Light is produced when it is oxidised. The action is brought about by the enzyme **luciferase**.

LUMB-, LUMBO-
the loins (L. *lumbus*).

lumb'ar—pertaining to the loins, e.g. **lumbar vertebrae**—the vertebrae (bones of the spine) in the 'small of the back'.

lumbo-abdominal—pertaining to the loins and the abdomen.

LUMEN, LUMIN-
light (L. *lumen*, *lumin-*).
This root is well-known in the words **luminous** and **illuminate**. (In science a luminous body is one which emits its own light, e.g. the Sun but not the Moon.)

lumen—a unit for measuring the flow of light from a source or on to a surface. (Also see below.)

lumin'escence—the giving out of light for reasons other than being at a high temperature, e.g. **chemo'luminescence** (as a result of chemical action), **electroluminescence** (as a result of an electric field), **cathodo'luminescence** (as a result of cathode rays).
Biologists use the word **lumen** to denote a cavity, e.g. an empty plant cell and, especially, the central cavity of a duct or gland. This strange use of the word may be traced to the fact that a cavity, in a section of tissue under a microscope, shines with the light which passes through it.

LUN(A)
the Moon (L. *luna*).
A **lunatic** was once supposed to be affected in the intensity of his ravings by the changes in the Moon.

lunar—pertaining to the Moon, e.g. a lunar eclipse.

cis-lunar space—"space on this side of the Moon"—the space between the Earth and the Moon's orbit.

lun'ate—crescent-shaped.

lun'arian—a supposed inhabitant of the Moon; one with a special interest in or knowledge of, the Moon.

LUTE-, LUTEO-
of an orange-yellow colour (L. *luteus*).

lute'ous—of an orange-yellow colour.

lute'in—a yellow colouring matter found in leaves and petals of various plants and also in egg-yolk.

corpus luteum—"a yellow body"—a yellowish nodule which develops from a Graafian follicle (in the ovary of mammals) after the expulsion of the egg.

luteo-cobaltic chloride — a complex chloride of cobalt and ammonia which is yellow in colour.

LYC-, LYCO-
a wolf (Gk. *lykos*).

lyc'anthropy—a form of madness in which a person imagines himself to be a wolf (or, loosely, an animal).

Lyco'podium — the genus of Club Mosses (plants botanically related to the Ferns but more like Mosses in appearance). The name ("wolf foot") probably refers to the resemblance of the root to a wolf's paw. (An alternative explanation is the supposed resemblance of a growing shoot to a wolf's paw.)

LYMPH, LYMPHO-

Lymph (Latin *lympha*, water) is a watery, colourless liquid which occurs in the tissues and organs of the body.

lymph'adenitis—inflammation of the lymph glands.

lymph'adenoma—a diseased growth of the lymph glands.

lympho'genic—produced in a lymph gland.

lymph'angi'ectasis—"a state of enlargement of the lymph vessels" (usually due to an obstruction).

lympho'cyte—a type of white cell formed in the lymph glands.

karyo'lymph—the liquid part of a cell nucleus.

LYO-

A prefix (derived from the Greek *lyō*, to loosen) used in modern terms to denote a solution or the liquid of a solution.

lyo'philic—"solution liking"—said of a colloid (e.g. starch solution) which is easily formed and is easily re-formed after the solid has separated out from it.

lyo'phobic—"solution fearing"—said of a colloid (e.g. a solution of gold) which is only formed with difficulty and is not easily re-formed after the solid has separated out from it.

lyo'sorption—the adsorption of a liquid on to a solid surface, especially of the solvent (liquid of a solution) on to particles suspended in it.

lyo'lysis—the breaking down of a dissolved salt (into an acid and a base) by the action of the solvent on it.

-LYSIS, -LYST, -LYTE, -LYTIC

loosening; hence breaking down, setting free (from Gk. *lyō*, to loosen).

An **analysis** is a "loosening up again" of (e.g.) a sentence or a chemical substance in order to examine its composition.

electro'lysis — the setting free of chemical substance by the passage of an electric current through a solution or a melted solid. The solution (or the substance which is dissolved to form the solution) is termed the **electrolyte**.

hydro'lysis—the decomposition of a compound by the action of water, especially the decomposition of a salt (or an ester) into an acid and base (or an alcohol).

cata'lysis—"a loosening down"—the speeding up of a chemical action by the presence of a substance which is not itself used up or changed by the action.

pyro'lysis—the decomposition of a substance by heat.

haemo'lysis, haemocyto'lysis — the breaking down and dissolving of red blood cells.

bacterio'lysis—the breaking down of bacterial cells by anti-bodies (**bacteriolysins**) formed to resist the bacteria.

para'lysis — "irregular (disordered) loosening"—loss of muscular power due to nervous disease.

-LYTIC is used to form adjectives, e.g. an **electrolytic** cell, a **paralytic** person.

-LYST is used to form the name of the agent which brings about the -LYSIS, e.g. an **analyst** is a person who analyses, a **catalyst** is a substance which brings about catalysis.

M

m-

(1) An abbreviation for *meso-* (chem.), e.g. ***m*-tartaric acid.**

(2) An abbreviation for *meta-* (chem.), e.g. ***m*-dinitrobenzene.**

-M, -MA, -ME, (-MATIC)

The addition of the suffix -MA to a Greek stem (e.g. a verbal stem) produces a noun (often expressing the result of a verbal action). Thus **drama**, from *dra-* (*draō*, to do), is a deed, an act, an action (including action on the stage). Similarly,

oidema (**oedema** in English)—a swelling (*oideō*—to swell).

sarcoma—a growth of flesh (*sarkoō*—to make into flesh, to become fleshy). This word has given rise to the 'artificial' suffix -OMA (q.v.) in medical terms.

stigma—a prick, a mark, a point (*stizō*—to prick, to mark).

plasma—a moulded or modelled thing (*plassō*—to mould, to form). (Hence **-plasm**, etc.)

zygoma—a yoke, especially the bridge of bone in front of the ear and under the eye (*zygoō*—to yoke).

exanthema—"an out-flowering" — a disease which results in an eruption on the surface of the skin.

eczema—"an out-boiling"—a general term for inflammations and irritations of the skin.

parenchyma—"that which is poured in at the side"—tissue consisting of soft cells forming the packing material, e.g. in softer parts of a plant.

(The Greek plural of such words ending in *-ma* is formed by the addition of *-ta*, e.g. *stigma, stigmata*. In science such plurals are sometimes used but the practice is declining.)

In English words derived from the Greek, the suffix often takes the form -ME or -M. The former is seen in **scheme, theme, rhizome** ("a thing which roots"); the latter is seen in **poem, chasm, symptom** ("that which befalls one"), **prism** ("a sawn thing"), **diaphragm** ("a fence across"), **axiom** ("that which one holds true or worthy").

Corresponding English adjectives often end in -MATIC, e.g. **prismatic, symptomatic, axiomatic, astigmatic.**

(The English word **cinema** has come to us in a roundabout way. The verb *kineō* (to move) gives *kinēma* (movement) and hence the combining-element *kinemat(o)*- or *cinemato-*. A **cinemato'graph** is an apparatus for producing moving pictures; this word, when abbreviated, gives **cinema**.)

MACRO-

large (Gk. *makros*).

macro'cephalic—having a large head.

macro'scopic—large enough to be seen by the unaided eye. (Contrast *microscopic*.)

macro'cyte—a large white cell in the blood.

macro'molecule — a large molecule, e.g. of diamond (built up from atoms), of cellulose, polymers, etc. (built up from simple molecules).

MAGNET, MAGNETO-

The Greek name for lodestone (the natural iron oxide with magnetic properties) was *Magnēs* (*lithos*), or *lithos Magnētis*, i.e. the Magnesian stone (from Magnesia in Thessaly). The term *magnes* (with a stem *magnet-*) passed into Latin and eventually gave rise to the English word **magnet.**

From early times, however, the mineral which we now call pyrolusite (manganese dioxide) was confused with lodestone. In the Middle Ages lodestone was called *magnes* (masculine) and pyrolusite *magnesia* (feminine). Further confusion resulted in the eighteenth century when another mineral (now called magnesium carbonate) was discovered and, for some strange reason, believed to be related to magnesia. It now became necessary to distinguish the two kinds of magnesia by calling pyrolusite *magnesia nigra* (black) and magnesium carbonate *magnesia alba* (white). From this tangle we have obtained not only the word **magnet** but also the names of two chemical elements: **magnesium** (directly from *magnesia*) and **manganese** (a corruption of *magnesia*).

The combining-form MAGNETO- relates to magnets or to magnetic fields.

magneto-electric generator—a simple electric generator which has permanent (not electro-) magnets. Now usually called a **magneto.**

magneto'meter — an instrument for measuring the strengths of magnets or of magnetic fields.

magneto-striction—the slight change in the size of a magnetic material (e.g. iron, nickel) when it is magnetised.

MAGNI-
large, great (L. *magnus*).
This root is well-known in **magnify** (to make larger or to make to appear larger) and **magnitude** (size, greatness, brightness (of stars)).

MALAC-, MALACO-, -MALACIA
soft; a softening (Gk. *malakós*, soft).

malaco'derm—a soft-skinned animal. Adj. **malacodermatous.**

malac'ostracous — (animal) having a soft shell.

malaco'phyllous — having soft, fleshy leaves.

osteo'malacia—a softening of the bone (e.g. due to a vitamin deficiency).

encephalo'malacia—a softening of the brain (by disease).

malaco'logy—the study of Molluscs. (The Molluscs may be regarded as soft-bodied animals although they often have hard shells; L. *mollis*, soft.)

MANIA
madness (Gk. *mania*).
A **mania**, in everyday speech, is a mental disorder marked by excitement and violence; a **maniac** is a person affected with a mania, i.e. raving mad.

klepto'mania—an irresistible desire to steal.

megalo'mania—the delusion of being "big", of being grand, of possessing great wealth, position, etc.

dipso'mania—an uncontrolled craving for drink.

pyro'mania—a mania for destroying things by fire.

MANO-
thin, rare (Gk. *manos*).

mano'xylic wood — wood of loose, "thin" texture (containing a good deal of parenchyma).

mano'meter — an instrument for measuring the pressure ("thinness") of gases and vapours.

MARI-
the sea (L. *mare* (n.); *marinus* (adj.)).
This root is seen in **mariner** and **submarine** (a boat which can travel under the surface of the sea).

mari'gram—a record, made by an instrument (a **marigraph**), of the varying height of the tide.

MARMOR-
marble (L. *marmor*).

marmor'eal—of, or like, marble.

marmor'ate—marked or coloured like marble.

MAST-, MASTO-
the breast (Gk. *mastos*).

mast'itis—inflammation of the female breast.

mast'odynia—pain in the breast.

Mast'odon—a large extinct Pliocene mammal (like an elephant) which had pairs of nipple-shaped projections on the molar teeth.

mast'oid—resembling a nipple (as the bony projection, the **mastoid process,** behind the ear).

masto-parietal—pertaining to the mastoid bone and parietal bone (forming side and top of skull).

(Note that MASTO- may refer to the breast or to the mastoid process.)

MASTIGO-
a whip (Gk. *mastix*, *mastig-*).

Mastigo'phora — a class of simple organisms (Protozoa) which bear one or more flagella (whip-like projections).

MAXILLA, MAXILLI-, MAXILLO-
the jaw-bone (L. *maxilla*); a feeding appendage of an arthropod animal.

maxilla—the jaw-bone (especially the upper) in vertebrate animals; in arthropod animals (e.g. insects), an appendage close behind the mouth used in feeding. Adj. **maxillary.**

maxillo'dental—pertaining to the jaws and the teeth.

maxilli'ped—an appendage, especially in Crustacea (Crabs, etc.) used to transfer food to the mouth.

MED-, MEDI-
middle (L. *medius*).
Medium means of middle quality; the

Mediterranean Sea is in the middle of the land known to the Romans.

medi'an—situated in the middle (e.g. a median nerve); a line drawn from a corner of a triangle to the mid-point of the opposite side; the middle number (or measure) of a series of numbers (or measures).

med'ulla—the soft middle part of an organ or tissue, e.g. the marrow inside a bone).

MEG-, MEGA-, MEGALO-, -MEGALY

large, great (Gk. *megas*, *megal-*).

mega'phone — an instrument for making one's voice sound louder and so carry farther.

mega'phyllous — having very large leaves.

mega'cephalic—having an unusually large head.

mega'lecithal—(egg) which contains a large quantity of yolk.

In scientific units, MEGA- denotes one million, e.g. **megohm**—one million ohms of electrical resistance, **megacycles**—millions of cycles of alternating current or of waves, **megawatt**—one million watts, **megaton**—an explosive power (e.g. of a nuclear weapon) equal to that of one million tons of T.N.T.

megalo'cyte—an unusually large red-cell in the blood.

megalo'mania—a delusion of being grand, of possessing great wealth, a high position, etc.

acro'megaly—a disease which causes the outer parts of the body (head, hands, feet) to become very large.

spleno'megaly—an unusual enlargement of the spleen.

MEIO-, MIO-

less (Gk. *meiōn*).

meio'sis—"a state of lessening"—the halving of the number of chromosomes in the formation of reproductive cells. (It is interesting to note that, for several centuries, meiosis denoted an understatement; it was not until the present century that the word acquired its specialised biological meaning.)

Mio'cene—"less recent (period)"—a geological period (part of the Cenozoic era) about 35 to 15 million years ago. (The Alps and the Himalayas were formed in this period but it is not represented by any known deposits in Britain.)

Note. *Miosis* (contraction of the pupil of the eye) is a corruption of *myosis* (from Gk. *myō*, to shut the eyes).

MELAN-, MELANO-

black (Gk. *melas*, *melan-*).

melan'in—a dark brown or black pigment in the hair and (sometimes) the skin. **melanism**—an abnormal condition caused by over-production of melanin.

melan'osis—an abnormal development or deposit of a black pigment in tissues of the body.

melan'ite—black garnet, a silicate of calcium and iron.

melano'sporous — producing black spores.

melan'aemia—a condition, associated with severe forms of malarial fever, in which the blood contains granules of a dark brown or black pigment.

For a note about **melancholy**, see CHOL-.

MELLI-

honey (L. *mel*, *melli-*).

melli'ferous — yielding or producing honey.

melli'vorous—honey-eating.

MENING-, MENINGO-

Pertaining to the **meninges** (the membranes covering the brain and spinal cord). (Gk. *mēnnix*, *mēning-*, membrane.)

mening'itis — inflammation of the meninges.

meningo'cele — a swelling of the meninges which pushes through a defective part of the skull.

meningo'myelitis—inflammation of the meninges and the spinal cord.

MENO-, MENSTRU-, -MENIA

Pertaining to the **menses**, the monthly discharge of blood by a female. (L. *mensis*, a month; *menstruus*, monthly; Gk. *mēn*, *mēno-*, a month.)

menstru'ation — the periodical discharge from the womb of a female mammal.

meno'pause—the natural ending of menstruation (because of age).

meno'rrhoea—excessive loss of blood during menstruation.

xeno'menia—"strange menstruation" —a condition in which, in the absence of normal menstruation, there is monthly bleeding from other parts of the body (e.g. the nose).

MER-, MERI-, MERO- (1)

a part (Gk. *meris*; *meros*). (Also see MERO- (2) below.)

penta'merous—consisting of five parts, e.g. (flower) with petals, sepals, etc., arranged in sets of five. So also **dimerous** (two), **trimerous** (three).

meta'meric animal—an animal, such as an Earthworm, which consists of a row of similar parts. Each part is a **metamere** or **merome.**

meta'meric compounds — chemical compounds of the same class, and built up from the same number and kinds of atoms, but differing in the arrangement of the atoms, e.g. diethyl ether $C_2H_5OC_2H_5$ and methyl propyl ether $CH_3OC_3H_7$.

mono'mer—a "unit part"—a simple chemical compound whose molecules can be made to join together (usually in chains), so producing **dimers** (two parts), **tetramers** (four parts), . . . **polymers** (many parts).

iso'mers—chemical compounds with "equal parts" — compounds built up from the same number and kinds of atoms but differing in the way the atoms are arranged in the molecules. The phenomenon is called **isomerism.**

meri'spore—one part of a multiple spore.

meristic variation—a variation from the usual number of parts, e.g. of petals in a flower, of parts in a metameric animal.

mero'blastic—said of an ovum (egg cell) in which only a part of it can undergo division.

mero'hedral — (crystal) having less than the normal number of faces.

Note. *Meridian* comes from L. *meridies*, midday.

MERO- (2)

the thigh (Gk. *mēros*). (Also see MERO- (1) above.)

mero'cele — a hernia (protrusion) of the thigh bone.

mero'sthenic—with "thigh strength"— having the hind legs exceptionally well developed (as in Frogs, Kangaroos).

mero'cerite—"thigh horn"—the fourth segment of an antenna of a Crustacean.

MES-, MESO-

middle, intermediate (Gk. *mesos*). **Mesopotamia** is the land "in the middle of (between) the rivers" (Tigris and Euphrates).

mes'encephalon—the mid-brain of a vertebrate animal.

meso'carp—the middle layer of a fruit wall.

meso'derm—the middle layer of a developing embryo.

meso'phyte—a plant which lives in conditions (especially with regard to water) which are not extreme.

Meso'zoic era—the "middle life" era of geological time (about 200 to 70 million years ago) following the Palaeozoic ("ancient life") era and preceding the Cenozoic ("recent life") era.

meso'tron, meson—an elementary particle of mass between that of an electron and a proton.

In chemical nomenclature, *meso-* denotes a "middle" form of a compound, i.e. one in which the optical* activities of the atom groups at the two ends of the molecule are equal but in opposite directions so that the compound (as a whole) is optically inactive, e.g. *meso-* **tartaric acid.**

MET-, META-

This prefix (Gk. *meta*, *meta-*) has a range

* Power of the solution to twist the plane of polarisation of a beam of polarised light.

of meanings which may be summarised as follows:

(a) with, among, between;

(b) after, next after, behind, following;

(c) a change, e.g. of place, of condition.

It is also used in several specialised senses (as described below) in the naming of chemical compounds.

meta'plasm—any body within a cell but not forming part of the protoplasm, e.g. a particle of food, a drop of fat.

meta'carpals—the bones of the hand between the fingers and the wrist, i.e. those which come "after the carpals (wrist bones)".

meta'meric animal—an animal, such as an Earthworm, which consists of a number of similar body parts one behind the other.

Meta'zoa—a sub-kingdom of animals (showing some degree of complexity) which comes after the sub-kingdoms Protozoa and Parazoa.

meta'bolism—the total of the chemical and physical changes which take place in a living organism.

meta'morphic rocks—rocks of "changed form"—rocks formed, e.g. by heat and pressure, from previous rocks, e.g. slate has been formed from a clay rock.

meta'morphosis—the process in which there is a marked change, taking place in a relatively short time, in the form of an animal, e.g. the changing of a caterpillar into a butterfly.

meta'chrosis—the changing of colour by an animal (e.g. a Chameleon).

meta'plasis—"a change in moulding" —the transformation of a tissue into one of another structure, e.g. the change of cartilage into bone.

The prefix is used in three different ways in naming chemical compounds.

(1) Many acids can be regarded as compounds of an oxide with water. When an oxide can combine with various proportions of water, the prefix META- is used to denote the acid which is formed when one molecule of the oxide combines with one molecule of water.

meta'silicic acid—SiO_2+H_2O, i.e. H_2SiO_3. The salts are **metasilicates,** e.g. calcium metasilicate $CaSiO_3$. (Contrast orthosilicic acid SiO_2+2H_2O.)

meta'boric acid—$B_2O_3+H_2O$, i.e. $H_2B_2O_4$ or HBO_2.

(2) It is used in the naming of certain kinds of compounds formed by the replacement of two of the hydrogen atoms of benzene (C_6H_6) by other atoms or groups. Fig. (a) represents a molecule of chlorobenzene (C_6H_5Cl) in which one of the hydrogen atoms of benzene has been replaced by a chlorine atom. A second chlorine atom could replace a hydrogen atom in an adjacent position (2 or 6), in the "next after" position (3 or 5), or in the "opposite" position (4). The "next after" position is the **meta-position** and Fig. (b) represents the molecule of *meta*-**dichlorbenzene** ($C_6H_4Cl_2$).

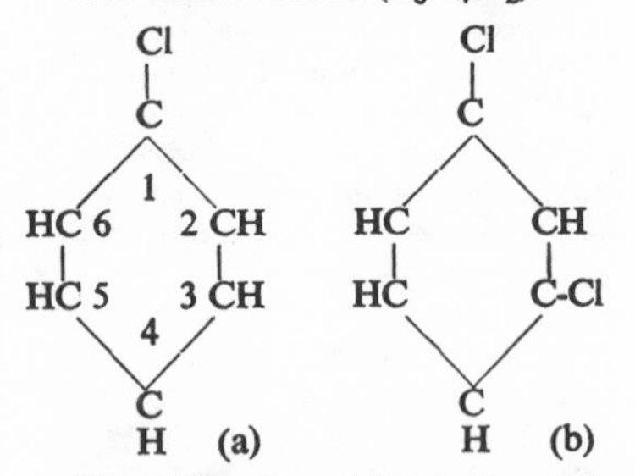

(The adjacent position is the *ortho*-position and the "opposite" position is the *para*- position.)

(3) The prefix is sometimes used to denote a polymer. Thus **met'aldehyde** is a polymer built up from three (acet)aldehyde molecules.

METALL-, METALLO-

a metal (L. *metallum*; Gk. *metallon*).

metall'oid—resembling a metal; a chemical element (e.g. arsenic) which has some of the qualities of a metal.

metall'isation—the changing of a substance (e.g. selenium) into a metallic form. **metallised valve**—a radio valve whose exterior has been coated with a conducting, metallic film.

metall'urgy—the science of extracting, refining and working metals.

metallo'graphy — that branch of metallurgy which deals with the structure of metals (and alloys) and the relation of this to their properties.

METEOR, METEORO-

high in the air; the atmosphere (Gk. *meteōros*, raised on high, lofty).

meteor—a small body which enters the Earth's atmosphere from space, a shooting-star.

meteor'ite—a mineral stone which has reached the Earth from space.

meteoro'logy—the study of the atmosphere, especially in relation to weather.

METER, METRE, METRO-, -METRY, (-METRICS)

Measurement is fundamental to science and it is therefore not surprising that the Greek *metron*, a measure, is the basis of a large number of scientific terms.

(Also see METRO-, the womb.)

(1) -METER, METRO-, a measure.

diameter—the "measure across" a circle.

perimeter—the "measure round" a figure (e.g. a rectangle).

iso'metric system (of crystals)—having "equal measures"—the system of crystal shapes based on three equal axes all at right-angles. (Also called the cubic system.)

metro'logy—the science of weights and measures.

metro'nome—an instrument which "arranges the measure"—an instrument which marks out time, e.g. by giving audible signals at regular intervals.

(2) METRE, the unit of length.

The **metre** is the standard unit of length in the **metric** system as used on the Continent and in scientific measurement. It is approximately equal to 39·37 inches. Multiples and sub-multiples are formed by adding the appropriate prefixes, e.g. **kilometre** (1,000 metres), **centimetre** (one-hundredth of a metre), **millimetre** (one-thousandth of a metre).

(3) -METER frequently denotes a measuring instrument. Thus a **thermometer** is an instrument for measuring temperature, a **barometer** is an instrument for measuring the "weight" (i.e. the pressure) of the air. From the very great number of instruments whose names end in -METER it is only necessary to quote a few typical examples.

(a) Names formed directly from Greek stems.

pyro'meter (instrument for measuring high temperatures), **photometer** (intensity of light or of sources of light), **hygrometer** (dampness of the air), **cyclometer** (turns of a wheel), **goniometer** (angles), **potometer** (intake of water by a plant).

(b) Names formed directly from Latin stems (i.e. hybrids).

calori'meter (instrument for measuring quantities of heat), **planimeter** (areas of plane figures), **pluviometer** (rainfall).

(c) Names formed from English or modern stems. (Note the frequent use of an *o* before -METER as if the suffix were -OMETER; this, of course, is a reflection of the *o* in names derived wholly from Greek elements.)

acidometer (instrument for measuring the density of an acid, e.g. in an accumulator), **galvanometer** (small electric currents), **refractometer** (refractive qualities of substances), **fathometer** (sea-depth in fathoms), **voltmeter** (electric pressure in volts), **wattmeter** (electric power in watts), **weatherometer**—an instrument used to determine the weather-resisting properties of paints.

(4) The suffix -METRY (or the modern alternative -METRICS) denotes the process of making the measurement or, more usually, the science of making the measurement. **Geometry** originated as the practical measurement of areas of the Earth's surface.

thermometry (the measurement of temperatures), **photometry** (of light intensities), **hydrometry** (of dampness of the air), **psychometrics** (of mental factors). **Tri'gono'metry** is "three angle measurement", i.e. the branch of mathematics which deals basically with calculations about the angles and sides of triangles; **telemetry** is "measurement from afar", i.e. the science of sending messages over a distance (e.g. from rockets, aircraft) to report on the measurements of instruments.

As with -METER, hybrids are common (often with *i* or *o* as a connecting vowel).

acidimetry (the measurement of strengths of acids), **alcoholometry** (the determination of the quantity of alcohol in a solution).

METH-, METHYL

'Wood spirit' (methyl alcohol) was discovered in wood-tar by Boyle in 1661 but it was not until 1812 that it was recognised as different from ordinary (ethyl) alcohol. It may be obtained by the destructive distillation of wood. The name **methyl** is derived from the Greek words *methy* (wine) and *hylē* (wood). **Methylated spirits** is alcohol which has been rendered unfit for human consumption by the addition of a proportion of methyl alcohol and colouring matter.

Methyl alcohol has the composition CH_3OH. The prefixes METH- and METHYL- refer to the characteristic group of atoms CH_3-.

meth'ane—$CH_3.H$, i.e. CH_4. A gaseous hydrocarbon, also known as marsh gas and fire damp.

methyl'ene—Methylene (CH_2-) stands in the same relation to methane (CH_4) as does ethylene (C_2H_4) to ethane (C_2H_6), but it only exists in combination with other atoms, e.g. **methylene chloride** CH_2Cl_2.

methyl'amine—the amine (q.v.) in which one of the hydrogen atoms of ammonia has been replaced by the methyl group, i.e. CH_3NH_2.

meth'oxy—the group of atoms CH_3O-, e.g. **methoxy-benzoic acid**—$CH_3O.C_6H_4.COOH$.

The origin of the word **amethyst** (a bluish-violet variety of quartz) is of interest. The ancients believed that a drinking cup made of amethyst prevented drunkenness; the name is derived from *a-* (not), *methystos* (drunken, from *methy*, wine).

METRE See METER.

-METRICS See METER.

-METRIUM, METRO-

the uterus (womb) (Gk. *mētra*).

metr'itis—inflammation of the uterus.

endo'metrium—the membrane lining the cavity of the uterus.

para'metrium — the joining-tissue round the lower part of the uterus.

metro'rrhagia — bleeding from the uterus (between the menstrual periods).

Also see METRO- (measure) under METER.

-METRY See METER.

MICRO-

small (Gk. *mikros*).

This prefix is well-known in **microscope** (an instrument for viewing small things) and **microbe** (a non-scientific name for a small organism especially one which causes a disease).

micro'analysis—the analysis of substances (determining their composition) by methods which permit the use of very small amounts of the substances.

micro'cephalic—having an unusually small head.

micro'climate—the climate (variations of light, temperature, dampness, etc.) of a very small region such as a field, a hedgerow, a flower-bed.

micro'phyllous—having very small leaves (as have plants living in very dry conditions).

micro'structure (of a metal or alloy)—the crystalline structure as revealed by a microscope.

micro'tome—an instrument for cutting very thin slices (e.g. of a rock) for examination under a microscope.

micro-film — a photographic reproduction of a document, etc., on a very small film.

In the names of units MICRO- denotes one millionth.

micro'farad—one millionth of a farad (unit of capacity of an electric condenser).

micr'ohm—one millionth of an ohm (unit of electrical resistance).

micro-second—one millionth of a second.

micron—a unit of length equal to one millionth of a metre, i.e. one thousandth of a millimetre.

MILL-, MILLE-, MILLI-

a thousand (L. *mille*).

Care is needed in interpreting this prefix. Occasionally it denotes a thousand (as in **mill'ennium,** a thousand years) but in scientific terms (especially the names of units) it more usually denotes one thousandth. (Note that it does *not* denote a million or a millionth.)

mille'pede, milli'pede — a kind of arthropod creature with numerous legs.

milli'metre—one thousandth of a metre.

milli'litre—one thousandth of a litre (very nearly the same as one cubic centimetre).

milli'ammeter—an instrument graduated to read electric currents in **milli'amps** (**milliamperes,** thousandths of an ampere).

milli'bar—one thousandth of a bar (unit for measuring atmospheric pressure).

MIM-, MIME-, MIMEO-

to copy, to imitate (Gk. *mimeomai*).

To **mimic** means to copy, to imitate; **mimicry** of people is often meant to bring ridicule. A **mime** was originally a simple drama in which there was much mimicry.

mimicry (*Zoology*)—the copying by one animal of the colour, habits or structure of another.

mime'sis—a state of close external resemblance between one animal and another or between an animal and an inanimate object.

mime'tic—pertaining to, or addicted to, mimicry or to mimesis.

mimeo'graph—an apparatus for making copies of written pages, etc.

mimet'ite—a mineral, an arsenate of lead, occurring in pale yellow or brownish hexagonal crystals. (So called because it resembles pyromorphite.)

MIO- See MEIO-.

MISS-, -MISSION, -MIT

to send; a sending (L. *mitto, miss-*, to send).

This root is seen in **emit, emission** (sending out) and **transmit, transmission** (sending across). A **missile** is an object or weapon suitable for throwing or discharging from some form of machine or engine.

MITO-

a thread (Gk. *mitos*).

mito'chondria—thread-like or rod-like bodies in living cells.

mito'sis—the normal process of cell division in the early stages of which the nucleus breaks up into thread-like chromosomes.

MNEMO-, MNEMON-, -MNESIA

memory (Gk. *mnēmē*).

mnemon'ic—a saying, formula or device for helping one to remember something else, e.g. All Cows Eat Grass (indicating the notes A, C, E, G in the spaces of the bass clef in music).

mnemo'taxis—the movement of an animal under the guidance (in whole or part) of the memory.

a'mnesia—loss of memory. (**Amnesty** is the intentional overlooking (and forgetting) of some act, hence a pardoning.)

ana'mnesia—the bringing "up again in the memory" of past things and events as a means of helping to indicate the cause or nature of a disease.

MODE, MODUL-, MODULUS

The Latin word *modus* has a range of meanings: a measure, a rhythm, a manner. The English word **mode** means the way or manner in which something is done; it is also a type of musical scale (e.g. the minor mode). To **modify** means "to make the proper measure" (usually to make partial changes or to make less severe).

The Latin word *modulus* means "a small measure". A **model** (which word has come through French from Latin) is a "small measure" or representation of a thing.

In scientific terms the roots usually have some reference to a measure or to the adjustment of a measure.

mode—This term is used in several ways in mathematics. In its simplest meaning it is the value of a quantity (e.g. height of a person, an examination mark) which occurs most frequently. In this sense it is one of the ways of indicating the central tendency of a set of quantities; it may or may not be the same as the average.

modulus—Again, this term is used in several ways. A **modulus of elasticity** is a measure of the elasticity of a substance; from it the deformation of any body made of the substance, when acted on by given forces, may be calculated. The **modulus of a complex number** (i.e. of a number which is partly real and partly imaginary) is the nearest approach to a 'real measure' of the number; if the complex number is $a+\sqrt{-1}.b$, the modulus is $+\sqrt{a^2+b^2}$.

modul'ate—"to adjust (or vary) the measure of." In radio communication, **modulation** is the impressing on a high frequency current (e.g. by varying its amplitude) of a lower frequency current to represent the speech, music, etc.

MOL-, MOLE-

a mass (L. *moles*).

mole'cule — "a little mass" — the smallest particle of a chemical substance which can exist independently, usually consisting of two or more atoms (like or unlike) joined together.

mol'ar—pertaining to that quantity of a substance whose weight in grammes is numerically equal to its molecular weight.

Note. *Molar* (one of the back grinding-teeth of a mammal) is derived from L. *mola*, a millstone.

MON-, MONO-

one, single, alone (Gk. *monos*).

(The prefix takes the form MON- before a vowel.)

This common prefix is seen in **monarch** ("sole ruler"). **monotonous** ("of one tone") and **monologue** (dramatic scene in which one person speaks by himself).

mon'aural—pertaining to the use of one ear instead of two.

mono'chromatic light—light of only one colour (wavelength).

mono'clinic system (of crystal forms) —the system in which one pair of the three (unequal) axes on which the forms are based intersect at an angle other than a right-angle.

mono'cotyledons—plants forming one of the two main classes of the Angiosperms (flowering plants), having, in addition to other features, only one cotyledon (seed-leaf) in the embryo, e.g. Grasses.

mono'dactylous—having one finger, toe, or claw.

mon'oecious—"with single houses"— having separate male and female flowers on the same individual plant (e.g. Hazel, Birch).

In the naming of chemical compounds the prefix denotes the presence of one atom (or group) of a stated kind in the molecule.

carbon mon'oxide—a compound of carbon and oxygen whose molecule contains one atom of oxygen, CO. (Contrast carbon dioxide CO_2.)

mono'chloro'methane—the compound formed when one hydrogen atom of methane (CH_4) is replaced by a chlorine atom, CH_3Cl. (Also called methyl chloride.)

mono'hydric alcohols—alcohols whose molecules contain only one hydroxyl group (-OH), e.g. ethyl alcohol (C_2H_5OH) but not glycol ($HO.CH_2.CH_2.OH$).

MORPH-, MORPHO-

form, shape (Gk. *morphē*).

morpho'logy—the study of the form and shape of plants and animals (or of crystals, etc.).

a'morph'ous—not having a definite shape; (*Chemistry*) not in crystalline form.

anthropo'morph'ous—of human form.

iso'morph'ism—the state of having "equal forms"—the formation of crystals of the same shape by different (though related) substances.

meta'morpho'sis — "a changing of form"—a marked change of form of an animal taking place in a short time, e.g. the changing of a caterpillar into a butterfly.

allelo'morph—in the "shape of one another"—one of a pair of contrasted inherited characters, e.g. tallness (or dwarfness) of peas, blue (or brown) colour of eyes.

Note that a few terms are derived from *Morpheus* (L.), the god of dreams and sleep.

morphia, morphine—the chief alkaloid of opium. Used extensively as a hypnotic for relief from pain.

morphino'mania—an uncontrollable craving for morphine or opium.

MUC-, MUCI-, MUCO-

Mucus (L.) is the sticky, slimy substance excreted by animals (e.g. from the nose) and a gummy substance found in plants.

muci'lage—a sticky substance such as gum.

muc'ins—protein substances occurring in mucus, saliva, etc.

muco'cele—a collection of mucus in a hollow organ (because the outlet is blocked).

muco'sanguineous — consisting of mucus and blood.

MULTI-

many (L. *multus*).

This prefix, which is seen in **multiply** and **multitude**, is less common in scientific terms than its Greek counterpart *poly-*.

multi'lateral—having many sides.

multi'cellular — consisting of many cells.

multi'parous—producing many young at a birth; (woman) who has borne several children.

mult'ungulate — having the hoof divided into three or more parts.

MUT-

to change (L. *muto*, *mutat-*).

mut'ation—a sudden change in character (e.g. in colour of petals) of an offspring which is not inherited but which is passed on to future generations.

per'mut'ations—"a changing through" the series—all the different ways of selecting and arranging a group (of given size) of numbers or objects from a given larger group.

trans'mut'ation—the changing of one chemical element into another.

com'mut'ator — an instrument for switching an electric current from one circuit to another, or for reversing the current, or for collecting current from several circuits.

muta-rotation—the gradual change in the optical* activity of a freshly prepared solution (e.g. of sugar).

MY- See MYO-.

-MYCES, -MYCETES, MYCETO-, MYCO-

a fungus (Gk. *mykēs*, *mykēt-*, a mushroom).

Saccharo'myces—the "sugar fungus" —yeast, which converts sugar into alcohol.

Asco'mycetes—the class of Fungi which bear their spores in asci (q.v.). So also other names, e.g. **Basidiomycetes**, **Eumycetes**, etc.

* Power of a solution to twist the plane of polarisation of a beam of polarised light.

myco'logy—the study of Fungi.

myco'sis—a disease of an animal caused by a parasitic fungus.

mycet'oma—a fungal disease of the foot or hand.

myceto'phagous—fungus-eating.

myco'rrhiza — a mutually beneficial association between a fungus and a root of a higher plant.

mycelium—the mass of threads which forms the body of many kinds of fungi. (Here the Greek root has been given a Latin ending.)

-MYCIN

A suffix used in forming the names (mostly proprietary) of substances derived from fungi (or fungus-like organisms) used as antibiotics in the treatment of disease, e.g. **Aureomycin, Chloromycin, Streptomycin.**

MYEL-, MYELO-

marrow (Gk. *myelos*).

(1) the marrow (soft inner part) of a bone.

osteo'myelitis — inflammation of the bone marrow.

myel'oma—a tumour composed of bone-marrow cells.

(2) the marrow of the spine, i.e. the spinal cord.

myel'itis — inflammation of the spinal cord. **polio'myelitis**—inflammation of the grey matter of the spinal cord.

myel'in—the white, fatty substance which forms the sheath of the spinal cord and of nerves.

a'myelin'ate — (nerves) lacking a myelin sheath.

myelo'cele—a pushing out of the spinal cord due to a defect of the spinal vertebrae.

myelo'malacia—a softening of the spinal cord by disease.

MYO-, MYOS-, MY-

The Greek *mys*, *myos* means (1) a mouse, (2) a muscle. The first meaning is seen in **Myosotis** (Forget-me-not), so called because of the 'mouse-ear' shape of the leaves. In scientific terms the root usually denotes muscle tissue.

my'algia—pain in the muscles.

my'asthenia—weakness in the muscles.

myo'carditis — inflammation of the muscular wall of the heart.

myos'in—the chief protein of muscle fibres.

electro'myo'graphy—the study of the electrical disturbances and variations in muscles in order to examine the condition of the muscles.

Note. *Myopia* (short-sightedness) and *myosis* (contraction of the pupil of the eye) are derived from the Gk. verb *myō*, to shut (the eye).

MYRIA-, MYRIO-

many, innumerable (Gk. *myrios*).

Myria'poda — Arthropod creatures with many legs—Centipedes and Millipedes.

myrio'sporous—producing very many spores.

MYRING-, MYRINGO-

the ear-drum (Late L. *myringa*).

myring'itis—inflammation of the ear-drum.

myringo'scope—an instrument for inspecting the ear-drum.

myringo'tomy—the surgical cutting of the ear-drum.

MYRMECO-

ant (Gk. *myrmēx*, *myrmēk-*).

myrmeco'phagous—feeding on ants.

myrmeco'chory — the spreading of seeds by ants.

MYX-, MYXO-

slime, mucus (Gk. *myxa*).

myx'oma—a tumour composed of clear jelly-like substance. **myxomat'osis**—a contagious disease of rabbits.

myx'oedema — "mucus swelling"— a condition caused by deficient action of the thyroid gland, characterised by thickening of the tissue just below the skin, loss of hair, gain of weight and slowness of mental and physical processes.

Myxo'mycetes—the "slime fungi"—a

group of very simple organisms mostly living on rotten wood or in the soil.

Myxo'phyceae—the "slime algae"—the group of Blue-green algae.

N

n-

(*Chemistry.*) An abbreviation for *normal*, indicating an unbranched chain of carbon atoms in the molecule, e.g. *n*-**butyl alcohol** $CH_3.CH_2.CH_2.CH_2OH$. Contrast *iso*-butyl alcohol

$$\begin{matrix} CH_3 \\ CH_3 \end{matrix} \rangle CH.CH_2OH.$$

NAN-, NANO-

small, like a dwarf (L. *nanus*).

nan'ism—the condition of being like a dwarf.

nano'soma (or **nanosomia**) **pituitaria**—dwarfism caused by under-activity of the pituitary gland.

nano'plankton — plankton (floating plants and animals) consisting of microscopic organisms.

The prefix NANO- is occasionally used in the metric system of units to denote one thousandth-millionth (10^{-9}).

NAPHTH-, NAPHTHA

Naphtha (in Latin and Greek) was the name given to inflammable liquids issuing from the earth. It is now used for one of the more volatile petroleum products consisting largely of heptane. In chemical names the element NAPHTH- indicates some relationship with naphthalene.

naphtha'l'ene — a crystalline hydrocarbon with a smell of 'moth balls'. The molecule, $C_{10}H_8$, consists of two benzene rings fused together.

naphth'ols—alcohols (phenols) formed by the replacement of a hydrogen atom of naphthalene by a hydroxyl (-OH) group, $C_{10}H_7OH$.

naphth'yl—the group of atoms $C_{10}H_7$-, e.g. **naphthylamine,** $C_{10}H_7NH_2$.

NARCO-, NARCOT-

numbness, loss of feeling, sleep (Gk. *narkē* (n.), *narkotikos* (adj.)).

narco'sis—a state of drowsiness, insensibility or unconsciousness produced by a drug.

narcot'ic—(drug) which produces narcosis.

narco'lepsy—a disease causing sudden, uncontrolled desires to sleep.

NASC-, NAT-

to be born (L. *nascor, nat-*).

nasc'ent—in the act of being born, e.g. **nascent hydrogen**—hydrogen in the state (probably in unjoined atoms) in which it is first formed in a chemical reaction.

nat'al—of, or from, one's birth.

ante'natal—before birth, e.g. antenatal care of pregnant women.

NASO-, NAS-

the nose (L. *nasus*).

nas'al—pertaining to the nose.

naso-labial—pertaining to the nose and the lips.

naso'sinusitis—inflammation of the air-containing bony cavities connected with the nose.

NAT- See NASC-.

NAT-, NATAT-

to swim (L. *nato, natat-*).

natat'orial—adapted for swimming.

super'nat'ant liquid—the clear liquid which is "swimming above" a sediment.

NECR-, NECRO-

a dead body, a corpse (Gk. *nekros*).

Necromancy is the art of predicting by supposed communication with the dead.

necr'opsy—the examination of a dead body (e.g. to discover the cause of death).

necro'phagous—feeding on the bodies of dead animals.

necro'phorous — carrying away the bodies of dead animals (as by certain Beetles who then bury them).

necro'genous—living or developing in the bodies of dead animals.

necro'biosis—the gradual death and decay of a cell in the living body; decay of body tissues.

NECT-, NECTO-, NEKT(O)-
swimming (Gk. *nēktos*).

necto'pod—an appendage adapted for swimming.

necton, nekton—actively swimming organisms (as contrasted with plankton which consists of floating and drifting organisms).

NEMA-, NEMAT-, NEMATO-
a thread, a filament (Gk. *nēma*, *nēmat-*).

Nemat'oda—a phylum of unsegmented worms: Thread Worms and Round Worms.

nemato'cyst—a stinging cell (e.g. of a Jellyfish) consisting of a bag of poisonous liquid and a pointed, thread-like sting.

nemato'parenchymatous thallus—"thread-like side-tissue body"—an algal body which is made of a mass of united threads.

proto'nema — "first thread" — a branched, filamentous plant-body produced when a moss spore germinates.

NEO-
new (Gk. *neos*).

neo'lithic—pertaining to the "new" (later) Stone Age when ground and polished stone implements were made.

neo'plasm—newly formed tissue in the body, especially a tumour.

neo'prene—a general name for artificial rubbers ("new" rubbers) made from chloroprene.

neo'pallium—"a new cloak"—that part of a brain of a mammal which is concerned with impressions from the senses (other than smell).

neo'dymium—a chemical element (a rare earth) discovered as one of the two elements in didymium (q.v.).

neon—one of the "new" gases discovered in 1898, an inactive gas present in small quantities in the atmosphere.

The prefix is sometimes added to an adjective or noun to give the notion of 'later, modern, recast', e.g. **Neo-Cambrian** (later Cambrian), **Neo-Darwinism** (a modern form of Darwinism).

NEPH-, NEPHO-
a cloud (Gk. *nephos*).

nepho'logy—the study of clouds.

nepho'graph — an instrument for photographing clouds.

nepheline, nephelite—cloudy, white minerals (silicates of sodium and aluminium) found in igneous rocks.

iso'nephs—lines drawn on a map to pass through places having equal amounts of cloud at a given time.

NEPHR-, NEPHRO-
the kidneys (Gk. *nephros*).

nephr'itis—inflammation of the kidneys.

nephr'ectomy—the cutting out of a kidney by surgery.

nephro'pathy — any disease of the kidneys.

nephro'ptosis—a dropped kidney, a 'floating' kidney.

hydro'nephro'sis—"a state of water in the kidney"—the swelling out of the kidneys because of urine which is held up by an obstruction in the urinary tract.

nephr'idium—"a little kidney"—an excretory organ of lower animals.

nephro'stoma—"kidney mouth" — a funnel-like opening leading to a nephridium.

NEUR-, NEURO-
a sinew, a cord, hence a nerve (Gk. *neuron*).

Probably the best known example of a word derived from this root is **neuralgia** —spasmodic pains along the course of a nerve.

neur'al—pertaining to nerves.

neur'on, neur'one—a nerve cell (with its branches).

neuro-muscular—pertaining to nerves and muscles.

neuro'logy—the study of the nervous system.

neuro'sis—any one of the diseases thought to be due to improper working of the nervous system.

The root is also used to denote the veins of a leaf or the vein-like ribs of an insect's wing.

neur'ation—the system of veins in a leaf or an insect's wing.

Neuro'ptera—the order (group) of insects which have two pairs of thin, transparent wings with a network of veins. Lace Wings, Scorpion Flies, and others.

NID-, NIDI-
a nest (L. *nidus*).

nidi'fy—to build a nest or nests.

nidi'fication—the process of nest-building; the manner of nest-building.

nidi'colous — "nest inhabiting" — (birds) which remain in the parents' nest some time after hatching.

nidi'fugous — "nest fleeing" — (birds) which leave the parents' nest soon after hatching.

nidamental—nest-forming; serving as a receptacle for eggs; (gland) which secretes material for the formation of a nest or for making a covering about an egg. (L. *nidamentum*, material for a nest.)

NIGR-, NIGRO-
black (L. *niger*).

nigr'escent — tending to black, becoming black.

nigro-punctate — marked with black dots.

NIMBUS
The rain cloud, a dense layer of dark, shapeless cloud with ragged edges. (L. *nimbus*, a cloud, especially a rain cloud.)

cumulo-nimbus—a great mass of piled-up cloud, ragged at the top, dark and irregular at the bottom; a thunder cloud.

NITR-, NITRI-, NITRO-, NITROSO-
nitrogen.
The parent name is **nitre.** Both the Latin *nitrum* and the Greek *nitron* referred to soda but nitre is now identified as potassium nitrate.

nitro'gen — "nitre producer" — a gaseous chemical element, forming about four-fifths of the air, the essential element of nitric acid and nitrates. (The name was proposed by Chaptal as an alternative to *azote* (q.v.).)

nitr'ic acid—HNO_3, a highly corrosive liquid. Its salts are called **nitrates,** e.g. sodium nitrate $NaNO_3$.

nitr'ous acid—HNO_2, a much weaker acid; the salts are called **nitrites.**

nitro'genous—containing, or yielding, nitrogen or nitrogen compounds.

nitri'fying bacteria — soil bacteria which are able to form nitrogen compounds from nitrogen in the air. **de'nitrifying bacteria** — soil bacteria which liberate nitrogen from nitrogen compounds.

A **nitro-compound** is strictly one which contains the nitro-group $-NO_2$ in place of a hydrogen atom.

nitro'benzene—a compound $C_6H_5NO_2$ in which the nitro-group replaces one of the hydrogen atoms of benzene C_6H_6.

tri'nitro'toluene — $C_6H_2(NO_2)_3CH_3$, from toluene $C_6H_5CH_3$. A powerful explosive commonly called T.N.T.

nitro'glycerine—an explosive in which three nitro-groups take the place of three hydrogen atoms of glycerine.

A **nitroso-compound** is one in which the nitroso-group $-NO$ takes the place of a hydrogen atom.

nitroso-benzene—C_6H_5NO (from benzene C_6H_6).

NOC-, NOCI-
harm, hurt (L. *noceo*).

in'noc'uous—not harmful.

noci'ceptive—sensitive to pain. **nociceptive reflex**—an unconscious action (e.g. shutting the eyes) which protects from pain.

For *in'oculate*, see OCUL-.

NOCT-, -NOX
night (L. *nox*, *noct-*).

noct'urnal—during the night; active at night.

noct'ambulation—walking "at night", i.e. in one's sleep.

equi'nox—a time of the year (about March 21st and September 22nd) when, at all places on the Earth, days and nights are equal in length (12 hours).

NOD-, NODE
a knot (L. *nodus*).

node—This word is used in various senses in different sciences, e.g. a knob or swelling on a root or branch, a small hard tumour, a point of least displacement in a set of stationary waves.

inter'node—the space between nodes, e.g. between those on plant stems.

nod'ule—"a small node"—a small, rounded lump, a small knotty tumour.

NOMEN-, NOMI-, NOMIN-
a name (L. *nomen, nomin-*).
To **nominate** a person (e.g. for an office) is to call or propose his name.

nomen'clature—"a calling of names"—the system of naming, e.g. of chemical substances, of rocks, of plants, etc.

bi'nomi'al—having "two names" or two terms. **binomial nomenclature**—"two-term naming"—the system of using two terms for naming a plant or animal, e.g. *Bellis perennis* (=Daisy). **binomial expression** — an algebraic expression which consists of two terms, e.g. $(2x+y)$. **binomial theorem**—a formula (rule) for working out expressions of the type $(a+b)^n$, i.e. a binomial expression raised to a power.

de'nomin'ator — "the named part under"—the figure under the fraction bar in a vulgar fraction.

-NOME, -NOMICS, -NOMY
management, arrangement, ordered study, hence systematised knowledge. (Gk. *nomos*, anything arranged or apportioned, a law.)
Economy is basically "the management of the house".

agro'nomy—the art of managing farms.

taxo'nomy—"the orderly management of arrangement"—the art and study of classifying, particularly of plants and animals.

metro'nome—instrument which "apportions the measure". It marks out time, e.g. by giving audible signals at regular intervals.

astro'nomy—the systematic study of the stars and other heavenly bodies.

ergo'nomics—the scientific study of human work, of the physical and mental problems of work, of the effects of conditions of work, etc.

NON-
This prefix, as in general speech, means "not", e.g. **non-conductor** (a substance which does not conduct electricity), **non-ferrous** (not containing iron), **non-luminous** (not luminous, not emitting its own light) (L. *non*).

NON-, NONA-
nine (L. *nonus*, ninth).

non'ane — a paraffin hydrocarbon whose molecule contains nine carbon atoms, C_9H_{20}.

non'ose—a sugar whose molecule contains nine oxygen atoms.

nona'gon—a plane figure with nine angles (and nine sides).

NOSO-
disease (Gk. *nosos*).

noso'graphy—the systematic description of diseases.

noso'logy—the systematic classification of diseases.

noso'phobia—fear of contracting diseases.

NOTO-
the back (Gk. *nōton*).

noto'nectal—(beetle) which swims on its back.

noto'branchiate—having gills on the back (upper side) of the body.

noto'chord—a stiffening rod present in the backs of a few animals (e.g. Lancelet) and present in embryonic stages of vertebrate animals (later replaced by a backbone).

NOV-, NOVA, NOVO-
new (L. *novus*).

Nova—a "new" star, a star which makes a sudden appearance in the sky and (often) then decreases in brightness. (Actually a faint star which suddenly flares up brightly.)

Novo'caine—"new cocaine"—a substance similar to cocaine widely used as a local anaesthetic.

-NOX See NOCT-.

NUCLE-, NUCLEO-, NUCLEUS
A **nucleus** is "a little nut" (L. *nux*, *nuc-*, a nut). In general it means a denser, central core round which other parts are assembled.

In astronomy the term is used for the denser part of the head of a comet but the more important uses of the term are in biology, where it denotes the denser, controlling part of a living cell, and in physics, where it denotes the positively charged central part of an atom.

nucle'ate — having a nucleus (or nuclei). **multi-nucleate**—(cell) possessing many nuclei.

nucleo'plasm—the denser protoplasm composing the nucleus of a cell.

nucleo'proteins—a group of protein-compounds which form an important part of the nuclei of living cells.

nucle'ic acid—an acid which forms a part of nucleoproteins.

nucle'olus—a small round body within the nucleus of a cell.

nucle'on — an elementary particle which forms a part of a nucleus of an atom, i.e. a proton or a neutron.

nucle'ar reaction—a reaction between atomic nuclei. Nuclei of light elements can be made to fuse (join into one) by firing one against the other under powerful electric forces. If the atoms are heated to a very high temperature (millions of degrees) they move fast enough to fuse when they hit each other. This is a **thermo'nuclear reaction.**

NUD-, NUDI-
nude, bare, uncovered (L. *nudus*).

de'nud'ation—the laying bare of rocks by chemical and mechanical action followed by the carrying away (e.g. by wind or water) of the particles.

nudi'branchiate — having the gills exposed (not within a gill-chamber).

nudi'caudate—having a bare tail, i.e. not covered by hair or fur (as have Rats).

NUT-, NUTATION
to nod, to sway (L. *nuto*, *nutat-*).

nut'ant (*Botany*)—hanging with the apex downwards.

nutation—a regular oscillation in the direction of the Earth's axis.

nutation, circum'nutation—the movement of the tip of a growing stem so that it traces out a spiral in space.

NYCT-, NYCTI-
night (Gk. *nyx*, *nykt-* (n.), *nyktios* (adj.)).

nyct'anthous—having flowers which open only at night.

nycti'tropic—(leaves, etc.) which turn in certain directions at night.

nyct'alopia — "night not-seeing" — night-blindness, abnormal difficulty in seeing objects in the dark (often due to lack of vitamin A).

O

o-
(*Chemistry.*) An abbreviation for *ortho-* (q.v.), e.g. *o*-**di'nitro'benzene.**

-O, -O-
The combining-form of a root often ends with the letter *-o*. Many Greek roots readily give rise to such combining-forms, e.g. *chlōros* (green) → *chloro-* (as in *chloro'phyll*), *lithos* (stone) → *litho-* (as in *litho'graphy*). On the analogy of Greek combinations and their adaptations and imitations in Latin, *-o* is now affixed to many Latin stems (which, in Latin compounds, would have ended in *-i*) and also to some English words. Examples are *concavo-*, *ferro-*, *linguo-*, *radio-*.

Some suffixes of Greek origin (e.g. *-logy*, *-meter*) are so frequently preceded by a combining-form ending in *-o* that the *o* tends to be regarded as forming a part of the suffixes. Hence have come about such curious words as **alcohol-ometer** and **speed'ometer**. The *-o-* in such words is sometimes described as a connecting link between the component parts but (as in the examples given) an effective join could often have been made without it.

OCT-, OCTA-, OCTO-
eight (L. *octo*; Gk. *oktō*, *okta-*).
October was the eighth month of the Roman year.

octo'pus—a marine creature (a cephalopod) which has eight long, muscular tentacles ("feet").

octa'gon—a plane figure with eight angles (and eight sides).

octa'hedron—a solid figure with eight faces.

oct'ane — a paraffin hydrocarbon whose molecule contains eight carbon atoms, C_8H_{18}.

oct'ode—a radio valve with eight electrodes.

OCUL-, OCULO-
the eye (L. *oculus*).

ocul'ar—pertaining to the eyes or to sight.

ocul'ist—a doctor who specialises in diseases of the eye.

oculo'nasal—pertaining to the eye and nose.

bin'oculars—an instrument (for seeing distant things) for use with both eyes.

in'ocul'ate—to inject into a person the bacteria or viruses of a disease in order to produce a mild form of the disease and so build up a resistance against it. (The Latin word *oculus*, besides meaning an eye, also means a bud—compare the 'eye' of a potato. To inoculate means "to place a bud in".) In chemistry inoculation is the putting of a small crystal into a solution in order to start crystallisation.

ocellus—"a small eye"—a simple eye or eye-spot in an invertebrate animal; an eye-shaped spot of colour (e.g. on a leaf).

-ODE
(1) A suffix, similar to -OID, meaning "(a thing) of the nature of". (Gk. *-ōdēs*, contracted from *-oeidēs*, like.)

phyll'ode—"a thing of the nature of a leaf"—a flattened leaf-stalk which acts as a leaf.

stamin'ode—"a thing like a stamen"—an imperfectly developed, or sterile, stamen.

nemat'ode—"like a thread"—pertaining to the Round Worms and Thread Worms (the **Nematoda**).

cest'ode—"like a ribbon"—pertaining to the Tapeworms (the **Cestoda**).

(2) a way, a path (Gk. *hodos*).
This element is commonly used to denote the plate or wire by which an electric current is led into or out of a liquid (as in electrolysis), a gas (as in a discharge tube), or a near vacuum (as in a radio valve).

electr'ode—a general name for such a plate or wire.

an'ode—"up way"—the positive electrode by which current is led in.

cath'ode—"down way"—the negative electrode by which current is led out.

di'ode—a radio valve with two electrodes. So also **triode** (three), **tetrode** (four), **pentode** (five), **hexode** (six), etc.

Also see HOD-.

ODONT-, ODONTO-, -ODON
a tooth (Gk. *odous*, *odont-*).

odont'oid—tooth-like.

odonto'geny—the origin and development of teeth.

odonto'stomatous—with a "toothed mouth"—having jaws which bear teeth.

orth'odontic—having to do with the proper growth and development of teeth (not with the repair and removal of diseased teeth).

rhynch'odont—having a toothed beak.

peri'odont'itis — inflammation of the membrane round that part of the tooth which is in the jaw.

Odonto'glossum—"toothed tongue"—a genus of orchids.

Mast'odon—a large extinct animal which had pairs of nipple-shaped projections on the molar teeth.

ODORI-
an odour, a smell (L. *odor*).

odori'ferous—bearing or producing a smell (usually pleasant).

odori'metry—the measurement of the strength and lasting-power of odours.

odori'phore — "odour bearer" — a group of atoms which give a compound an odour.

-ODYNIA, -ODYNE
pain (Gk. *odynē*).

dors'odynia—pain in the muscles of the back.

pleur'odynia—pain in the muscles of the ribs, pain in the chest. So also **glossodynia** (tongue), **scapulodynia** (shoulder blade), etc.

an'odyne—(drug) which eases pain.

-Œ-, -OE-
The ligature (tied letters) œ has almost disappeared from English spelling. As ligatures are difficult to write, they are usually 'untied' and written as separate letters. Among the roots affected are **homoeo-, oedema, -oecious,** and **-rrhoea.** Having thus separated the letters, the *o* then appears to be unnecessary and the modern tendency is to omit it. In American spelling *-œ-* and *-oe-* are seldom seen but in English spelling the tendency to omit the *o* has not yet gained much ground. (This comment does not apply, of course, to words such as **coefficient** in which the juxtaposition of the *o* and *e* arises for a different reason.) It is interesting to note, however, that **ecology,** like **economy** (without an initial *o*), is now the standard spelling.

-OECIOUS, (-ECIOUS), -OECIUM
Literally, pertaining to "a house, a dwelling" (Gk. *oikos*, *oikion*, a house). The same root is seen in **eco'nomy** ("the management of a house") and **eco'logy** (formerly **oecology**)—the study of a plant or animal in relation to the environment in which it lives.

heter'oecious — having "different houses" — (parasite) which lives on different hosts at different stages of its life.

di'oecious—having "two houses"—having male and female flowers on separate plants of the same species, e.g. Willow, Stinging Nettle. Hazel and Birch have separate male and female flowers on the same plant; they are **mon'oecious.**

andr'oecium—"the male house"—the group of stamens in a flower; the male organs in a Moss.

The word **gynaeceum** (L. from Gk.) basically denotes the woman's part of a house. The use of the word to denote the female organs of a flower dates from about the middle of the last century. The spelling **gynoecium** (corresponding to **androecium**) is not infrequently seen but is incorrect.

OEDEMA, EDEMA
a swelling (Gk. *oidēma*).
(The *o* is often omitted from this root.)

oedema—a state of swollen tissue and accumulation of liquid in the body; dropsy.

erythr'(o)edema—a disorder of infants in which there is swelling and redness of the face, fingers, etc.

papill'(o)edema—a swelling of the head of the optic nerve within the eye.

OESOPHAG-, OESOPHAGO-
the **oesophagus** (food-pipe) (Gk. *oisophagos*).

oesophag'ectasis—enlargement of the oesophagus.

oesophago'scope—an instrument for viewing the inside of the oesophagus.

oesophago'tomy—the cutting into the oesophagus by surgery.

The Greek word *oisophagos* apparently contains the root *phag-* (to eat) but its exact derivation has not been satisfactorily explained.

OESTR-, OESTRO-, (ESTRO-)
The **oestrum** (or **oestrus**) is the period of sexual desire in a female animal. (L. *oestrus*; Gk. *oistros.*)

oestro'gens—a group of substances, produced (as an internal secretion) by the ovary, which cause the series of changes during the **oestrous cycle** and (particularly **oestrone**) the development and maintenance of the secondary female characteristics (e.g. texture of hair and skin, nature of voice).

-OID, -OIDAL, -OIDEA

A suffix (from Greek *-oeidēs*) forming adjectives and nouns with the meaning of "(thing) having the form of, resembling, like". It is chiefly used with Greek stems (as in **sphenoid**) but is also used with Latin stems (as in **fibroid**) and with stems of other origins (as in **alkaloid**).

Words ending in -OID are usually adjectives but are also sometimes used as nouns, e.g. **thyroid** (adj. and noun), **colloid** (noun). When the word ending in -oid is clearly a noun, the corresponding adjective is formed by adding *-al*, e.g. **colloidal**.

This suffix is very common, and it is only necessary to give a small, representative list of examples.

anthrop'oid—(ape) resembling a man.

arachn'oid — like a spider, like a spider's web.

schiz'oid — having a tendency towards a "split" personality (schizophrenia).

thyr'oid—"shield-shaped", as is the thyroid cartilage of the larynx (voice-box). Hence, the thyroid gland (or the thyroid) which is situated on the larynx.

aster'oid—having the form of a star —a minor planet, one of the many small planets which have their orbits round the sun between those of Mars and Jupiter.

alkal'oid — a substance "like an alkali" — a naturally-occurring basic ('alkaline') substance, containing nitrogen, found in plants (often in combination). Examples are nicotine, quinine, morphine.

In mathematics the suffix is used:.

(a) in the names of plane curves.

astr'oid—a star-shaped curve.

cardi'oid—a heart-shaped curve.

troch'oid—the curve traced out by a point fixed relatively to a circle (e.g. a point on a radius) which rolls along a straight line. (Also see **cycloid, epicycloid.**)

(b) in the names of solids formed by the revolution of a plane figure about a central axis.

ellips'oid—the solid formed by the revolution of an ellipse about a diameter—the shape of a Rugby football. So also **paraboloid** (parabola), **catenoid** (catenary), etc.

The suffix -OIDEA is sometimes used, somewhat inconsistently, to form the names of classes of organisms, especially a class which comprises a group of families, e.g. **Aster'oidea**—"the star-like animals" — the Starfish.

-OL

(1) an alcohol.

The word **alcohol,** from which this suffix is derived, has a curious history. It comes from the Arabic *al koh'l* (from Hebrew) which denoted the fine metallic powder (antimony sulphide) used for staining the eyelids.

Later this word became associated with any powder but especially with one which could be obtained or purified, as was the original powder, by sublimation (converting the solid into a vapour and then condensing it). Even as late as the beginning of the nineteenth century flowers of sulphur was sometimes referred to as alcohol of sulphur.

The word also gradually came to be used for a liquid which could be obtained by distillation. Hence the essence or spirit of wine, obtained by distillation, was called 'alcohol of wine'. The words 'of wine' are now dropped and the liquid just called alcohol.

Ordinary alcohol (spirit of wine) is ethyl alcohol, C_2H_5OH. It may be regarded chemically as derived from ethane (C_2H_6) by the replacement of one hydrogen atom by the hydroxyl group (OH) or as a compound formed by the joining of the ethyl group (C_2H_5-) with a hydroxyl group.

The chemist uses the term alcohol in a general way for any compound which consists of a hydrocarbon group joined to one or more hydroxyl groups. (Some chemists prefer not to use the term for compounds in which the hydrocarbon is based on benzene but to call them **phenols.**)

methan'ol—CH_3OH—the alcohol corresponding to methane (CH_4), i.e. methyl alcohol.

ethan'ol — C_2H_5OH — the alcohol corresponding to ethane (C_2H_6), i.e. ethyl alcohol.

glyc'ol—$C_2H_4(OH)_2$—an alcohol with two hydroxyl groups, a thick liquid with a somewhat "sweet" taste.

phen'ol—C_6H_5OH—a compound in which one hydrogen atom of benzene (C_6H_6) is replaced by a hydroxyl group. Also known as carbolic acid.

cres'ols — compounds with the formula $CH_3.C_6H_4.OH$. A solution of cresols in soft soap is known under the trade name of **Lysol** and is used as a disinfectant.

thym'ol—$C_{10}H_{13}OH$—an important antiseptic found in oil of thyme.

quin'ol—$C_6H_4(OH)_2$, *p*-dihydroxybenzene—a compound in which two hydrogen atoms of benzene are replaced by hydroxyl groups. Similarly, **pyro'gall'ol**—$C_6H_3(OH)_3$, containing three hydroxyl groups, used as a developer in photography.

naphth'ol — $C_{10}H_7OH$ — a compound in which one hydrogen atom of naphthalene ($C_{10}H_8$) is replaced by a hydroxyl group.

thi'ols — thio-alcohols — compounds similar to the alcohols but in which the oxygen atom is replaced by a sulphur atom, e.g. ethane'thiol, C_2H_5SH (compare ethyl alcohol C_2H_5OH).

ster'ols—a group of alcohols of complex chemical structure, found in nature with fatty acids. The best-known is **chole'sterol** which is found in nerve tissue, gall-stones and other body tissues.

(2) an abbreviation for -OLE (=oil).

The suffix is not used regularly in this way.

benz'ol—This was the name proposed by Liebig for benzene. Both **benzol** and **benzole** are now trade names for crude benzene as used for motor-spirit.

fur'ol—furfuraldehyde, a colourless oil obtained by distilling bran, wood, etc., with sulphuric acid.

lan'ol'in — a substance obtained from sheep's wool used as a basis of ointments.

-OLE

(1) A diminutive suffix, indicating a small specimen of the thing named in the main part of the word (L.).

arteri'ole—a very small artery.

bronchi'ole—one of the small end-divisions of the bronchia (the branches of the wind-pipe).

vacu'ole—a "small empty" space—a small space or cavity, usually containing liquid, in the protoplasm of a cell.

(2) an oil (see next entry).

The suffix is not used regularly in this way.

benz'ole—a trade name for crude benzene as used for motor-spirit.

pyrr'ole—a colourless liquid obtained from bone-oil (a product obtained by the dry distillation of fatty bones). See PYRR-.

ind'ole—a compound derived from pyrrole and benzene which forms the basis of the indigo molecule. (It is not, however, an oil but a colourless, crystalline solid.)

OLE-, OLEO-, -OLEUM

oil (L. *oleum*).

The term **oleum** is still used commercially for fuming sulphuric acid. **Linoleum** is made from canvas ("flax") and oxidised linseed "oil"; **petr'oleum** is "rock oil".

ole'fiant—"oil making", e.g. olefiant

gas (the old name for ethylene C_2H_4) which makes oily substance with chlorine and bromine. Hence, **olefines**—hydrocarbons of the ethylene series.

ole'ic acid—the commonest of the acids which, when combined with glycerine, forms fatty oils (e.g. olive oil.) The compound with glycerine is called **glyceryl oleate** or **olein.**

oleo'some—a fatty body in the protoplasm of a cell.

oleo'resin—a natural or artificial mixture of oil and resin, e.g. balsam.

OLIG-, OLIGO-

little, few (Gk. *oligos*).

oligo'carpous—bearing only few fruit.

olig'uria—the passing of abnormally little urine.

oligo'cyt'haemia—a deficiency of red cells in the blood.

oligo'siderite—a stony meteorite containing only a small proportion of iron.

Oligo'chaeta—a class of segmented worms which have relatively few bristles, i.e. the Earthworms.

Oligo'cene period—"little recent"—the geological period, about 45 to 35 million years ago, which was earlier than the Pliocene ("more recent") and the Miocene ("less recent") but came after the Eocene ("dawn of recent").

-OLOGY See -LOGY.

-OMA

As is explained more fully under -M, the addition of the suffix -MA to the stem of a Greek verb produced a noun which expressed the result of the action of the verb. Thus *sarkoō* (to make flesh) was converted into *sarkoma* (a growth of flesh). By analogy, a *karkinoma* (from the noun *karkinos*, a crab) denoted a crab-like growth, and so on. -OMA has now become a specialised suffix which is freely used in medical terms to denote an unhealthy or diseased growth, i.e. a tumour.

sarc'oma—a tumour (diseased growth) of connective tissue, e.g. of muscular tissue.

carcin'oma—a form of cancer, a disorderly growth of epithelial (surface) cells which invades the tissues and also spreads (e.g. by the blood) to other parts of the body.

angi'oma, haem'angi'oma—a tumour composed of blood vessels.

fibr'oma—a tumour composed of fibrous tissue. (Note that in this word the suffix has been added to a Latin stem.)

ec'chondri'oma—a tumour composed of cartilage and growing out from the surface of a bone.

OMBRO-

rain (Gk. *ombros*, rain-storm).

ombro'meter — an instrument for measuring rainfall, a rain-gauge.

ombro'phyte—a plant which flourishes in rainy places.

-OMETER See -METER.

OMNI-

all (L. *omnis*).

An **omnibus** (now called a 'bus') is a vehicle "for all".

omni'vorous—eating all kinds of food (e.g. both flesh and vegetable).

omni'directional aerial—a radio aerial for sending or receiving in all directions in a horizontal plane. (Compare unidirectional.)

OMPHAL(O)-

the navel (the small hollow in the middle of the front of the body where the baby was attached to the mother) (Gk. *omphalos*).

omphal'oid—navel-shaped.

omphal'itis—inflammation of the navel.

omphal'ectomy—the surgical cutting out of the navel.

-ON, -TRON

(1) -ON is a termination of Greek neuter nouns. It is found in scientific terms which have been taken, virtually without change of spelling, though possibly with restriction of meaning, from the Greek or have been formed from Greek roots on the Greek pattern.

Examples of such words are **amnion, cotyledon, enteron, encephalon, tetrahedron, plankton, ganglion.**

The Greek plural is formed by substituting *-a* for *-on*, e.g. **ganglion/ganglia,** but the practice is not always followed in scientific terms (e.g. **cotyledon/cotyledons**).

In recent years a number of new words and names have been invented, some of them by the French, apparently in imitation of Greek. Examples are **aileron, fluon, longeron, nylon, rayon.**

(2) -ON occurs as the termination of the names of some non-metallic chemical elements but the practice has not been followed to the same extent as -IUM (q.v.) for metals.

The first name of this type was **carbon** (from L.). **Boron** and **silicon** followed. With the discovery of the rare, inactive gases a series of names was invented, most of them being neuter forms of suitable Greek adjectives: **argon** (idle), **krypton** (hidden), **neon** (new), **xenon** (strange) and **radon** (formed from radium). The first member of the chemical group, helium, was discovered (in the Sun) and named before the others were discovered; a more suitable name would have been *helion.*

(3) -ON also occurs as the termination of the names of elementary atomic particles.

The parent of these names is **electron,** the name given in 1891 to the negatively charged sub-atomic particle of very small mass. Although *ēlektron* (amber) is the root from which *electric* and related words are derived, it is probable that the name electron was re-formed from the English root ELECTR(O)-.

When it was found that there was another fundamental particle, positively charged and much heavier than an electron, the name **proton** (Gk. *prōtos*, first, chief) was given to it. (A normal hydrogen atom consists of a proton with one electron in an orbit round it.) From then on the fashion was set for using -ON as a termination for the names of elementary particles.

neutron—a neutral particle; **positron**—a positively charged particle similar in mass to an electron; **meson**—a particle of "middle" mass, i.e. between that of an electron and a proton; **deuteron**—the nucleus of deuterium (heavy hydrogen); **hyperon** — a particle of greater mass than a proton; **photon**—a particle of light.

(The *r* of *positron* may appear to be strange as there is no *r* in *positive* nor in the corresponding Latin root. It is probable, however, that the name originated as a contraction of *posi(tive elec)tron.* A similar explanation can be given of *mesotron,* the older name for *meson.*)

-TRON

A **cyclotron,** invented in 1931, is an instrument for producing high-speed particles which travel within the instrument in approximately circular paths. The name was built up from *cyclo-* and *(elec)tron.* It started the use of -TRON as a suffix for the names of other instruments, e.g. **betatron, synchrotron,** used for producing high-speed particles.

The element -TRON, originating apparently from the word *electron,* is now a fashionable ending for the names of various instruments in which electrons or charged particles are caused to move in particular ways. A **thyratron** is a special type of gas-filled radio valve; a **pyrotron** is an apparatus for forcing very hot gas into a small space by reflecting it back by means of a magnetic field when it tries to escape; an **Emitron** is a form of television camera.

ONCO-, (ONCHO-)

(1) a barb, a hook (Gk *onkos*, barb of an arrow).

(This root is sometimes spelt ONCHO-.)

onco'sphere, onchosphere—a larval form of cestode worm which has hooks.

oncho'cerci'asis — an infection of the skin tissue by a hooked nematode worm (**Onchocerca**).

Onc'idium — "a little barb" — a genus of American orchids (including the Butterfly plant) so called from the form of the lower petal of the flower.

(2) mass, bulk, hence (in modern terminology) a tumour (Gk. *onkos*).

onco'meter — an instrument for measuring variations in the size of an organ.

onco'logy — the medical science which deals with tumours.

onco'genous—inducing, or tending to induce, the formation of tumours.

onco'tomy — the surgical cutting into (or cutting out of) a tumour.

-ONE

The Greek suffix *-ōnē* was used to form a (feminine) name from that of an ancestor. Thus **anemone** may be freely translated as "daughter of the wind".

The suffix -ONE is used in forming the names of a number of chemical (especially organic) compounds. The name may be formed from (a) that of a parent compound, or (b) a root representing the occurrence of the compound or one of its characteristic properties. Examples are **hydrazone** (derived from hydrazine), **lactones** (derived from lactic acid), **histones** (relatively simple proteins, see HIST-), **peptones** (products formed in digestion from proteins, see PEPS-), **ionones** (which give the smell to violets and iris roots) and **ozone** (Gk. *ozō*, to smell).

The name **acetone** for the compound $CH_3.CO.CH_3$ is based on *acetic acid.* The molecule shows the characteristic features of the **ketones**: a -CO- group in association with two hydrocarbon groups. In modern naming, the suffix -ONE is usually reserved for the names of compounds of this kind. In **aceto'phen'one,** $C_6H_5.CO.CH_3$, the hydrocarbon groups are methyl (CH_3-) and phenyl (C_6H_5-). In **quinone** (q.v.) two -CO- groups form part of a benzene ring.

The **sulphones** are somewhat similar to the ketones in structure; an SO_2 group links two hydrocarbon groups, e.g. **(di)ethyl-sulphone** $C_2H_5.SO_2.C_2H_5$. In a **sulphonic acid** an $-SO_2.OH$ group is joined to a hydrocarbon group, e.g. **ethyl sulphonic acid** $C_2H_5.SO_2.OH$.

The **silicones** were misnamed. It was anticipated that when compounds of the type $R_1R_2SiCl_2$ (in which R_1 and R_2 represent hydrocarbon groups) were hydrolysed, the product would be a compound of the type $R_1.SiO.R_2$. Such a compound, being similar in structure to a ketone but with a silicon atom in place of a carbon atom, was called a silicone. In fact, however, the products of hydrolysis are long chain polymers which would be more correctly described as polysiloxanes.

ONTO-

The combining-form of a Greek participle (*ōn, ont-*) meaning "being, existing".

onto'genesis—the origin and development of an individual being.

onto'logy — (1) the history of the development of an individual being. (2)=ontogenesis (as above).

ONYCH-, -ONYCHIA, ONYCHO-

a talon, a claw, a finger-nail (Gk. *onyx, onych-*).

onychia—inflammation of the bed of a finger-nail.

par'onychia — "at the side of the finger-nail"—an infected inflammation of the tissue near the finger-nail, a whitlow.

onych'ium—a pad on an insect's foot.

onycho'genic — (substance) which forms or produces nails.

onycho'crypt'osis — "a (unhealthy) state of hidden nails"—an in-growing toe-nail.

Onycho'phora—"the claw-bearers"—a sub-phylum of the Arthropoda with two genera of which Peripatus is the better known. The creatures, which are found in warm countries, have soft bodies composed of segments each of which bears a pair of fleshy legs ending with a pair of claws.

OO-
egg (Gk. *ōon*).

oo'genesis—the origin and development of ova (eggs).

oo'gonium — "egg generator" — the female organ in Algae and Fungi.

Oo'mycetes—"egg fungi"—one of the two sub-divisions of the lower Fungi which includes the fungi responsible for Potato Disease and the 'damping-off' of Mustard and Cress. Egg-like **oospores** are formed in the oogonia.

oo'lite—"egg stone"—a form of limestone consisting of egg-like grains.

oo'phoro'salping'ectomy—the surgical removal (*-ectomy*) of an ovary (*oo'phoro-* = "egg-bearer") and Fallopian tube (*salping-*). This horrible word is deliberately included to show how a long medical term is built up and how it can be analysed.

OPHI-, OPHID-, OPHIO-
a snake, a serpent (Gk. *ophis; ophidion*).

ophid'ian—pertaining to, or a member of, the **Ophidia**—an order of reptiles including snakes.

ophi'cephalous—with a snake-like head.

ophid'iasis—snake-poisoning.

Ophio'glossum — the Adder's-tongue Fern, so called from the appearance of its spikes.

OPHTHALM-, OPHTHALMO-
the eye (Gk. *ophthalmos*).

ophthalm'ic—pertaining to, or situated near, the eye.

ophthalm'ia—inflammation of various parts of the eye (especially of the conjunctiva).

xer'ophthalmia—a form of ophthalmia in which the eye is dry. (Due to lack of vitamin A.)

ex'ophthalmic goitre — a condition, caused by over-activity of the thyroid gland, in which the eye-balls stand out.

ophthalmo'scope—an instrument for inspecting the inside of the eye.

-OPIA
eyesight (Gk. *-ōpia*, from *ōps*, eye).

my'opia—short-sight. Light from distant objects cannot be focused on the retina without the help of a suitable spectacle lens.

hyper'metr'opia — ("over measure") long-sight. The eye-ball is too short so that the eye-lens cannot be made strong enough to focus light from near objects.

presby'opia — long-sight which is brought on by old age.

ambly'opia—"dull (blunt) eyesight"—general dimness of vision.

OPISTHO-
behind, back, backwards (Gk. *opisthen, opistho-*).

opistho'glossal — having the tongue free at the back (and fixed in front) as in a Frog.

opistho'tonos—"backward stretching" —an extreme bending of the body backwards by contraction of the muscles as in a seizure.

OPS-, -OPSY, OPT-, OPTO-, -OPTRE
the eye; seeing (Gk. *ōps; opsis; optikos* (adj.)).
This root is seen in **optical** (pertaining to seeing, e.g. an optical illusion) and **optician** (a maker or seller of spectacles). (Note. *Option* (a choice) and *optimism* (expecting the best) do not come from this root.)

optics—the study of light. **Physical optics** deals with the nature of light and with its properties when considered as waves; **geometrical optics** deals with its properties when considered as rays.

syn'optic chart—a chart for "seeing all together"—a weather map on which are shown all the elements of weather (e.g. temperatures, air pressures, amount of cloud) at a given time.

rhod'ops'in—"red seeing-substance"—a purple pigment found in the rods of the retina of the eye, required for seeing in weak light, e.g. at night. Lack of it, which may be caused by lack of vitamin A, causes night-blindness.

erythr'opsin—a pigment found in the elements of the retina of certain night-flying insects.

aut'opsy—"seeing for one's self"—inspecting a dead body, e.g. to find the cause of death.

opto'gram—the image on the retina.

opto'coele—the cavity of one of the optic lobes of the brain.

di'optre, diopter—a unit for expressing the power of a lens. A lens of focal length 3 metres has a power of one-third dioptre.

OPSON-
cooked, ready for eating (Gk. *opson*, cooked meat).

opson'ic—having the effect of making bacteria in the blood more easily 'eaten' (destroyed) by white cells.

opson'in—a substance produced in the blood which increases the opsonic power of the white cells.

OPT- See OPS-.

ORCH(I)-, ORCHID-, ORCHIDO-
a testicle (testis), one of the two male reproductive glands (Gk. *orchis*; *orchidion*).

orch'itis—inflammation of a testicle.

orch'ectomy, orchid'ectomy—the surgical removal of a testicle.

orchido'pexy — the surgical fixing in place of a displaced testicle.

The familiar flower **Orchis (Orchid)** and its family **Orchidiaceae** were so named because of the twin nature of the tubers.

ORGAN-, ORGANO-
The basic meaning of the Greek root *organon* is "an instrument, a tool, that which works". An **organ**, besides being a musical instrument, is a part of an animal or plant body (e.g. the heart, the liver) which has special, vital work to do. To **organise** (a word which has come through Latin) means "to give an orderly structure so that the parts work together". An **organism** is a body which is organised, i.e. a living animal or plant.

organo'graphy—the descriptive study of the outward form of organisms (especially plants) with reference to the work they do.

organo'therapy—the treatment of a disease by the giving of extracts from animal organs.

Organic chemistry — At one time chemical compounds were divided into Organic (those which occur in the animal and plant worlds) and Inorganic (those which occur in the mineral world). The first to adopt this division was Lémery in 1675. It was thought that, whereas inorganic compounds could be made in the laboratory, organic compounds could only be made in living organisms under the influence of a vital force. This assumption was shown to be untrue when Wöhler, in 1828, made urea from substances which were considered to be inorganic. Shortly afterwards, acetic acid was made from substances which were clearly inorganic. Now Organic Chemistry may be described as the chemistry of the carbon compounds.

organo-metallic compound—a compound consisting of one or more hydrocarbon groups joined to an atom of a metal. e.g. $Zn(C_2H_5)_2$, a compound of zinc and two ethyl groups. Similar compounds but including an atom of bromine or iodine, e.g. $C_2H_5.Mg.I$, are called Grignard reagents.

organo'sol—a colloidal solution in an organic liquid.

ORNITH-, ORNITHO-
a bird (Gk. *ornis*, *ornith-*).

ornith'ic—pertaining to birds.

ornith'oid — somewhat resembling birds in form or structure (as some reptiles).

ornitho'logy—the study of birds.

ornitho'lite—a fossil of a bird (or part of a bird).

Ornitho'poda—a group of extinct reptiles with feet like those of a bird.

OR-, ORO-
the mouth (L. *os*, *oris*)
This root is seen in the well-known word **oral.**

ab'oral—opposite to, away from, or leading away from, the mouth.

oro'nasal—pertaining to the mouth and the nose.

ORO-
a mountain (Gk. *ōros*).

oro'graphy—the branch of physical geography which is concerned with mountains.

oro'graphic rain—rain caused by damp winds driving against, and rising up, mountains.

oro'genesis—the formation of mountains.

ORTHO-
straight, hence upright, at right-angles, correct (Gk. *orthos*, *ortho-*). This prefix also has special meanings in the names of chemical compounds; see below.

ortho'cladous—having long, straight branches.

Ortho'ptera — the order of insects which have straight, narrow fore-wings —Cockroaches, Crickets, Grasshoppers, etc.

ortho'clase — a common feldspar mineral which splits in two directions at right-angles.

ortho'gonal—at right-angles, e.g. 'lines cutting orthogonally'.

ortho'rhombic system (of crystals)—The shapes of the crystals of this type are based on three unequal axes at right-angles to each other.

ortho'cephalic — having a "correct head", i.e. not abnormally broad or narrow.

ortho'paedics—the branch of medical science which deals with the putting 'right' of body deformities of children (and others).

ortho'chromatic film—a photographic film which responds to colours in their 'correct' brightness as seen by the eye, i.e. yellow (which is bright to the eye) is shown bright (white) on a black and white photograph.

Special uses of ORTHO- in chemical names.

(a) A mineral acid can usually be regarded as a compound of an oxide and water, e.g. sulphuric acid H_2SO_4 is a compound of sulphur trioxide SO_3 and water H_2O. If an oxide forms a series of acids by combining with different proportions of water, the **ortho-acid** is the one in which the oxide has combined with the greatest number of water molecules.

ortho'silicic acid—H_4SiO_4, i.e. SiO_2+2H_2O. (Contrast meta'silicic acid $SiO_2.H_2O$.)

ortho'phosphoric acid — H_3PO_4, i.e. $P_2O_5+3H_2O$ ($=2H_3PO_4$). (Contrast pyro'phosphoric acid $P_2O_5+2H_2O$ and meta'phosphoric acid $P_2O_5+H_2O$.)

(b) An *ortho*-compound is formed when two atoms (or groups) replace two hydrogen atoms of benzene which are next to each other, e.g. *ortho-***di'nitro'benzene** (as shown). Also see META-.

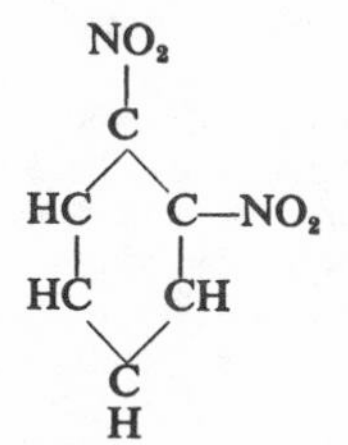

OSCILL-, OSCILLO-
The Latin root (*oscillo*, *oscillat-*, to swing), which is familiar in **oscillate,** needs little explanation.

oscillo'graph—an instrument for producing a curve to represent the waveform of an alternating current (and hence of a sound, etc.).

oscillo'scope — a low voltage cathode ray oscillograph.

-OSCOPE See -SCOPE.

-OSE
(1) A suffix (L. *-osus*), more commonly seen in the form -OUS, used in forming adjectives. Basically it means "full of, abounding in".

pil'ose — abounding in hairs, covered with hairs.

com'ose—bearing silky hairs or down.

foli'ose—leafy, like a leaf.

(2) As a suffix in a chemical name, it denotes a sugar or a related carbohydrate. This 'artificial' suffix seems to have been taken from the parent name **glucose** (q.v.).

sucr'ose (cane sugar), **lact'ose** (milk sugar), **malt'ose** (malt sugar), **hex'ose** (a sugar containing six oxygen atoms), **cellul'ose** (the carbohydrate which forms the walls of plant cells), **poly'ose** (any complicated carbohydrate, such as cellulose, built up from many simpler molecules), etc.

-OSIS

As is explained more fully under -SIS, the addition of the suffix *-sis* to the stem of a Greek verb produces a noun. Thus the verb *morphoō* (to form, to shape, from *morphē*, a form) gives rise to *morphōsis* (a shape, a shaping). Such nouns were also formed by the addition of *-sis* directly to the stem or combining-form of nouns and adjectives, e.g. *exostōsis* (an outgrowth of bone, from *osteon*, a bone), *sklērōsis* (a hardening, from *sklēros*, hard). Many words of this type have passed into English.

meta'morpho'sis — "a changing of form"—a marked change of form and structure taking place in an animal within a fairly short time, e.g. the change of a caterpillar into a butterfly.

sym'bio'sis — "a state of living together", as an Alga and a Fungus in a Lichen. (From *bioō*, to live.)

sclero'sis—a hardening, a state of hardening, e.g. of tissues. (From *sklēros*, hard.)

thrombo'sis—"a curdling, a clotting" —the formation of a blood-clot in a vessel or organ. (From *thrombos*, a lump, a clot.)

On the analogy of *sclerosis*, *thrombosis*, and others, -OSIS has become a suffix in itself. (Note that the *o* really belongs to the stem to which *-sis* is added.) It is freely added to Greek and (less often) Latin stems, but in medical terms, in which it is common, it denotes an *unhealthy*, *diseased* or *damaged* state or condition. Only a sample of the many such terms are given here in illustration.

narc'osis—a state of unconsciousness produced by a narcotic drug.

psych'osis—a state of serious mental disorder.

cyan'osis—blueness of the skin due to insufficient oxygen in the blood.

acid'osis—an acid condition of the blood.

tubercul'osis—an infection of the body, especially the lungs, lymph glands and joints, by the tubercle bacillus. Tubercles develop in the body tissues.

ec'chym'osis—"a (damaged) state of juice out"—a discoloured patch due to an escape of blood under the skin.

fibr'osis — the formation of fibrous tissue as a result of injury or inflammation.

silic'osis—a lung disease caused by breathing in particles of sand (as by stone-workers).

ana'stom'osis—"a state of mouthing up"—a join between two blood vessels; an artificial joining up of two parts of the intestine.

chlor'osis — "an unhealthy state of greenness"—an unhealthy condition of a plant in which there is insufficient chlorophyll (green matter).

OSM-, OSMO-

There are two roots here which are liable to be confused.

(1) a smell, an odour (Gk. *osmē*).

osm'ium — a metallic chemical element (the densest substance known), so called because of the pungent smell of the oxide which it forms with oxygen.

osmo'phore — a group of atoms which are responsible for the smell of a compound. (Also called an odoriphore.)

(2) a push, a thrust (Gk. *ōsmos*).

osmo'sis—When two solutions of different strengths (or one solution and water) are separated by certain kinds of membrane, e.g. pig's bladder, the solvent (the water of the solution) diffuses through the membrane from the weaker solution to the stronger. This process is called osmosis. The extra pressure which builds up in the stronger solution is the **osmotic pressure.**

osmo'meter — an apparatus for measuring osmotic pressure.

OSS-, OSSI-, OS

a bone (L. *os, ossis*).

Note that the Latin *os, oris* means a mouth. Thus the *os uteri* is the mouth of the uterus. The combining-form of this root is OR- (q.v.).

os calcis—the heel bone.

oss'eous—bony, consisting of bone.

ossi'fy—to turn into bone, to become rigid and bony.

ossi'cles—"small bones", e.g. those in the ear.

ossein—the chief organic substance of bone.

OSTE-, OSTEO-, -OSTEUM

a bone (Gk. *osteon*).

oste'itis—inflammation of a bone.

osteo'myel'itis—inflammation of the marrow of a bone.

osteo'logy—the study of bones.

osteo'pathy — a system of treating certain diseases by skilled handling of the bones.

osteo'malacia — a softening of the bones as a result of the loss of calcium salts from them.

osteo'clasis—the breaking of a bone by a doctor so as to set it again in a better way.

peri'osteum—the covering tissue on a bone.

OSTRAC-, OSTRACO-, -OSTRACUM

a piece of earthenware, the hard shell of an animal (e.g. of a tortoise, a snail, a crab) (Gk. *ostrakon*).

To **ostracize** a person means to banish him from society. In ancient Athens a person was banished by popular vote; the name of the person was written on potsherd (a broken piece of earthenware) or an earthen tablet. In scientific terms the root refers to the hard shell of an animal.

Ostrac'oda—a sub-class of the Crustacea (comprising minute creatures which have a shell in two parts).

malac'ostrac'ous—having a relatively soft shell. **Malacostraca**—another subclass of the Crustacea; among the many members are Crabs and Shrimps.

Ostraco'derma—animals with "shell skins"—fish-like creatures of Palaeozoic times (found as fossils) which were heavily armoured with bony scales.

peri'ostracum—the horny, outer layer of the shell of a Mollusc.

OT-, OTO-

the ear (Gk. *ous, ōtos*).

ot'itis—inflammation of the ear.

oto'logy — that branch of medical science which deals with diseases of the ear.

ot'algia—pain in the ear, ear-ache.

oto'scope—an instrument for inspecting the inside of the ear.

par'ot'id—situated near ("at the side of") the ear, e.g. the parotid gland (a salivary gland) at the angle of the lower jaw in front of the ear.

-OUS

This suffix of Latin origin (*-osus*) is adjectival. Its normal use, as in **carnivorous, fibrous, anhydrous, luminous,** needs no further explanation.

It is used in specialised senses in the names of chemical compounds.

(a) In the names of acids. The suffix *-ous* indicates an acid which has a smaller proportion of oxygen than another similar acid, e.g. **nitrous acid** HNO_2 (nitric acid HNO_3), **sulphurous acid** H_2SO_3 (sulphuric acid H_2SO_4).

(b) In the names of oxides, salts, etc. The suffix *-ous* indicates that the metal has a lower valency (chemical joining-power) than in another similar compound, e.g. **ferrous chloride** $FeCl_2$ (in which iron has a valency of 2), but ferric chloride

$FeCl_3$ (in which iron has a valency of 3); **stannous (tin) oxide** SnO, but stannic oxide SnO_2.

OV-, OVI-, OVO-, OVUL-

an egg (L. *ovum*, pl. *ova*).

An **oval** is a closed curve like the outline of an egg. (The term is often applied, wrongly, to an ellipse.)

ov'ate—(leaf) which is oval in shape.

ov'oid—(solid) having the shape of an egg.

ov'ary—an organ in a female animal in which eggs (female cells) are produced; the lower part of the female structure of a flower, made up of carpels, containing the ovules (see below).

ovi'duct—the tube which leads from an ovary and through which eggs pass out.

ovi'position — the act of depositing eggs.

ovi'parous—egg-laying (as are birds, snakes, etc.).

ovo'vivi'parous—"bringing forth eggs alive" — producing eggs which are hatched within the body of the mother.

ov'ule—"a little egg"—the female cell, in the ovary of a flower, which becomes a seed after it has been fertilized.

ovul'ation—the formation of eggs; in mammals, the escape of an ovum (egg cell) from the ovary.

OX-, OXAL-, OXY-

(1) sharp (in shape, to the taste, to the feeling, etc.), keen, pointed (Gk. *oxys*).

oxy'carpous—having pointed fruit.

oxy'dactylous — having narrow, pointed fingers (or toes).

oxy'cephaly—a deformity of the skull in which it has a high forehead and a pointed top.

Amphi'oxus — the small water creature (also called the Lancelet) which is pointed at both ends.

par'oxysm—a sharp fit (spasm) of a disease.

Ox'alis — the Wood Sorrel, so called because of the sharp (acid) taste of the leaves.

(2) related to, or derived from, oxalic acid.

oxal'ic acid—$(COOH)_2$, a poisonous acid which is found in Wood Sorrel and other plants; its salts are called **oxal'ates**.

oxal'uria—the presence of crystals of oxalates in the urine.

oxal'yl—the 'stem' of oxalic acid, -OC.CO-, e.g. oxalyl chloride ClOC.COCl.

ox'amide — the amide (q.v.) of oxalic acid, $NH_2.CO.CO.NH_2$.

(3) pertaining to, or containing, oxygen. The word **oxygen** means "acid-producer". In 1777 Lavoisier described the gas as *la principe oxygine*, a name which he changed to *oxygène* in 1786. He was led to this description by the fact that when certain substances, e.g. carbon, sulphur, burnt in oxygen the resulting oxides formed acidic solutions. Unfortunately the name is a wrong one; Lavoisier did not realise that not all oxides are acidic and that not all acids contain oxygen. He would, in fact, have been nearer the mark if he had given the name to the gas now called hydrogen.

oxy'haemoglobin — the loose compound formed by the action of oxygen on the haemoglobin of the blood. Oxygen is carried in this form over the body and given up where required.

an'ox'ia, an'ox'aemia — a state of having insufficient oxygen in the blood.

oxy-acetylene flame—the very hot flame, used (e.g.) in welding, produced by the burning of acetylene in oxygen.

ox'ide—a compound of oxygen and another element (or group of atoms), e.g. zinc oxide ZnO, sulphur dioxide SO_2, ethylene oxide C_2H_4O.

oxid'ation—basically, the addition of oxygen to an element or compound, e.g. copper can be **oxidised** to copper oxide. For a more specialised and extended use of this term a chemistry textbook should be consulted. (This word is often misspelt and mis-said by non-scientists as 'oxidisation'.)

oxid'ase—an enzyme in a plant or

animal cell which brings about oxidation.

hydr'oxide—a compound of a metal (usually) with one or more hydroxyl (-OH) groups, e.g. sodium hydroxide NaOH, calcium hydroxide $Ca(OH)_2$. The word comes from *hydr(ated) oxide*; calcium oxide, for example, is formed by the addition of water (H_2O) to calcium oxide (CaO).

OZONE, OZON-

Ozone is formed by the action of ultraviolet rays or radium radiations on oxygen or by an electric discharge through oxygen. The molecule consists of three oxygen atoms. The gas has a distinctive smell—hence its name (Gk. *ozō*, to have a smell).

ozon'iser—an apparatus in which oxygen is changed into ozone by an electric discharge.

ozon'ides—compounds formed by the addition of ozone to certain organic compounds (hydrocarbons, alcohols, etc., which contain a double bond between two carbon atoms).

P

p-

(*Chemistry.*) An abbreviation for *para-* (q.v.), e.g. ***p*-di'chlor'benzene**.

PACHY-

thick, massive (Gk. *pachys*).

pachy'derm—an animal with a thick skin, e.g. an elephant. *Adj.* **pachydermatous.**

pachy'phyllous—having thick leaves.

pachy'carpous—having large, thick fruit.

pachy'glossal—(lizard) with a short, thick, fleshy tongue.

pachy'meter — an instrument for measuring the thickness of glass, metal, paper, etc.

PAED-, PED-

a child, the rearing of children (and hence education) (Gk. *pais*, *paid-*).

paed'iatrics, ped'iatrics, pediatry—the branch of medical science which deals with the study of childhood and of the diseases of children.

ortho'paedics—the branch of medical science which deals with the correction of deformities of children (and of others).

dasy'pedes—"hairy children"—birds which have a complete covering of down when hatched.

ped'agogy—"the leading of children" —the art of teaching.

An **en'cyclo'paedia** gives an "all-round education".

PALAEO-, (PALEO-)

ancient, old (Gk. *palaios*).

palaeo'botany — the study of fossil plants.

Palaeo'lithic—pertaining to the "Old Stone" Age when early Man made crude stone tools.

palae'ontology—the study (by means of fossils, etc.) of plant and animal life in the past and of the history of life on the Earth.

Palaeo'zoic era—the "ancient life" era —one of the great divisions of geological time (from 500–400 million to 200 million years ago).

PALAT-, PALATO-

Pertaining to the palate (the roof of the mouth) (L. *palatum*).

palat'ine—pertaining to the palate.

palato'plegia—paralysis of the palate.

palato'gram — a record of the use made of the palate in producing a sound.

PALIN-

again, back again (Gk. *palin*).

A **palindrome** ("running back again") is a word, such as *level*, which reads the same backwards and forwards.

palin'dromic disease—a disease which apparently ceases and then comes on again.

palin'genesis—the re-making of rock such as granite by melting and solidifying again; the reproduction, in a plant or animal, of some ancestral character.

PALLI-, PALLIO-, PALLIUM
a cloak, a mantle, a covering (L. *pallium*).
A **pall** is a cloth which is spread over a coffin or tomb.

pallium—the mantle (a fold of covering tissue of the body parts) of a Mollusc or Brachiopod; part of the wall of the hemispheres of the brain of vertebrate animals. **neo'pallium**—"the new pallium"—that part of the brain of mammals which deals with sense impressions (other than smell).

palli'ative — (that) which eases or lightens the effects of a disease (i.e. puts a "cloak" over it) but does not cure it.

pallio'branchiate — (mollusc, etc.) having breathing-tubes in the mantle.

Note. The name of the chemical element *palladium* was given in 1803 after the name of the newly discovered minor planet Pallas.

PAN-, PANTO-
all (Gk. *pan-*; *pas*, *pant-*).
A **panorama** is an all-round view, a **panacea** is a supposed remedy for all ills, and a **pantomime** was originally a performance consisting of all forms of mime (dumb show).

pan'chromatic film—a photographic film which is sensitive to all the visible colours.

pan'demic—(disease) which spreads over "all the people", i.e. over a whole country or continent.

pan'carditis — inflammation (at the same time) of all of the heart, i.e. of the outer covering, the muscular wall, and the lining of the cavities.

pan'hyster'ectomy—the removal by surgery of the whole of the womb.

panto'phagous—eating all kinds of food.

panto'graph — "all drawings" — an instrument by which a point (e.g. a pen) is constrained to mark out, to any required scale, the path traced out by another point (e.g. being moved round the drawing to be copied).

PAPAVER-
a Poppy (L. *papaver*).

papaver'ous—of, like, or related to, a Poppy.

Papaver'aceae—the Poppy family.

papaver'ine—a complicated substance (an alkaloid) which occurs in opium (prepared from a kind of Poppy).

PAPILL-
A **papula** (or **papule**) is a pimple, a small projection from the skin (as in chickenpox), or a small fleshy projection on a plant. A **papilla** is "a little papule". (L. *papula*, a pimple, a pustule.)

papill'ose—covered with papillae.

papill'itis—inflammation of the head of the optic nerve (at the back of the eye). **papill'oedema**—a swelling of this head.

papill'oma — originally, a diseased growth of a papilla, now (generally) a harmless tumour produced by new growth of the skin.

PAR-, PARA- (1)
This commonly used Greek prefix has several meanings, including specialised meanings in the names of chemical compounds.

(1) by the side, at the side of, to one side.
Par'allel lines run "by the side of one another"; a **paragraph** was originally indicated by a short, horizontal mark at the side.
A **parasite** in ancient Greece was one who "fed beside" another, i.e. by his side at a table. Then the idea grew that the parasite fed at his companion's expense (for which end he indulged in flattery). The biological use of the word, meaning an organism which lives on or in another organism and obtains its food from the host, dates from the beginning of the eighteenth century.

para'mastoid—by the side of the mastoid (a bony projection behind the ear).

par'axial—lying alongside, or on each side of, the axis of the body.

para'biotic twins—"living side by

side"—two persons living joined together (Siamese twins).

para'physes — "growths at the side" — sterile (non-reproductive) hairs found among the reproductive parts of some lower plants.

par'otid—situated "by the side of (or near) the ear", e.g. the parotid gland.

par'enchyma—"juice poured in at the side"—soft, non-specialised tissue which forms the 'packing material' of plants or animals.

para'meter—"a side measure"—(1) a measure or quantity (e.g. electrical resistance) which is fixed for a particular case (e.g. a given circuit) but which may be different in other cases. (2) (*Mathematics.*) The relation between two quantities X and Y may be given in terms of two **parametric equations** which state how X and Y each depend upon an "outside", third quantity P. P is the parameter.

Occasionally the idea of "side by side" may be better interpreted as "to be compared with, similar to". A **parable** (which is "thrown by the side") is essentially a story for comparison; a **parabola** (see -BOLA) is similarly a curve with which others are compared. (This idea of "similar to" is also reflected in the first of the specialised chemical meanings below.)

para'typhoid — a kind of fever similar to typhoid but caused by different bacteria.

(2) wrong, faulty, amiss, disordered.

para'lysis — "faulty loosening" — the loss, in a part of the body, of the power of movement due to disease of the nerves or brain.

para'noia—"disordered mind"—a serious form of mental disorder in which the sufferer has fixed delusions (wrong ideas), e.g. that others are trying to harm him.

par'aesthesia — a state of "disordered feelings (sensations")—the state of having any abnormal sensations on the surface of the body, e.g. tingling, tickling.

(3) Specialised meanings in the names of chemical compounds.

(a) It is used occasionally (and inconsistently) to denote an alternative form of a substance, especially an isomer (alternative arrangement of the atoms in the molecule) or a polymer (large molecule built up from a number of smaller molecules).

para'lactic acid—the form of lactic acid which is found in the juice of flesh. (Also called sarco-lactic acid and *d*-lactic acid.)

par'aldehyde—a polymer built up from three molecules of acetaldehyde, a colourless liquid used as a sleep-producing drug.

par'cyanogen—a polymer of the gas cyanogen (C_2N_2), a brown powder.

(b) It is used rigidly to indicate that two atoms (or groups of atoms) which replace hydrogen atoms in a benzene ring are situated at opposite points on the ring. The diagram represents a molecule of *para*-dichlorbenzene; the two chlorine atoms are at opposite points on the ring. (See META-.)

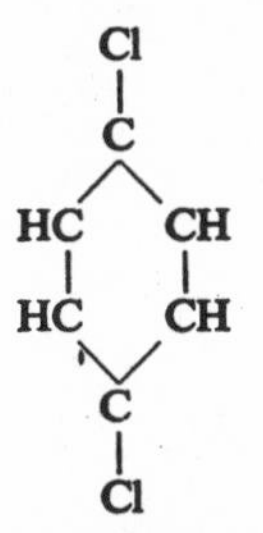

Note. The term *paraffin* is derived from the Latin *parvum affinis* ("little affinity"). These hydrocarbons, many of which are obtained from petroleum, do not readily combine with other substances.

PARA- (2)
to ward off, to protect from (Italian *parare*).
A **parasol** is to provide protection from the sun; a **parachute** is to ward off the effects of a fall.

In some modern words the root is used without a proper indication of what is warded off. Thus the word **paratroops** does not mean 'to ward off troops' but is a contraction of 'parachute troops'. A **paravane** is not an instrument 'to ward off vanes' but an instrument towed by a ship, at a depth regulated by vanes, to cut the moorings of explosive mines.

PARIET-, PARIETO-
Pertaining to a wall (L. *paries, pariet-*), especially a wall of the body or of any of its cavities.

pariet'al—pertaining to a wall, e.g. the parietal bones which form part of the sides and top of the skull.

parietes—the walls of an organ or of a body cavity.

parieto-occipital—pertaining to the parietal and occipital (back part) bones.

-PAROUS
bringing forth young (L. *pario, parit-*, to bring forth, to create).

primi'parous—bearing a child for the first time. The woman who is doing so is a **primipara.**

multi'parous—bringing forth many young at a birth; (woman) who has given birth to more than one child.

vivi'parous — producing the young alive (not laying eggs).

PARTHENO-
unfertilized (Gk. *parthenos*, a virgin).

partheno'genesis—reproduction from an egg-cell which has not been fertilized by a male cell.

partheno'carpy—the production of a fruit without the fertilization of the egg-cell so that the fruit does not contain seeds.
The **Parthenon** on the Acropolis at Athens was named after the goddess Athene, the virgin goddess.

PATH-, PATHO-, -PATHY
a feeling, a suffering, hence a disease (Gk. *pathos*).
Sympathy is "a feeling with" another person; **apathy** is "lack of feeling". A **pathetic** appearance is one which indicates suffering and so calls for pity. In scientific terms the root usually relates to the suffering from a disease.

patho'logy—the branch of medical science which deals with the causes, nature and body effects of diseases.

patho'gen—an organism or substance which produces a disease; **pathogenesis**—the production of a disease.

patho'gnomonic—"fit to give judgment on disease"—(signs, etc.) clearly indicating a particular disease.

psycho'path—"a sufferer in the mind" —a person who is mentally abnormal, especially one who lacks a normal social conscience, but is not suffering from a true mental disorder.
The element -PATHY may often be interpreted as 'the treatment of a disease'.

electro'pathy—the treatment of disease by electrical means. (Also called electrotherapy.)

homoeo'pathy—a system of medicine based on the idea that a disease may be cured by small amounts of medicines which produce, in a healthy person, similar effects to those of the disease. (The system was founded by Dr. Samuel Hahnemann (1755–1843).)

osteo'pathy—a method of healing by the skilful manipulation of parts of the human frame (bones, muscles, etc.).

PECT-, PECTO-
made thick or solid, congealed (Gk. *pēktos*).

pect'isation—the formation of a jelly.

pect'ins—soluble carbohydrates found in the cell walls of fruit and vegetables. They form jellies with fruit juices and so cause jams to set.

pecto-celluloses—complex forms of cellulose, found (e.g.) in flax fibres, which can be broken down to pectins.

PECTIN-
a comb (L. *pecten, pectin-*).

pecten—any comb-like structure in the body of an animal.

pectin'ate, pectin'eal—comb-like, e.g. a pectinate leaf, the pectineal ridge on the thigh-bone to which the **pectineus muscle** is attached.

pectines—comb-like structures on the under surface of the second abdominal segment of Scorpions.

PECTOR-, PECTORI-
the breast (L. *pectus, pector-*).

pector'al—pertaining to the breast or chest, e.g. **pectoral girdle**—the collar bones and shoulder blades; **pectoral fins**—the front pair of fins of a fish.

pector'ales—muscles connecting the upper part of the fore-limb with the front of the pectoral girdle.

pectori'loquy — "breast-speaking" — the carrying of the sound of words, spoken by a person, to the chest-wall so that they can be clearly heard through a stethoscope. (Indicating a disorder of the lungs.)

PED-, -PED, -PEDE
a foot (L. *pes, ped-*). (Also see PAED- child.)
This root is familiar in **pedal** (which is worked by the foot), **pedestrian** (who goes on foot) and **pedestal** (which forms a foot or base).

ped'ate—(leaf) which is divided like the toes of a foot (or the claws of a bird).

ped'uncle—"a little foot"—the stalk of a flower, fruit, or a cluster.

ped'ometer—an instrument for counting the number of paces which a walker takes and so indicating how far he walks.

bi'ped—an animal (e.g. Man) which has two feet.

quadru'ped—an animal which has four feet.

fissi'ped—having "split feet"—having the digits separate and free.

centi'pede — "a hundred feet" — a worm-shaped, crawling creature (a member of the Chilopoda) which has many feet.

-PEL, -PULSE, -PULSION
to drive (L. *pello, puls-*).
This root is seen in **expel** ("to drive out"), **impel** ("to drive in or on") and **repel** ("to drive back"). A **propeller** is the instrument which drives the ship or aeroplane along.

The **pulse** is the throbbing of the arteries as the blood is driven along them. Words with -PULSE are sometimes used as verbs (e.g. **to repulse**) but more often as nouns (e.g. **a repulse, an impulse**). Words with -PULSION are nouns (e.g. **expulsion**).

PELAG(O)-
the sea (Gk. *pelagos*).
The **Archipelago** ("the chief sea") was the name given to the Aegean Sea. Hence the word has come to mean a sea studded with small islands.

pelag'ic—pertaining to the open sea, e.g. pelagic sediments (formed under deep-water conditions).

pelag'ian—inhabiting, or an inhabitant of, the open sea.

PEND-, -PENSION
to hang (L. *pendo, pens-*).
This root needs little explanation. It is seen in **pendant, pendulum, suspend** ("to hang under"), **suspension.** Both **appendage** and **appendix** mean a thing "hung on", i.e. an attachment, an addition.

PENE-, PEN-
almost (L. *paene*).
A **peninsula** is "almost an island" (*insula*); **penultimate** means "almost last", i.e. last but one.

pene'plain—a gently rolling region of low land produced by the wearing down of higher land over a long period.

pen'umbra—the region of part-shade which bounds the true shade (umbra) when an object obstructs the light which is coming from an extended (not point-like) source.

-PENIA
poverty, lack of, less than usual (Gk. *penia*).

erythro'penia—the state of having less than the normal number of red cells in the blood.

leuco'penia, leuco'cyto'penia — the state of having less than the normal number of white cells in the blood.

PENICILL-

like a painter's brush, tufted (L. *penicillum*, a small brush).

A **pencil** was originally an artist's paint-brush and the term is still sometimes used for certain small brushes. A **pencil of light-rays** is a set of rays spreading out from a point like the hairs of a brush.

penicill'ate—having, or forming, little tufts.

Penicill'ium—one of the common moulds which forms blue-green patches on bread, jam, leather, etc. So called because it forms chains of spores arranged in a brush-like head.

penicill'in—a substance produced by Penicillium moulds (especially *P. notatum*). Discovered by Sir Alexander Fleming in 1920 and since developed as an important antibiotic drug.

PENNI-, -PENNATE

a feather (L. *penna*), hence winged (L. *pennatus*).

penni'form — having the form or appearance of a feather.

brevi'pennate—having short wings.

longi'pennate—having long wings.

PENT-, PENTA-

five (Gk. *pente, penta-*).

penta'gon—a plane figure with five angles (and five sides).

pent'ode—a radio valve with five electrodes.

penta'grid—a radio valve with five grids (and hence seven electrodes in all).

pent'ane — a paraffin hydrocarbon whose molecule contains five carbon atoms, C_5H_{12}.

pent'ose—a sugar whose molecule contains five oxygen atoms.

pent'oxide—an oxide which contains five oxygen atoms, e.g. phosphorus pentoxide P_2O_5.

penta'valent — having a valency (chemical joining-power) of five.

penta'merous — having parts (e.g. petals, stamens) arranged in a set (or in sets) of five—as in many well-known flowers.

penta'dactyl—having five digits, e.g. five fingers on the hand.

PEPS-, -PEPSIA, PEPT-

digestion (Gk. *pepsis*).

dys'pepsia — "bad digestion" — indigestion.

brady'pepsia, bradypepsy — slow digestion.

eu'peptic—having a good digestion.

peps'in—a substance (an enzyme) contained in the stomach juices which begins the digestion of proteins.

The meaning of the root has been extended to cover the substances which are formed by the breakdown of proteins in the process of digestion. Thus proteins are broken down into simpler **peptones**; these are forms of **poly'peptides** (chains of amino-acids).

PER, PER-

As a word on its own (a preposition), **per** means "through, by means of" (as in **per post**) or "for each" (as in **per second** and **per cent.**) (L. *per*).

As a combining-element it means "through, all over, completely". Familiar examples are **permanent** ("remaining (lasting) through"), **perforate** (having holes "bored through"), **perspire** ("to breathe through"), **percolate** ("to strain through"). **Perturb** means to disturb completely, to throw into complete confusion.

per'ennial—a plant which lasts "through the years" (until it dies or is dug up).

im'per'vious—"not having a way through" — (substance) not allowing a liquid, gas, etc., to pass through it.

per'meable—allowing a liquid or gas to "pass (diffuse) through".

per'mut'ations—"changes all through" —the different arrangements that can be made by taking a given number of items from a given (larger) group of items.

The prefix PER- is used in specialised senses in chemical names.

(a) When used loosely, it indicates that the compound contains the maximum (or an unusually large) number of atoms, or groups of atoms, of a certain kind.

lead per'oxide—brown oxide of lead, PbO_2. (Contrast the normal oxide PbO.)

per'chloric acid—$HClO_4$; the salts are **perchlorates** (e.g. sodium perchlorate $NaClO_4$). Contrast chloric acid $HClO_3$ and sodium chlorate $NaClO_3$. (The corresponding acid of iodine is **per'iodic** acid, HIO_4: it must not be thought that it has anything to do with *period*.)

per'manganates—salts of permanganic acid $HMnO_4$, e.g. potassium permanganate $KMnO_4$. Compare this with potassium manganate K_2MnO_4; the permanganate contains a relatively larger proportion of the manganate group $-MnO_4$.

(b) When used strictly, it indicates an oxide (or acid, or salt) in which two oxygen atoms are joined together.

hydrogen peroxide — H_2O_2, i.e. H-O-O-H—a liquid, usually met as a solution in water, used for bleaching.

barium peroxide—BaO_2. With a dilute acid, hydrogen peroxide is formed.

per'sulphuric acid — one of the higher acids of sulphur whose molecule may be represented by

$$\begin{array}{l} O\text{-}SO_2\text{-}OH \\ | \\ O\text{-}SO_2\text{-}OH \end{array}$$

The salts are **persulphates.**

PERI-

round, around (Gk. *peri, peri-*).

peri'scope—an instrument for "seeing round" a corner, over a wall, out of a submarine, etc.

peri'meter—"the measure round"—the line which bounds a closed figure; the length of this line.

peri'helion—"around the Sun"—the point in the orbit of a planet or comet when it is nearest to the Sun.

peri'anth—"around the flower"—the ring (or rings) of petals and sepals which surrounds the reproductive parts of a flower. Used especially when the sepals (calyx) and petals cannot be distinguished.

peri'carp—the wall of a fruit (formed from the wall of the ovary).

peri'cardium—the space surrounding the heart; the membrane enclosing the heart.

peri'osteum—the membranous tissue covering a bone.

peri'toneum, peritonaeum — a thing "stretched round"—the double membrane which lines the cavity of the abdomen.

A **peri'od** (from Gk. (*h*)*odos*, a way) is "a going round, a circuit". In its elementary sense it is an interval of time which comes round and round again, e.g. a week, the time of swing of a pendulum. The word is also used for a less clearly marked interval of time (e.g. the time in which a disease runs its course, an indefinite portion of history, of life, etc.) and in other senses.

-PETAL

seeking (L. *peto*, to seek).

centri'petal—tending, or acting, towards the centre, e.g. the centripetal acceleration of a body which is moving in a circle, a centripetal inflorescence (or raceme) in which the flowers of the flower-head develop in turn from the outside towards the centre.

acro'petal — "seeking the tip" — (leaves, flowers, etc.) developing in turn from the bottom of the stalk towards the top.

Note. The word *petal* is not related to this root; it is derived from the Gk. *petalon*, a leaf, a thin plate (of metal).

PETR-, PETRI-, PETRO-

a rock (Gk. *petra*).

The name **Peter** is derived from this word.

petri'fy—to convert into stone. The process is **petrification.**

petro'logy—the study of rocks.

petr'oleum—"rock oil"—mineral oil found in rock layers (or on the surface of water) used as a source of oils for lighting, fuel, lubricating, etc. **Petrol,** as used in motor-car and aeroplane engines, is obtained from it.

salt'petre—"salt of stone" (L. *salpetra* or *sal petrae*), i.e. salt found as a stony crust—potassium nitrate (or nitre) KNO_3.

-PEXY

a fixing, a making firm or solid (Gk. *pēxis*).

rheo'pexy—making "a flow set solid" —the making of certain colloidal solutions (suspensions of very small particles) quickly set into a jelly or solid by gentle rhythmic shaking.

The root appears more commonly in medical terms in which it denotes "the fixing in place" by surgery.

nephro'pexy—the fixing in place of a kidney which is unusually movable (a floating kidney). So also **hepato'pexy** (liver), **gastro'pexy** (stomach), etc.

PHAC-, PHACO-

a lens (Gk. *phakos*, a lentil, anything of lentil shape).

phaco'lite—"lens stone"—a mineral (a zeolite) which forms lens-like crystals.

phaco'lith—a small lens-shaped mass of igneous rock which has been forced into the top of an upward curve of rock layers.

phac'oidal structure—a rock structure which includes lens-shaped minerals or fragments.

phaco'malacia—a softening of the lens of the eye.

PHAEO-, (PHEO-)

dusky, dark-coloured, dark brown (Gk. *phaios*).

phaeo'sporous—having dark-coloured spores.

Phaeo'phyta, Phaeo'phyceae — the large group of Algae which consists mainly of the Brown Seaweeds.

PHAGO-, -PHAGE, -PHAGIA, -PHAGOUS, -PHAGY

to eat, to swallow (Gk. *phagō*).

geo'phagous—"earth-eating" (as are Earthworms). The phenomenon is **geophagy.** So also **hylo'phagous** (wood), **entomo'phagous** (insects), etc.

dys'phagia—difficulty in swallowing.

phago'cyte—"an eating cell"—a white blood cell which "eats" (absorbs) foreign bodies such as bacteria.

bacterio'phage—"bacteria eater"—an agent (possibly a virus) which causes the breakdown of bacteria.

The **oesophagus** is the canal leading from the mouth to the stomach, i.e. the food-pipe. The word is derived from the Gk. *oisophagos*; this is clearly related to the root *phag-* but the exact origin is obscure.

PHALANG-

The Greek word *phalanx*, *phalang-*, was used in various senses—a line of battle or battle-array, a spider, a spider's web. English scientific terms derived from this root reflect the idea of ordered rows or of webs.

phalanx—any one of the bones of the finger or toe. Plural **phalanges.**

phalang'eal—pertaining to a phalanx.

phalanger—an Australian animal (e.g. the flying squirrel) with webbed toes.

PHAN-, -PHANE, PHANERO-

showing, visible, easily seen (Gk. *phainō* (v.); *phaneros* (adj.)).

dia'phanous — "showing (appearing) through"—transparent. (The idea of transparency is reflected in the trade name Cellophane.)

hydro'phane—"water transparent"—a form of opal which is opaque when dry but becomes transparent when soaked in water.

phanero'crystalline—(igneous rock) in which the crystals of all the essential minerals can be seen with the naked eye. (Contrast cryptocrystalline.)

Phanero'gams—"visible marriage"—the great class of higher plants, including the Gymnosperms (e.g. Fir, Pine) and

Angiosperms (flowering plants), in which the reproductive parts can easily be seen and identified. (Contrast Cryptogams.)

PHARMAC-, PHARMACO-
a drug (Gk. *pharmakon*).

pharmaco'logy—the study of drugs and of their effects.

pharmacy—the preparation and selling of drugs; the shop where drugs are sold (called a chemist's shop in Britain).

pharmaceutical—of, or engaged in, pharmacy. (Gk. *pharmakeutēs*, a druggist.)

pharmaco'poeia — "the making of drugs"—a book which gives a list of drugs with instructions for their making and use.

PHARYNG-, PHARYNGO-
The **pharynx** (Gk. *pharynx, pharyng-*) is the cavity, with muscular walls, between the back of the mouth and the top of the food-pipe and wind-pipe.

pharyng'itis — inflammation of the pharynx.

glosso-pharyngeal—pertaining to the tongue and the pharynx.

pharyngo'plegia — paralysis of the muscles of the pharynx.

pharyngo'tomy—the surgical cutting into the pharynx.

pharyngo'scope—an instrument for viewing the pharynx.

-PHASIA
speech (Gk. *phasis*, a saying, speech).

a'phasia—loss of the power of speech due to brain injury.

dys'phasia—"bad", imperfect speech due to a brain injury.

PHELLO-
cork (Gk. *phellos*).

phello'gen—"cork producer"—a layer of cells just inside the surface of a root or stem which forms cork (**phellem**) on its outside and **phelloderm** ("cork skin") on its inside.

PHEN-, -PHENE, PHENO-, PHENYL
The Greek verb *phainō* means to appear, to show, to give light, to shine. Scientific words derived from this root are conveniently considered in two groups.

(a) Words relating to an appearance.

A **phenomenon** is anything that appears (or is perceived by the senses) especially when the cause of the thing is in question.

pheno'crysts—large, usually perfect, easily seen crystals in an igneous rock.

pheno'type—any one of a group of individuals which appear to be alike although, in fact, possibly differing in hereditary factors.

phos'phene—"an appearance of light"—an appearance of rings of light produced by pressure on the eye-ball. (This word must not be confused with *phosphine*.)

pheno'logy—(originally) the study of the times of re-happenings of natural phenomena especially in relation to climate; (now) the science which deals with the relation between living things and the physical conditions (e.g. climate, altitude) of their environment.

(b) Terms relating to phenol or other coal-tar products.

Coal-tar, now an important source of chemical substances, was originally a by-product in the manufacture of gas for illuminating purposes. The element PHEN(O)-, from Gk. stem *phain-*, to shine, was therefore used in forming the names of some of these substances.

phen'ol—carbolic acid. The molecule, C_6H_5OH, consists of a benzene ring (C_6H_6) in which one of the hydrogen atoms has been replaced by a hydroxyl (-OH) group. **Phenolic resins**, which form the largest group of artificial plastics, are built up from phenol (or a closely related compound) and an aldehyde (e.g. formaldehyde).

phenyl—the group of atoms C_6H_5-, i.e. a benzene ring less one hydrogen atom. **phenyl acetate**—the compound derived from phenol and acetic acid, $CH_3.COOC_6H_5$.

phenyl'amine — the amine (q.v.) formed by the replacement of one hydrogen atom of ammonia (NH_3) by the phenyl group, i.e. $C_6H_5NH_2$. (Also called aminobenzene or aniline.)

tri'phenyl'methane dyes—Triphenyl-methane is a compound in which three of the hydrogen atoms of methane (CH_4) have been replaced by phenyl groups, i.e. $CH(C_6H_5)_3$. It is the parent of a number of dyestuffs including malachite green, rosaniline and aurine.

phen'acetin—a complex substance (aceto-*p*-phenetidine), derived from phenol, used as an anti-pyretic (for easing or preventing fever).

thio'phene — a colourless liquid, closely resembling benzene, found in coal tar. The molecule (as shown) includes a sulphur atom (*thio-*) in the ring.

```
HC=CH
|     >S
HC=CH
```

PHEO- See PHAEO-.

PHIL-, PHILO-, -PHILIA, -PHILIC, -PHILOUS

loving, liking, hence preferring, seeking (Gk. *philō*, to love, to like).

A **philo'sopher** is "a lover of wisdom"; a **phil'anthropist** is "a lover of mankind".

hydro'philous—(plant) which lives in and flourishes in water. (Also one which is pollinated by water.)

anemo'philous — (plant) which "likes the wind" and is pollinated by it.

luci'philous—(plant) which seeks the light.

psammo'philous—(plant) which flourishes in sandy soils.

hydro'philic—(colloid) which readily forms a colloidal solution with water.

haemo'philia—"a liking for blood"—a hereditary tendency to bleed very easily (even from slight wounds) and to continue bleeding.

basi'phil—(cell, etc.) which readily stains with basic dyes.

PHLEB-, PHLEBO-

a vein (Gk. *phleps, phleb-*).

phleb'itis—inflammation of the walls of a vein.

phlebo'tomy—the surgical cutting of a vein to let out blood.

phlebo'lith—"vein stone"—a blood-clot, in a vein, which has turned chalky.

PHOBO-, -PHOBIA, -PHOBE

fear, fright, hence dislike (Gk. *phobos*).

claustro'phobia—fear of being shut up in a small, closed space.

agora'phobia—dread of public places.

hydro'phobia—an aversion to water, especially as a sign of rabies (dog madness) in Man; hence, rabies itself.

lyo'phobic colloid—"loosening (dissolving) disliking"—a colloidal solution (e.g. of gold) which is only formed with difficulty and from which the solid is readily deposited.

phobo'taxis—"arrangement to fear"—the movement made by an organism to draw back from that which is unpleasant or dangerous.

PHON-, PHONO-, -PHONE

sound (Gk. *phōnē*).

This root is seen in **telephone** (which brings "sound from afar") and **microphone** (which picks up "small sounds" and changes them into electrical impulses). The word **gramophone** is probably an inversion of **phonogram** ("sound writing") which name was given to the sound-record of a **phonograph** (early form of gramophone). The musician will recognise the root in **polyphony** ("many sounds") and in **symphony** ("sounds together").

phon—a unit for measuring the loudness of a sound.

a'phonia—a state of being "without sound"—total loss of the voice (due, e.g. to hysteria, laryngitis, or paralysis of the vocal cords).

phono'chemistry—the study of the effects of sound waves (and of similar waves of higher frequencies) on chemical reactions.

phono'lite — "sound stone" — a volcanic rock which rings when struck by a hammer.

phon'endo'scope—an instrument for "seeing sounds within" — a doctor's stethoscope fitted with a device for strengthening the sounds heard through it.

-PHORE

that which bears or carries (Gk. *phoros*). A **semaphore** (as a piece of apparatus) is "that which bears the signs".

sporo'phore—that part of a fungus which bears the spores.

carpo'phore—"fruit bearer"—a general name for the stalk of a fruit-structure, especially in lower plants.

chromato'phore—"colour bearer"—a colour-producing substance in a cell of a plant.

chromo'phore — This also means "colour bearer". It is a group of atoms (e.g. -N:N-) which, when combined with hydrocarbon groups, is responsible for the colour of a dye.

pyro'phoric—"life bearing"—(powder) which takes fire when exposed to the air. The name **phos'phorus** means "light bearer"; the substance slowly burns in air at ordinary temperatures and so glows in the dark.

-PHORESIS

the act of bearing or carrying (Gk. *phorēsis*).

(1) In the physical sciences the root is used to denote the movement of small particles (or of ions) in a liquid.

electro'phoresis—the movement of the small particles in a colloid when an electric field is applied. Such movement towards the anode (positive electrode) is **ana'phoresis**; that towards the cathode (negative electrode) is **cata'phoresis.**

photo'phoresis—the movement of particles in a liquid under the influence of light.

(2) In medical terms the root denotes sweating.

dia'phoresis—"carrying through" —sweating. A **diaphoretic** drug produces sweating.

PHOS-, PHOSPH-, PHOSPHAT-, PHOSPHOR(O)-

The chemical element **phosphorus** slowly oxidises (burns) in the air at ordinary temperatures and so glows in the dark. Such glow is called **phosphor'escence** (though the term is also used for the glow of certain substances after they have been in the light). A substance which produces phosphorescence is **phosphoro'genic.**

Phos'gene ("produced by light") is a very poisonous gas with the chemical formula $COCl_2$. The name refers to its production by the action of sunlight on a mixture of chlorine and carbon monoxide.

In most scientific terms, the root (in its various forms) refers to phosphorus.

phosph'ates—salts of **phosphor'ic acid** (H_3PO_4).

phosphat'ic deposits—rock beds containing calcium phosphate (formed especially in dry regions).

phosph'ine—a colourless, bad-smelling gas (PH_3) which usually catches alight in the air.

phosphor-bronze—a tough, hard form of bronze which contains a small amount of phosphorus.

phosphat'uria—the presence of excess phosphates in the urine.

PHOT-, PHOTO-

light, by light (Gk. *phōs*, *phōt-*).

This root is seen in the familiar word **photograph** ("a drawing made by light").

photo-chemistry—the study of the chemical effects produced by light and of the production of light by chemical means.

photo-electric cell—in general, any instrument which responds to light (or to infra-red or ultra-violet rays) by producing an electric current (as in a

photo-emitter) or by producing a change in an electric current.

photo'meter—an instrument for comparing the powers of two sources of light.

photo'sphere—the visible surface-layer of the Sun from which light and heat are radiated.

phot'on—a particle of light. (In some effects, e.g. the photo-electric effect, light behaves as if it were a series of particles and not waves.)

photo'synthesis—"putting together by light"—the building up of simple carbohydrates (sugars) from carbon dioxide and water by the green cells of a plant in the presence of sunlight.

photo'tropism — "light turning"—the response (e.g. by curved growth) of a part of a plant to light. A shoot, for example grows towards the light; it is positively **phototropic**.

photo'phobia — "dreading light" — suffering when bright light falls on the eye.

Photogenic originally meant "producing light" but in these days of the cinema and television it has acquired another meaning. Using *photo* as an abbreviation for *photograph*, it now more commonly means 'having the qualities for producing a good photograph'.

PHRAGMA, -PHRAGM, PHRAGMO-

a hedge, a partition (Gk. *phragma*; *phragmos*).

phragma—a partition or septum in a body.

dia'phragm—"a partition across"—the midriff, the sheet of muscle which separates the chest cavity from the abdominal cavity.

phragmo'basidium—a basidium (q.v.) in which septa (partitions) form and is then divided into four cells.

PHREN-, PHRENI-

the diaphragm (as above) (Gk. *phrēn*).

phren'ic — pertaining to the diaphragm, e.g. the phrenic nerve.

phrenic'otomy—the surgical cutting of the phrenic nerve in order to paralyse the diaphragm on one side.

phreni'costal—pertaining to the diaphragm and the ribs.

This meaning of the root must be distinguished from that which follows. The Gk. word *phrēn* was extended in meaning to include the parts near the heart, hence the seat of passions, and hence the mind or understanding.

PHRENO-, -PHRENIA

the understanding, the mind. (Gk. *phrēn*. See the note above.)

phreno'logy—the old-fashioned study of the shape of the outside of the skull as a means of indicating something of a person's mental abilities.

brady'phrenia—a "slow mind" state—general slowness of the mental processes.

hebe'phrenia—a "dull mind" state—a form of mental disorder in which the person has outbursts of excitement and periods of weeping and depression.

schizo'phrenia—a "split mind" state—a disordered state of mind in which thoughts, feelings and actions are not consistent with each other.

PHTHAL-

Related to, or derived from, phthalic acid.

phthalic acid—a colourless, crystalline acid, $C_6H_4(COOH)_2$, formed by the oxidation (e.g. by nitric acid) of naphthalene. The name was formed as an abbreviation of *naphthalic* (of naphthalene).

phthal'ate—a salt of phthalic acid.

phthalic anhydride—a white crystalline solid (with formula as shown) formed by

$$C_6H_4 \begin{matrix} \diagup CO \\ \diagdown CO \end{matrix} \!\!> O$$

heating phthalic acid (in which process water is eliminated).

phthal'eins—a series of dyes formed by the combination of phthalic anhydride with phenols (with the elimination of water). Ordinary phenol gives **phenolphthalein**, a substance used as an indicator for acids and alkalis.

tere'phthalic acid—a form of phthalic acid (the *para*-form) obtained by the oxidation of oil of turpentine.

PHTHIS-, PHTHISO-, PHTHISIS
a wasting away (Gk. *phthisis*).

phthisis — a wasting away of body tissue (as a result of disease).

phthis'ergate—a worker ant (ergate) which, because of a parasitic infection, fails to develop. Similarly, **phthiso'gyne** —a queen ant which does not develop.

PHYCO-, -PHYCEAE
a seaweed (Gk. *phykos*), hence Algae in general.

phyco'logy—the study of Algae.

phyco'cyanin — the pigment of the Blue-green Algae.

Chloro'phyceae — the Green Algae. Similarly, **Phaeo'phyceae** — the Brown Algae, **Rhodo'phyceae**—the Red Algae.

Phyco'mycetes—the "algae-like fungi" —the great class of the lower Fungi. The body threads do not have septa (cross-walls).

PHYLACT-, -PHYLAXIS
guarding against (Gk. *phylax* (n.), *phylaktikos* (adj.)).

pro'phylact'ic—"guarding before"— (drugs, measures taken, etc.) which help to prevent a disease. Such preventive treatment is **prophylaxis.**

ana'phylaxis—"a guarding up against" —extreme sensitiveness to a foreign protein which has previously been introduced into the blood. The protein promotes the formation of an antibody; if, later, a second injection of the protein is given, there may be a violent reaction between it and the antibody. This is an **anaphylactic shock.**

PHYLLO-, -PHYLL, -PHYLLOUS
a leaf (of a plant) (Gk. *phyllon*).

phyllo'phagous—(animal) which feeds on leaves.

phyllo'taxis, phyllotaxy—the arrangement of the leaves on a stem or shoot.

phyllo'podium — "a leaf foot" — the stalk and main axis of a leaf such as that of a Fern.

phyllo'clade — "a leaf-like young branch"—a flattened stem which looks like and acts like a leaf.

meso'phyll—the soft middle tissues, between the upper and lower skins, of a leaf.

chloro'phyll — the green colouring-matter of leaves (and some other parts of a plant).

micro'phyllous — having very small leaves.

dasy'phyllous—having hairy leaves.

PHYLO-, PHYLUM
a tribe, a race (Gk. *phylē*).

phylum—one of the major divisions of the plant and animal kingdoms, e.g. *Bryophyta* (the Mosses and Liverworts), *Arthropoda* (animals, such as Insects, Crabs, Spiders, which have "jointed feet").

phylo'genesis—the evolution of a tribe or race or of any characteristic feature of the race.

phylo'geny—(1) phylogenesis, as above. (2) the history of the development of a race.

phylo-ephebic — pertaining to the period of greatest strength and vigour of a race. (*Ephebic*—pertaining to the period of greatest vigour of an individual, i.e. to 'the prime of life'.)

PHYMA
A **phyma** (Gk.) is a growth on the body, an inflamed swelling, a boil.

rhino'phyma—an outgrowth on the nose.

-PHYSEMA See PHYSO-.

PHYSIC(O)-, PHYSIO-
The Greek word *physikos* (from *physis*, nature, inborn quality) means "natural, pertaining to nature, produced by nature". It has given rise to a range of English words with rather divergent meanings. The adjective **physical** means natural (as in physical geography), material (as in physical properties) and bodily (as in physical exercise). A **physician** (who is skilled in the medical arts) is really one who has a knowledge of nature. **Physics** is that branch of

science which deals with the general properties of matter and with the various forms of energy.

physio'logy—"study of nature", i.e. Natural Science. By the end of the sixteenth century this all-embracing subject was beginning to be limited to the study of Man's body. Gradually other branches of science, e.g. chemistry, geology, broke away from the parent science and physiology thus became that which was left. It is now the science which deals with the working of the living body (especially that of Man).

physio'gnomy — "judgment by outward nature (features)"—the characteristic features of a thing, e.g. a person, an area of land, a group of vegetation, by which it may be recognised and its general nature judged.

physio'therapy—medical treatment by massage and electrical means.

-PHYSIS

The Greek word *physis*, which usually means the nature, inborn quality, *or* property of a thing, is derived from the Gk. verb *phyō*, to bring forth, to grow, i.e. it really means the nature of a thing resulting from its production and growth. The element -PHYSIS, as occurring in scientific words, is often better interpreted as "a growth, a growing".

sym'physis—"a growing together"—the union (growing together) of two corresponding bones (or other parts) of the body, e.g. the pubic symphysis in the pelvis.

para'physes—"growths at the side"—non-reproductive hair-like growths among the reproductive structures of many lower plants.

hypo'physis — "a growth under" — a general term for a downward growing structure in a plant or animal; in vertebrate animals the term particularly denotes the pituitary body.

apo'physis—"a growth away from"—This term is used in many connections, e.g. a growth from a bone (usually for the attachment of a muscle), the swollen end of the seta (stalk) of a Moss capsule, an enlargement of the outer end of a scale of a Pine cone, a side-branch or offshoot of a larger vein of igneous rock. (For derivatives of this word, e.g. **hyp'apophysis**, see APOPHYSIS.)

PHYSO-, -PHYSEMA

a bladder (Gk. *physa*, bellows, a bladder, a bubble, from *physaō*, to blow, to puff).

physo'stomous—(fish) having the air bladder joined to the alimentary canal (food tract) by a duct.

Physo'phora—"the bladder bearers"—a genus of oceanic Hydrozoa which float by means of numerous bladder-like organs.

Physo'pod—a mollusc with a kind of sucker on its foot.

Physo'stigma—a genus of leguminous (Pea and Bean family) plants; in the flower the style is continued into a hood ("bladder") above the stigma. The genus includes the Calabar Bean from which the alkaloid **physostigmine** is obtained.

em'physema—a swelling due to the presence of air in the body tissues, e.g. the formation in the lung of air-containing spaces.

PHYTO-, -PHYTA, -PHYTE

a plant (Gk. *phyton*).

phyto'genesis — the generation or evolution of plants.

phyto'benthon—the plants living at the bottom of the sea.

Bryo'phyta—the phylum (large group) of plants comprising the Mosses and Liverworts.

Pterido'phyta—the phylum of plants comprising the Ferns, Club-mosses and related plants.

Spermato'phyta — the phylum of plants which bear seeds, i.e. the Gymnosperms (Conifers, etc.) and the flowering plants.

hydro'phyte—a plant which lives in, or on the surface of, water.

epi'phyte—a plant which grows on another plant (e.g. on a leaf or stem) but is not parasitic upon it.

sapro'phyte—a plant which obtains its

food from the dead remains of living things. (Many Fungi are **saprophytic.**)

geo'phyte—a plant which spreads by means of underground buds.

gameto'phyte—that form of a plant (e.g. of a Fern) which reproduces by means of gametes (sex cells).

sporo'phyte—that form of a plant which reproduces by means of spores.

Phyto'thora — "plant corruption" — the fungus which causes Potato disease. The term **phyto'logy** ("the study of plants"), unlike the corresponding term *zoology* ("the study of animals") is not often used; the science is more commonly called *botany*.

PIC-, PICI-

(1) a Woodpecker (L. *picus*).

Picus—the genus of Woodpeckers.

pic'oid—resembling, or pertaining to, Woodpeckers.

Pici'formes—the order of birds which includes the Woodpeckers, Toucans, and others.

(2) pitch (L. *pix*, *pic-*), hence derived from coal-tar.

pic'ene — one of the complex hydrocarbons, $C_{22}H_{14}$, which has been isolated from that part of coal-tar which boils above 360°C.

pic'ol'ine — an organic base, $C_5H_4N(CH_3)$, related to pyridine, obtained from bone-oil and coal-tar naphtha.

PICO-

one millionth-millionth (10^{-12}).

pico-farad—one millionth-millionth of a farad (a unit of electrical capacitance). Also called a micro-micro-farad.

PICR-, PICRO-

bitter (Gk. *pikros*). Used especially in the names of (a) minerals containing bitter salts of magnesium, and (b) compounds derived from picric acid.

picro'toxin—the bitter, poisonous constituent of the seeds of *Cocculus indicus*, resembling strychnine in its action.

picr'amin—either of two bitter constituents of Jamaica quassia.

picr'ite — a general name for coarse-grained igneous rocks which consist essentially of silicates of magnesium and iron with a small amount of plagioclase.

picro'mer'ite — "bitter in part" — a white crystalline mineral consisting of sulphates of magnesium and potassium, $MgSO_4.K_2SO_4.6H_2O$.

picr'osmine—a greenish or greyish, fibrous silicate of magnesium which emits a bitter odour when moistened.

picr'ic acid — tri'nitro'phenol, $C_6H_2.(NO_2)_3OH$—a very bitter, bright yellow crystalline compound, formerly used as a dye, now for making explosives. The salts are called **picrates.**

picr'yl—the group of atoms $C_6H_2.(NO_2)_3$-, e.g. picryl chloride $C_6H_2.(NO_2)_3Cl$.

-PIESIS, (-PIESIA)

pressure (Gk. *piesis*). Used especially to denote blood pressure.

hyper'piesis—abnormally high blood pressure.

hypo'piesis—abnormally low blood pressure.

PIEZO-

pressure, caused by pressure (Gk. *piezō*, to press).

piezo-chemistry — the study of the effects of very high pressures on chemical reactions.

piezo-electric effect—the production of opposite electric charges on the faces of certain crystals (e.g. quartz) when the shape is changed by outside forces; conversely, the expansion of a crystal in one direction and the contraction in another when an electric field is applied. The effect is used in certain instruments, e.g. the **piezo-microphone.**

PIL-, PILI-, PILO-

a hair (L. *pilus*).

This root is seen in the **pile** of a carpet.

pil'ose—hairy.

pili'ferous—bearing hairs, e.g. the surface layer of a root (which bears root-hairs).

de'pil'atory — (substance) used for removing hairs.

horri'pil'ation—a bristling (a standing up) of the hair as when cold or frightened. (Also called goose-flesh.)

pilo'motor — (nerves) which bring about the movement of hairs.

Note. *Pilocarpine* is an alkaloid obtained from the leaves of various species of *Pilocarpus* (from Gk. *pilos*, wool, felt+*karpos*, fruit).

PILE-, PILEUM, PILEUS

a cap (L. *pileum*, *pileus*, a felt cap).

pile'ate — shaped like a cap; having a pileus (as below).

pileum—the top of the head of a bird.

pileus—the cap-shaped part of a mushroom or similar fungus.

pil'idium—"a small cap"—the helmet-shaped larval form of some Nematode worms.

PINAC-, PINACO-

a slab, a tablet (Gk. *pinax*, *pinak-*, a board, a tablet).

pinac'oid—"slab-like"—said of a plane (or group of planes) in a crystal shape which intersects one of the axes but is parallel to the other two.

pinaco'cyte—a flattened cell of the outer layer of the body wall of Sponges.

pinac'one—a complex glycol, $(CH_3)_2$: $C(OH).C(OH):(CH_3)_2$, formed by the reduction of acetone. It crystallises to form large white tablets. Hence **pinac'ol'in(e)**—a colourless oily liquid, $CH_3.CO.C(CH_3)_3$, produced by the action of dilute sulphuric acid on pinacone.

PINN-, PINNATI-, PINNI-

a feather, a wing, a fin (L. *pinna*; *penna*).

pinna—a leaflet (part of a compound leaf); a fin of a fish; a feather of a bird; in mammals, the outer part of the ear.

pinnate — "feathered" — bearing branches on each side of a main axis, e.g. a compound leaf bearing a row of leaflets on each side of the main stem.

pinnati'fid—"pinnately cleft"—(leaf) which is cut into about half-way towards the middle so making a number of lobes.

pinn'ule—"a small pinna"—one of the small parts of a leaflet of a pinnate leaf when the leaflet itself is pinnate.

pinnati'ped, pinni'ped—"wing (or fin) footed"—having the toes joined together by flesh or membrane (as has a Seal or a Walrus).

pinni'grade—moving by means of fins.

PISCI-

a fish (L. *piscis*).

Pisces — the group of vertebrate animals which includes the Sharks and the Bony Fishes; the star constellation which forms the twelfth sign of the zodiac.

pisci'culture—the artificial rearing of fish.

pisci'vorous—fish-eating.

pisci'colous—living inside a fish (as do certain parasites).

PISI-, PISO-

a pea (L. *pisum*; Gk. *pisos*).

pisi'form—pea-shaped.

piso'lite—"pea stone"—a form of limestone built up of rounded particles (which are bigger and less regular than those of oolite).

PITHEC-

an ape (Gk. *pithēkos*).

pithec'oid—ape-like.

Pithec'anthropus — "ape man" — an early form of Man who may be the 'missing link' between apes and true Men; the fossil remains were found in Java in 1891.

The root is used in naming various possible ancestors of apes and men, e.g. **Australo'pithecus** ("southern ape"), **Neo'pithecus** ("new ape").

PLAC-

a plate (Gk. *plax*, *plak-*).

plac'oid—plate-shaped, e.g. placoid scales of some fish.

plac'ode—any plate-like structure.

plac'enta—the flattened, round organ to which the unborn child is attached, by the umbilical cord, within the womb;

that part of the carpel (container) to which the ovules are joined. (The word is derived, through Latin, from Gk. *plakous*, a flat cake.)

PLAGIO-
oblique, slanting, at an angle (Gk. *plagios*).

plagio'cephalic — having a "slanting head"—having the front part of the skull more developed on one side and the back part more developed on the other side so that the head is not symmetrical.

plagio'clase — "oblique splitting" — a type of feldspar mineral, containing sodium or calcium silicates, which splits in two directions at an angle. (Contrast orthoclase.)

Plagio'stomi—cartilaginous fishes, including the Shark and Dogfish, which have a crosswise mouth beneath the snout.

PLAN-, PLANI-, PLANO-
flat, level (L. *planus*).
(This root must not be confused with that which follows.)
In carpentry a **plane** is a tool for making a surface flat and smooth. In mathematics a **plane** is a flat surface (real or imaginary), i.e. one such that a straight line joining any two points of it lies wholly in the surface. (Note that a plane need not be level.) Lines, etc., which lie in the same plane are **co-planar.**

Plan'arians—a group of free-swimming forms of Flatworms.

plani'petalous — (flower) having flat petals.

plani'meter — an instrument for measuring the area of an irregular plane figure.

plano-convex—(lens) having one surface plane and the other convex (curved outwards). **plano-concave**—having one surface plane and the other concave (curved inwards).

-PLANE, PLANET, PLANO-
a wanderer; wandering, free-moving (Gk. *planētēs* (n.); *planos* (adj.)).
(This root must be distinguished from the preceding one.)
The ancients noticed that, while the great number of stars remained fixed in their positions relative to each other, a few appeared to have a motion of their own and to change their positions among the other stars. These are the **planets** ("wanderers"). They are not true stars but bodies which, like the Earth, move round the Sun.

planet'arium—a model of the Sun and planets; an instrument which shows the motions of the planets on a rounded roof (representing the sky); the building in which this is done.

planet'oid—a minor planet (an asteroid).

plano'gamete — a free-swimming gamete (sex cell).

An **aero'plane** is "a wanderer through the air". The element -PLANE has now almost become accepted as an alternative to aeroplane or aircraft. Hence a **converti'plane** is "a convertible aeroplane", i.e. an aircraft which can take off and land like a helicopter but act as an ordinary aeroplane when in forward flight.

PLANKTON
that which is drifting (Gk. *planktos*, drifting, roaming).

plankton—animals and plants (usually small) floating and drifting in the waters of seas, lakes, etc., as distinct from the forms of life, which are free-swimming or are attached to, or crawl on, the bottom.

zoo'plankton—plankton consisting of animals.

phyto'plankton — plankton consisting of plants.

sapro'plankton — plankton living in foul water.

PLANTI-
the sole of the foot (L. *planta*).

planti'grade—walking on the soles of the feet (as does Man).

It is probable that the name **Plantain** for a well-known weed with flat-lying leaves is derived from this root.

-PLASIA, PLASMA, PLASM-, PLASTI-, and similar forms

Gk. *plassō*—to mould, to form, to shape. *plasma* — anything moulded, modelled, or formed. *plastos, plastikos*—moulded, formed.

These Greek roots appear in scientific terms in a variety of spelling forms. The interpretation of one form may differ from that of another form (although all forms carry the basic idea, literally or figuratively, of moulding, shaping or forming). For convenience of explanation and of reference, the various forms are treated in separate groups.

-PLASIA

moulding, formation (especially of tissue) (Gk. *plasis*).

hyper'plasia—excessive multiplication of the cells of the body so producing an overgrowth of tissue.

hypo'plasia—under-development.

meta'plasia — "following growth" — the period of maturity (complete development) in the life-cycle of an individual.

a'chondro'plasia — "lacking cartilage growth"—a dwarf condition in which the arms and legs are small but the head and body are of normal size.

PLASMA, PLASMO-, -PLASM

that which is formed or moulded (Gk. *plasma*).

The term **plasm** is used with a number of different meanings.

(a) the watery liquid of the blood in which the various blood cells are carried.

(b) the region of gas in an electric discharge tube where there are equal numbers of positive and negative ions.

(c) a gas so hot, as in a star, that all its atoms have become ions.

(d) a bright green, transparent form of quartz.

neo'plasm—a new formation of tissue in the body, a tumour.

proto'plasm—"the first moulded substance", is the semi-liquid, semi-transparent, almost colourless substance, of complicated composition, which forms the basis of life in all plants and animals. The elements PLASMO- and -PLASM usually refer to the protoplasm or living substance of cells.

nucleo'plasm—the dense protoplasm which forms the nucleus of a cell.

cyto'plasm—the protoplasm of a cell other than the nucleoplasm.

plasmo'lysis—"protoplasm loosening" —the removal of water from a cell (by osmosis, so that the protoplasm shrinks.

plasmo'gamy — "plasm marrying" — the joining of the cytoplasm of two cells without the joining of the nuclei.

plasmo'tomy — "plasm cutting" — the splitting of the cytoplasm (of a Protozoan cell containing several nuclei) without the splitting of the nuclei.

plasmo'dium—a mass of naked protoplasm formed by the joining of amoeba-like bodies without the joining of their nuclei. (The word is also used in other senses.)

PLASTI-, -PLAST

A combining-element used particularly to denote a small mass of dense substance in a cell (from Gk. *plastos*, formed, moulded).

plast'ids—very small bodies in the protoplasm of a cell which often develop and give rise to chromoplasts.

chromo'plast—"a coloured body" in a cell, e.g. a **chloro'plast**—such a body which contains the green substance chlorophyll, **leuco'plast**—such a body which is "white" (colourless).

PLASTIC

able to be moulded (Gk. *plastikos*).

A **plastic** material is one such as clay or wax which can be shaped and moulded. **Plasticine** (trade name) is a plastic substance used, especially by children, as a substitute for clay for modelling purposes.

Plastics are resin-like substances which at some stage (usually under heat and pressure) can be moulded. Some occur naturally; many are made artificially by polymerisation or other processes. Those which are **thermo'plastic** can be made plastic (at any time) by heating; those which are *thermosetting* cannot be re-moulded after heating.

-PLASTY
the shaping or moulding of a part of the body by surgery (Gk. *plassō*, to shape).

kerato'plasty—the grafting of a new cornea (*kerato-*) on to an eye.

hetero'plasty—"different moulding"—the grafting on one person of body-tissue (e.g. skin) removed from another person.

thoraco'plasty — the operation of collapsing a diseased lung (causing it not to fill with air) so as to rest it.

PLATY-
broad and flat (Gk. *platys*).

platy'cephalic—having a flattened or broad head.

platy'kurtic—said of a distribution of measures which are more spread out from the average than in a 'normal' distribution so that the distribution curve is broader and flatter. (For further explanation, see KURTO-.)

Platy'helminthes—the class of Flat-worms.

Platy'pus — "broad footed" — the Duckbill or Duck-mole, an egg-laying mammal of Australia with a duck-like snout and broad, webbed feet.

-PLEGIA
a stroke, paralysis, loss of power because of brain injury (Gk. *plēgē*).
(This root also appears in the form -PLEXY (q.v.).)

hemi'plegia — paralysis of one side ("half") of the body.

para'plegia — paralysis of the lower part of the body and the legs.

palato'plegia—paralysis of the palate (roof of the mouth).

ophthalmo'plegia—paralysis of one or more muscles of the eye.

PLEI-, PLEIO-, PLEO-, PLIO-
more, more than usual (Gk. *pleiōn*, or *pleōn*, the comparative of *polys*, many).

plei'androus—(flower) having a large, indefinite number of stamens.

plei'on—an area in which temperature (or other weather element) is above usual.

pleio'merous—(flower) "having more parts" (e.g. petals) than the basic number.

pleo'phagous—(parasite) which "eats" (attacks and lives on) more than one species of host plant.

pleo'cyto'sis—"a state of more cells" —an increase in the number of white blood cells (especially in the fluid which bathes the surfaces of the brain and spinal cord).

Plio'cene period—"more new (recent)" —the geological period from about 15 to 1 million years ago in which there was much mountain building. It is the last period of the Tertiary era and came immediately before the Pleistocene period (see below).

PLEISTO-
most (Gk. *pleistos*, the superlative of *polys*, many).

Pleisto'cene period — "most new (recent)"—the geological period from about 1 million years ago. It includes the Great Ice Age.

PLESIO-
near, close to (Gk. *plēsios*).

plesio'morphous—very near in form, crystallising in forms which (though not identical) are very similar.

Plesio'saur—"very near a lizard"—an extinct marine reptile.

PLETHORA, PLETHYSMO-
an increase, a fullness (Gk. *plēthō*, to be or become full; *plēthōrē*, fullness; *plēthysmos*, an enlargement).

plethora—an increase above normal of the volume of the blood (with or without an increase in the number of red cells).

plethysmo'graph—an instrument for measuring changes in the size of body parts and of the flow of blood through them.

PLEUR-, PLEURO-
The Greek words *pleura*, *pleuron* mean "a rib, the side of the body". The element PLEUR(O)- is used to denote (a) the chest of the body, and (b) "a sideways position".

pleura—the membrane lining the **pleural cavity** which contains the lungs.

pleur'isy, pleur'itis—inflammation of the pleura.

pleuro-pneumonia — inflammation of both the pleura and a lung. (Used especially for a contagious disease of cattle.)

pleur'odynia—pain in the muscles of the ribs and chest.

pleuro'carpous—bearing the fruit in a sideways position. So also **pleurosporous** (spores).

pleur'odont—having the teeth fastened to the side of the bone which bears them (as in some lizards).

pleuro'genous—produced in a sideways position.

pleur'acro'genous—produced both at the tip and on the sides of stems.

-PLEX, PLEXUS
a network, a tangle (L. *plecto*, *plex*-, to interweave).
Complex means "well tangled".

plexus—a network of fibres, nerves, etc. in the animal body, e.g. **solar plexus** — "sun network" — the tangled nerve centre in the pit of the stomach from which nerves radiate in all directions.

-PLEXY
a stroke, paralysis, loss of power because of brain injury (Gk. *plēsso*, to strike).

apo'plexy—"a striking away (completely)" — sudden paralysis, usually with loss of consciousness, caused by an escape of blood into the brain.

cata'plexy — "a striking down" — an attack of weakness in which the person falls to the ground and remains motionless.
Also see -PLEGIA.

PLIC-
to fold (L. *plico*, *plicat*-).
This root is familiar in **duplicate** ("twofold"), **multiplication** ("a folding many times") and **complicated** ("well folded").

plica — a fold of tissue. **plic'ate** — folded.

ex'plic'it—"folded out, unfolded". An **explicit function** is a relationship between two mathematical quantities (e.g. x and y) which is openly stated (e.g. $y=2x^2-3$). In an **im'plicit** ("folded in") **function** the relationship is not openly stated (e.g. $x^2+y^2-3xy=5$).

PLIO- See PLEIO-.

-PLOID, PLOIDY
When a plant or animal cell is about to divide, the chromatin particles of the nucleus mass together to form chromosomes (heredity bodies). Each species has its characteristic number of chromosomes. A body cell has a double set of chromosomes (e.g. 2×23 in Man). It is **dipl'oid** ("of double form"). A reproductive cell has a single set (e.g. 23 in Man). It is **hapl'oid** ("of single form"). (The double set is re-made by the joining of two sex cells.)

A variation in the chromosome number sometimes occurs in plants, less commonly in animals, and may be produced artificially. By continuing the 'shape' of the words haploid and diploid, cells may be described as **triploid** (3 times the haploid number), **tetraploid** (4 times), and so on.

Hence has come about the extraordinary word **ploidy** to denote the number of chromosomes in relation to the haploid number. **Tetra'ploidy** is common among cultivated plants. **Eu'ploidy** ("good", true ploidy) is the state of having an exact multiple of the haploid number. Even numbered **euploids** breed true; odd numbered euploids often produce **an'euploid** ("not euploid") plants in which the actual number of chromosomes is a few more than the proper diploid number.

The crossing of one species with

another (e.g. radish with cabbage) gives curious combinations. The combination of a haploid set of each kind usually results in a sterile plant but the combination of a diploid set of each (**amphi'diploidy**) results in hybrids which can breed.

PLUM-

a feather (L. *pluma*).

A **plume** is a feather, especially one used for ornament.

plum'age—the feathers of a bird considered as a whole.

plum'ule—"a little feather"—the tiny bud and shoot in a seed and immediately after it grows out from the seed.

PLUMB-

the metal lead (L. *plumbum*).

A **plumber** is one who fits and mends lead pipes, tanks, etc. A **plumb-line** is a lead ball (or other heavy object) on the end of a string, used for testing if a wall (etc.) is vertical. **Pb** is the chemical symbol for lead.

plumb'ism—lead poisoning.

plumb'ite—a kind of salt (e.g. potassium plumbite) formed from an alkali and lead hydroxide.

In the early days of chemistry the name **plumbago** was applied to a variety of substances, including graphite, molybdenum sulphide, lead sulphide and even litharge. Up to 1779 graphite and molybdenum sulphide were often confused; in that year Scheele demonstrated that they were quite different chemically and that graphite could be made to produce carbon dioxide. It was then thought that graphite was a carbide of iron. Some years later it was shown to be a pure form of carbon. The old names plumbago and black-lead are still sometimes used for it.

PLUR-, PLURI-

more, several (L. *plus*, *plur-*).

This root is immediately recognised in the English word **plural**.

pluri'cellular—composed of a number of cells.

pluri'glandular — pertaining to, or affected by, or affecting, several (ductless) glands.

pluri'locular — having "several little places"—said of a spore-container or an ovary which is divided by cross-walls into several compartments.

PLUTO, PLUTON-

Pluto or *Pluton* (L.), *Ploutōn* (Gk.), was the god of the Underworld.

pluton'ic rocks or **intrusions** — large masses of igneous rocks which have cooled at great depths below the Earth's surface.

Pluto—the ninth planet out from the Sun (lying beyond Neptune), which was discovered in 1930. Named after the god.

pluton'ium—a radioactive chemical element derived from neptunium (which is produced by neutron bombardment of uranium). These three elements are named to correspond with the sequence of planets Uranus, Neptune, Pluto.

Note. The word *pluto'cracy* (government by the wealthy classes) is derived from Gk. *ploutos*, wealth.

PLUVI-

rain (L. *pluvia*).

pluvi'al—pertaining to rain.

pluvi'ometer — an instrument for measuring rain-fall, a rain-gauge.

PNEUMAT-, PNEUMATO-, PNEUMO-

air (Gk. *pneuma*, *pneumat-*, air, breath).

pneumat'ic—of, or worked by, air, e.g. a pneumatic tyre on a motor-car, a pneumatic drill worked by compressed air.

pneumato'cele—a swelling containing air; a pushing out of a lung through a weak place in the wall of the chest.

pneumato'cyst—an air-sac in the body of (e.g.) a bird or a fish.

pneumat'uria—the presence of air or gas in the urine.

pneumo'taxis—"arrangement to air" —the response of an organism to a gas (e.g. to dissolved carbon dioxide).

PNEUMON-, PNEUMONO-, PNEUMO-
a lung (Gk. *pneumōn, pneumon-*).

pneumonia, pneumon'itis—inflammation of the lungs. It is caused by bacteria of the type **pneumono'coccus.**

pneumon'ectomy—the cutting out of lung tissue by surgery.

pneumono'koni'osis, pneumo'koni'osis—a disease of the lungs caused by dust (e.g. from a coal-mine).

pneumo'gastric — pertaining to the lungs and the stomach, e.g. the pneumogastric nerves.

-PNOEA
breathing (Gk. *pnoē*).

dys'pnoea—difficult breathing.

tachy'pnoea—excessively fast breathing.

hyper'pnoea—an increase in the speed and depth of breathing.

a'pnoea—"absence of breathing"—the stopping of breathing.

Di'pnoi — animals with "double breathing"—the Lung Fish which have, in addition to gills, air-sacs which can act as lungs.

-POD, -PODA, -PODIAL, -PODIUM
a foot (Gk. *pous, pod-*).
A **tripod** is a stool, table, support, etc. which has "three feet".

deca'pod—an animal which has ten legs, e.g. Shrimp, Lobster, Crab. They form the order **Decapoda.** (Cuttlefish and similar animals are also sometimes called decapods because they have eight true arms and two other arm-like projections.)

arthro'pod—an animal such as a Crab, Spider, Insect, Centipede, which has "jointed legs". They form the large group (phylum) **Arthropoda.**

cephalo'pod — an animal such as the Octopus in which the mouth is apparently in the middle of the foot. They form the class **Cephalopoda.**

gastro'pod—an animal such as a Snail which moves about on a muscular outgrowth ("foot") on the underside ("stomach") of the body. The class **Gastropoda** includes Limpets, Whelks, Snails and Slugs which have a single (often conical) shell.

necto'pod—a foot-like part which is adapted for swimming.

-PODIUM ("a small foot") may sometimes be better interpreted as "a foot-like thing".

pseudo'podium—"a false foot"—a foot-like piece of protoplasm pushed out by some simple animals (e.g. Amoeba) as a means of moving or of enclosing a particle of food.

para'podium—one of a pair of fleshy flaps borne on each segment of some annelid worms (especially the Polychaeta) carrying the bristles. They serve as organs of locomotion (and sometimes of sensation).

phyllo'podium—"leaf foot-like thing"—the main stalk and axis of a leaf such as that of a Fern.

mono'podium—"single foot"—a system of branching in a plant which results in a cone-shaped plant on one foot. (Each branch continues to grow lengthwise, side branches develop in turn with the youngest nearest the tips of the branches.) The branching is said to be **monopodial.**

-POEIA, -POIESIS
making, the making of (Gk. *poieō* (v.), *poiēsis* (n.)).

pharmaco'poeia—a book which gives a list of drugs with instructions for their making and use.

haema'poiesis—the making of blood.

leuco'poiesis—the making of white blood cells.

chylo'po(i)esis—the making of chyle (milky liquid formed by the digestive processes in the intestine).

POIKIL-, POIKILO-, (POECILO-)
many coloured, hence varied, not all of one kind (Gk. *poikilos*).

poikil'itic rock—an igneous rock with small crystals of one mineral mixed with larger crystals of another.

poikilo'cyte — a badly formed

("varied") red blood cell. Their presence causes **poikilo'cyto'sis.**

poikilo'thermous—having the body temperature vary with the temperature of the surroundings. (Commonly called cold-blooded.)

POLAR, POLARI-

Gk. *polos*, a pivot, an axis on which a thing turns. L. *polus*, an end of an axis, a pole (in the sky). L. *polaris*, of a pole. The **poles of the Earth** are the two points at the ends of the axis about which the Earth turns. The **poles of a magnet** may be compared with the poles of the Earth although a magnet does not naturally turn about its axis. Similarly, a molecule can have poles, i.e. a positive charge towards one end and a negative charge towards the other; it is then a **dipole.** The **poles of a cell** (or battery) are the two terminal points.

Polarised light is light which has been modified so that the vibrations of the light waves are all in one plane; a ray then has a particular axis at right-angles to its direction of travel.

The element POLARI- may refer to poles (ends of an axis) or to polarised light.

polar—of, or pertaining to, poles.

polari'locular spore—with "pole-like places"—a spore with two compartments and a small connection between them.

polari'scope—an instrument for observing effects caused by polarised light.

polari'meter — an instrument for measuring the twisting of the plane of polarisation of light by (e.g.) certain solutions.

-POLATE, -POLATION

This root (from L. *polio*, to smooth, to polish, to clean up; *-polo*, *-polat-* in compounds) occurs in two important scientific terms.

Consider a series of measures, e.g. the temperatures at hourly intervals throughout a morning. One can fill in measures between those of the series (e.g. temperatures at each half hour) with a reasonable chance of their being reliable. This is the process of **inter'polation.** One can add measures outside the range of the series, e.g. temperatures during the afternoon, by following the general trend of the series. This is the process of **extra'polation.** It is clearly less reliable than interpolation.

POLIO-

grey (Gk. *polios*).

polio'plasm — grey, granular protoplasm.

polio'myel'itis—inflammation of the grey matter of the spinal cord.

polio'encephal'itis — inflammation of the grey matter of the brain.

POLY-

many (Gk. *polys*, *poly-*).

This is a common prefix. A **Polyanthus** produces "many flowers"; a **polytechnic** is a place where "many arts" are taught. It is not necessary to give more than a small, representative list of the many scientific words which contain this prefix.

poly'gon—a plane figure with many angles and sides.

poly'hedron—a solid figure with many faces.

poly'androus—having "many males" —(flower) having a large, indefinite number of stamens: (female animal) having many 'husbands'.

Poly'chaeta — the class of marine Bristle Worms. The parapodia (foot-like stumps) bear "many bristles".

poly'dactyl—having more than the usual number of digits (fingers or toes).

poly'ploid—having more than twice the haploid number of chromosomes. (For further explanation, see -PLOIDY.)

poly'hydric—(an alcohol) whose molecule contains more than one hydroxyl (-OH) group.

poly'saccharides, poly'saccharoses, poly'oses—complex carbohydrates, such as starch, cellulose, which may be regarded as built up from many simple sugar molecules.

polyp See -PUS.

poly'mer—a substance built up from

"many parts" — a complex molecule built up from a number of simple molecules of the same kind, e.g. paraldehyde $(CH_3CHO)_3$ is a polymer of acetaldehyde CH_3CHO. (Some 'plastics' are formed by the process of **polymerisation.** Others, e.g. Bakelite, are formed by the joining together of molecules with the elimination (setting free) of simple molecules such as water, H_2O.)
The prefix POLY- is sometimes used to denote 'a polymer of', e.g. **poly'ethylene** (or **polythene**) is a polymer of ethylene C_2H_4. So also **poly'styrene, poly-esters,** etc.

POME, POMI-, POMO-
an apple, a fruit (L. *pomum*).
A **pome** is a fruit like an apple, pear or quince. A **pome'granate** is a "seeded fruit".

pomi'culture—the cultivation of fruit, fruit-growing.

pomo'logy—the study and science of fruit-growing.

pomi'form—apple-shaped.

-PONIC
labour, toil (Gk. *ponos*).

geo'ponic — "earth-toiling" — pertaining to agriculture.

hydro'ponics—the culture of plants in water (not soil) containing the necessary chemical substances.

POR-, -PORE, PORI-, PORO-
a passage, a pore (Gk. *poros*).
This root is seen in the English words **pore, porous** and **porosity.**

poro'plastic—both porous and plastic, e.g. poroplastic felt.

poro'meter — an instrument for measuring the rate at which air can be drawn through the pores of a leaf.

Pori'fera—the "pore bearers"—a phylum (group) of Parazoa which includes the Sponges. Water can enter the single cavity of the body through the numerous pores in the body wall.

por'encephaly—the presence of small cavities ("pores") in the substance of the brain (due to imperfect development).

blasto'pore—an opening in a gastrula (hollow developing embryo).

PORPHYR-
purple (Gk. *porphyros*).
A certain rock was quarried in ancient Egypt for building and ornamental purposes. It consisted of large white or red crystals set in a purplish mass of small crystals. The Greeks named it *porphyritēs* (*lithos*)—the purple stone. This name became *porphyrites* in Latin and hence **porphyrite** or **porphyry** in English.

porphyry—a general name for an igneous rock which, like the original porphyry, consists of large crystals set in a ground-mass of fine grains (though not necessarily of a purple colour). **porphyritic,** adj.

porphyr'ins — complex compounds (based on pyrrole groups) many of whose derivatives are of a reddish colour. Haematin (part of the red haemoglobin of blood) and chlorophyll (the green substance of plants) are complex porphyrins.

POST, POST-
after, behind (L. *post*).
Some Latin expressions have been taken into the English language unchanged, e.g. **post meridiem** (p.m.)—after midday, **post mortem**—an examination of a body "after death".
POST- is used in English as a prefix, e.g. **postpone**—"to place after".

post'humous—"after" being put in the "ground"—(child) born after the death of the father; (book) published after the death of the author.

post'costal—situated behind a rib.

post'hepatic — situated behind the liver.

post-glacial—after the glacial period (Ice Age).

POT-, POTO-
to drink (L. *poto*, *potat-*).

potable — suitable for drinking purposes; **potion**—a dose of liquid (medicine or poison) to be drunk.

poto'meter — "drink measurer" — an instrument for measuring the rate at which a plant takes in water.
There are other scientific words beginning with POT- which are not derived from this root. For POTAM- and POTENTI- see below. **Potash** (which gave rise to the name **potassium**) means "pot ash"—a reference to the alkaline substance obtained by evaporating a solution of vegetable ashes in an iron pot.

POTAM-, POTAMO-
a river (Gk. *potamos*).
A **hippo'potamus** is "a horse of the river"; **Mesopotamia** is the land "in the middle of (between) the rivers" Tigris and Euphrates.

potam'ous — living in rivers and streams.

potamo'plankton — plankton (floating plants and animals) of rivers and streams.

POTENTI-
to have power (L. *potens*, *potent-* having power).
The English word **potential** means 'capable of coming into being or action'. It implies powers which are hidden and not actual. The word is used in this sense in **potential energy** — the energy possessed by a body because of its position (e.g. a raised weight) or its distorted shape (e.g. a wound-up clock spring) and which becomes available when the body returns from that position or shape.
The term **potential** is also used, mainly as a noun, to denote what may be simply described as 'electric pressure'. A **potenti'ometer** is an instrument for the accurate measurement of potential differences (differences of electric pressure). (The term has also been applied to a variable resistor from which various potential differences may be obtained.)

PRAE- See PRE-.

PRASE, PRASEO-
green like a leek (Gk. *prasios*).

prase—a transparent, dull green form of chalcedony.

praseo'dymium—one of the rare earth elements whose oxide is pale green in colour. (See DIDYM-.)

PRE-, PRAE-
before (in time, place, importance, etc.) (L. *prae*).
The original Latin form PRAE- is retained in only a few words and even in these PRE- is gradually taking its place.
The prefix PRE- occurs in a large number of familiar words, e.g. **predict** ("to say beforehand"), **precede** ("to go before"), **prepay** ("to pay in advance"), **prevent** ("to come in front of"). When the prefix is added to an existing English word a hyphen is sometimes inserted.

prae'cocial—(birds) whose young have a complete covering of down when hatched and are able at once to seek their own food. (Compare the word **precocious**.)

pre'cordial—situated or occurring in front of the heart.

pre'frontal—in front of the frontal bone of the skull.

pre'molars — the grinding teeth in front of the true molars.

pre'gnant—"before birth"—(woman, female animal) who is carrying an unborn child (or offspring).

pre-Cambrian—before the Cambrian times (i.e. before about 500 million years ago).

pre-cast cement—cement cast into blocks before use in building.

pre-ignition—the ignition (firing) of the petrol vapour in an engine before the normal ignition by the electric spark. Caused, e.g., by a red-hot piece of carbon.

-PRENE
Isoprene is a colourless hydrocarbon, with the formula $CH_2{:}C(CH_3).CH{:}CH_2$, obtained by the destructive distillation of rubber. The name was apparently invented by C. G. Williams in about 1860.

The 'artificial' word-element -PRENE, taken from *isoprene*, has been used in forming the names of certain substances from which synthetic rubbers are made, e.g. **chloro'prene** (CH_2:C(Cl). CH:CH_2 from which the synthetic rubber **Neo'prene** is made.

PRESBY-
old age (Gk. *presbys*, an old man).
A **Presbyterian Church** is governed by elders of equal rank.

presby'opia—a form of long-sightedness which comes on with advancing years.

PRIM-, PRIMI-, PRIMO-
first (and hence early), chief (L. *primus*). This root is seen in **primary** (first, chief), **prime** (chief, most important, of first quality) and **primitive** (early, ancient). A **primrose** (*prima rosa*) is the "first rose" but why it was so called has not been satisfactorily explained.

prime mover—an engine by which energy derived from a natural source is converted into mechanical energy, e.g. a steam engine (which uses coal) but not an electric motor (which must be provided with electrical power).

prim'ordial—"first beginning"—existing at, or from, the beginning.

Prim'ates—the highest class of mammals, including Monkeys, Apes and Man.

primi'gravida — a woman who is carrying her first unborn child.

primi'para—a woman who has given birth to a child for the first time.

primo'genitor—one's earliest ancestor.

PRIMUL-
Related to the Cowslip or Primrose.
The name **Primula** (a feminine diminutive of L. *primus*, first) was originally applied to the Cowslip. It is now the name of the genus of flowering plants which includes the Cowslip (*P. veris*), Primrose (*P. vulgaris*) and other species.

Primul'aceae—the family of flowering plants of which *Primula* is the typical genus.

primul'ine — a primrose-yellow dye stuff. (A derivative of thiazole.)

PRO-
L. *pro*, in front of, for, forth.
Gk. *pro*, before (in time, place, order, etc.).
This common prefix, especially in its Latin senses, is found in a large number of general words, e.g. **proceed** ("to go forth"), **produce** ("to lead forth"), **progress** ("to walk forth"), **project** ("to throw forth").

Latin

pro'lapse—a slipping forward (or out of place) of an organ or part of the body.

pro'geny—"a begetting, a bringing forth" — the offspring of a person, animal or plant.

pro'strate—"laid forth flat"—(person) laid out flat on the ground; (plant stem) which is **pro'cumbent**—lying on the ground for much of its length.

Occasionally the Latin prefix means "on behalf of, a substitute for", e.g. **pro-leg** —one of the fleshy projections on the body of larvae of butterflies, etc.

Greek

pro'boscis—"a feeder in front"—the trunk of an elephant; the sucking organ of some insects; similar parts of other animals.

pro'dromal — "running before" — (symptoms) forewarning of a coming disease.

pro'gnathous—having projecting jaws; having the mouth-parts directed forwards.

pro'gnosis—"a knowing beforehand" —a forecast of the probable course of an illness.

pro'phylactic — "guarding before" — (drug) which helps to prevent a disease.

pro'ptosis—"a falling in front"—a forward displacement or protrusion of a part of the body (especially of the eye).

pro'thallus—a small plant-body which grows from a fern spore and which reproduces sexually to give rise to the familiar form of fern.

PROCT-, PROCTO-
the anus (external opening of the intestine), the rectum (final part of the intestine) (Gk. *prōktos*).

proct'itis—inflammation of the rectum.

proct'algia—pain in the rectum.

procto'scope—an instrument for inspecting the rectum.

a'proct'ous—(kind of animal) without an anus.

PROP-, PROPION-, PROPYL

The fatty acids have the general formula $CH_3.CH_2.CH_2 \ldots COOH$. Formic acid (H.COOH) and acetic acid ($CH_3.COOH$) really begin the series but the first acid which may be considered fatty or oily is that with the formula $CH_3.CH_2.COOH$. It is called **pro'pion'ic acid** (Gk. *pro* + *pion*, fat). This name (and especially its first four letters) has formed the basis of the naming of related compounds.

propion'aldehyde—the aldehyde (q.v.) $CH_3.CH_2.CHO$, corresponding to propionic acid.

prop'ane—the corresponding paraffin hydrocarbon, $CH_3.CH_2.CH_3$.

prop'yl — the group of atoms $CH_3.CH_2.CH_2$-, i.e. C_3H_7-. (See -YL.) Hence **propyl alcohol** C_3H_7OH, **propylamine** $C_3H_7NH_2$. (The atoms of the propyl group can be re-arranged to form the *iso*-propyl group $\begin{matrix}CH_3\\CH_3\end{matrix}\!\!>CH$-.)

propyl'ene, prop'ene—the corresponding hydrocarbon with a double join between two of the carbon atoms, $CH_2{:}CH.CH_3$. (See -ENE.)

PROPER, PROPRIO-

one's own, one's self (L. *proprius*).

The word **proper** basically means belonging to, or relating exclusively to, a particular person or thing. A **proper name** is used to indicate an individual person, place, etc. (e.g. John Jones, Birmingham); the **proper motion** of a star is its own, actual movement relative to the other stars.

Property is that which one owns. The **properties of a substance** are its own particular qualities and characteristics. That which is **ap'propri'ate** is suited to, or set apart for, a particular person or purpose.

proprio'ceptor—a nerve-ending which receives stimuli from within one's own body.

proprio'spinal — (nerve fibres) which arise within the spinal cord itself.

PROPYL See PROP-.

PROS-

to, towards, in addition (Gk. *pros*).

pros'thesis—"a placing to (in addition)"—the supplying of an artificial part of the body in place of a part which is inadequate or missing (e.g. supplying false teeth or a false leg).

pros'enchyma—plant tissue composed of long, pointed cells arranged with their ends towards each other.

PROTER-, PROTERO-

before (especially in time) (Gk. *proteros*).

Protero'zoic era — the era "before life"—the second of the geological eras (following the Archaean and preceding the Palaeozoic ("old life") eras). The only undisputed remains of living things are those of lowly marine Algae.

proter'androus, protero'gynous — these terms are now more commonly spelt *protandrous*, *protogynous*. See PROTO- below.

PROT-, PROTO-, PROTEO-

first, original, hence chief (Gk. *prōtos*, *prōto-*).

Protista—"first of all"—the group of very simple organisms which are not clearly distinguished as plants or animals.

Proto'phytes — "first plants" — the most simple plants (consisting of only one cell).

Proto'zoa—"first animals"—the large group (phylum) of simple animals, mostly microscopic, consisting of a single mass of protoplasm and one/or more nuclei.

prot'androus—"male first"—(flower, animal) in which the male cells ripen before the female cells ripen. So also **proto'gynous**—"female first".

prote'ins — complex organic substances, built up mainly from amino-acids (q.v.), which form essential parts of living materials. (The word is derived from *prōteios* (primary, fundamental) + *-in* (q.v.).)

proto'nema — "first thread" — a branched, thread-like plant produced when a moss spore germinates.

proto'plasm — "the original (fundamental) moulded substance"—the semi-liquid, semi-transparent, almost colourless substance, of complicated composition, which forms the basis of life in all plants and animals.

proto'xylem—the first xylem (woody tissue) to be formed in a plant stem.

proto'type—the first type or model (e.g. of an aeroplane) as distinct from any copies or improved types which may be made later.

proto'actinium—a radioactive chemical which breaks down to form actinium.

prot'oxide of iron—the "first" oxide of iron, FeO. (The prefix PROT(O)- is occasionally used, as here, in the name of a compound in which an element combines with the smallest proportion of another element.)

prot'on — the "first" (fundamental) particle—the positively charged elementary particle which forms the nucleus of a hydrogen atom.

Special uses of these elements.

(a) PROTO- —pertaining to a proton.

proto'philic—"proton liking"—capable of joining with a hydrogen ion (a proton).

proto'genic—(originally) pertaining to the first period of formation; (now) capable of producing (supplying) a hydrogen ion (a proton).

(b) PROTE(O)- —pertaining to proteins.

prote'ase—an enzyme (e.g. pepsin in the stomach) which brings about the breakdown of proteins.

proteo'lytic—"protein loosening"—(enzyme) which causes proteins to break down into simpler substances.

PSAMM-, PSAMMO-

sand (Gk. *psammos*).

psammo'phile — "a sand lover" — a plant which flourishes in sandy soils. **psammophilous**—adj.

psamm'oma — a tumour (diseased growth) of the brain tissues containing sand-like grains.

psamm'itic gneiss—gneiss (a striped rock formed by heat and pressure from earlier rocks) formed from sandy rocks.

PSEUD-, PSEUDO-

false, imitating, as if but not really (Gk. *pseudēs, pseudo-*).

pseudo'carp—"a false fruit"—a fruit, such as an apple, which is formed from other parts than the ovary.

pseud'odont — "false teeth" — having horny jaws or ridges instead of true teeth.

pseudo'podium — "a false foot" — a piece of protoplasm pushed out by a simple animal (e.g. an Amoeba) as a means of locomotion or for enclosing a particle of food.

pseudo'parenchyma—a mass of closely tangled fungal threads which looks like parenchyma (soft cellular tissue of higher plants).

pseudo-alums—Alums are double sulphates of an element of valency (joining-power) 1 and an element of valency 3, e.g. a double sulphate of potassium (1) and aluminium (3); in pseudo-alums the place of the element of valency 1 is taken by an element of valency 2 (e.g. manganese).

PSIL-, PSILO-

bare, smooth (Gk. *psilos*).

psilo'paedic — "bare children" — (birds) hatched naked or without down.

psil'osis—"a (diseased) state of being bare"—a disease (also called Sprue) in which there is loss of energy, loss of weight, anaemia, and the passing of pale, frothy matter from the body.

psilo'melane — a "smooth black" mineral consisting mainly of an oxide of manganese.

PSITTAC-

a parrot (L. *psittacus*).

psittac'ine — pertaining to parrots, parrot-like.

Psittaci'formes—the order (class) of birds which includes Parrots, Parakeets, Macaws and Cockatoos.

psittac'osis—a disease of parrots (probably due to a virus) which can also be given to Man.

PSORA, PSORI-
an itch, a skin disease (Gk. *psōra*).

psori'asis—a skin disease in which red, scaly patches are formed.

PSYCH-, PSYCHO-
the mind (Gk. *psychē*, *psycho-*, the soul, the reason, the mind).

psycho'logy—the science which deals with the nature of behaviour and the working of the mind (normal and abnormal).

psycho'metrics—the science which deals with the measurement of mental abilities and processes.

psycho'path—"a mind sufferer"—a person who is mentally abnormal, especially one who lacks a normal social conscience, but is not suffering from a true mental disorder.

psycho'sis—a serious disorder of the mind.

psych'iatry, psychiatrics—"healing the mind"—the branch of medical science which deals with disorders and diseases of the mind.

psycho-analysis—a method of investigating and treating mental disorders (introduced by Freud and based on a certain system of psychology) in which the sufferer is induced, by various means, to bring unconscious conflicts into the conscious mind.

PTERIDO-
a fern (Gk. *pteris*, *pterid-*).

pterido'logy—the study of Ferns.

Pterido'phyta—the great group of plants which includes the Ferns, Horsetails and Club-mosses.

Pterido'spermae—"the seed ferns"—an ancient group of plants (known from fossils) which were similar to Ferns but produced nut-like seeds.

-PTER, -PTERA, -PTEROUS, PTERO-
a wing, a winged creature, anything like a wing (Gk. *pteron*).

ptero'paedes—"winged children"—young birds which are able to fly as soon as they are hatched.

Ptero'dactyl—"winged fingers"—an extinct reptile with a bird-like skull, long jaws, and wings.

ptero'pod—"winged foot"—a marine mollusc (shellfish) whose foot is formed into wing-like parts which are used for swimming.

ptero'stigma—"wing mark"—a dark spot on the wing of an insect.

Chiro'ptera, Cheiroptera—"hand wings"—the order of mammals, including Bats, in which the forelimbs are specially modified for flying.

helico'pter—"screw (spiral) wings"—a form of aircraft, with horizontally turning blades, which can take off and land vertically.

coleo'pter—"sheath wing"—a modern form of aircraft, with a 'wing' in the shape of a barrel round the body, which can take off and land vertically and turn to fly horizontally.

-PTERA (plural of *pteron*) is much used in naming sub-classes of insects.

Lepido'ptera ("scaly wings")—Butterflies and Moths; **Coleo'ptera** ("sheath wings")—Beetles; **Iso'ptera** ("equal wings")—White ants or Termites; **Di'ptera** ("two wings")—Flies, Gnats, etc.; **Hymeno'ptera** ("membrane wings")—Ants, Bees, Wasps.

-PTEROUS is used in forming adjectives.

a'pterous—without wings.

lepido'pterous—pertaining to the Lepidoptera.

-PTERIN
The **pterins** (or **pteridines**) are complex chemical compounds (with a characteristic molecular structure) which were first detected in the pigments of insects' wings. They are of great biological importance because they appear to have marked effects on the rate of growth of cells in an animal body.

leuco'pterin—the pigment of the wings of the Cabbage White Butterfly.

xantho'pterin — a yellow pigment found in the wings of the Lemon Butterfly and in the bodies of wasps.

metho'pterin—a related compound, also similar to folic acid, which appears to slow down the quick growth of cancer cells.

PTERYG-, PTERYGO-, -PTERYX
a wing, anything like a wing (e.g. a fin) (Gk. *pteryx, pteryg-*).

pteryg'ial—pertaining to a wing or a fin.

pteryg'oid — wing-shaped, especially the two projections from the sphenoid bone of the skull in Man and forming part of the framework of the face in some other vertebrate animals.

pteryg'ium—(in zoology) a limb or wing-like part; (in medical science) an encroachment on to the cornea of the eye of a wing-like part of the conjunctiva.

A'pterygo'ta—the class of primitive insects which show no signs of wings.

Archaeo'pteryx — "ancient wings" — the oldest known bird, retaining some characteristics of reptiles, found as a fossil in the Jurassic rocks of Bavaria.

PTILO-, -PTILE
a feather, the feathers of a bird (Gk. *ptilon*).

ptilo'sis—the plumage, or the arrangement of the feathers, of a bird.

coleo'ptile — "a sheathed feather" — the first leaf of a grass seedling.

Note. The word *reptile* is not derived from this root but from the Latin verb *repo, rept-*, to crawl.

-PTOSIS
a falling, a downward displacement (Gk. *ptōsis*).

ptosis—a dropping of the upper eyelid because of a loss of muscular power.

hepato'ptosis—a state in which the liver is fallen or displaced downwards. So also **gastro'ptosis** (the stomach), **viscero'ptosis** (the viscera), etc.

In mathematics an **a'sym'ptote** is a line to which a given curve gets nearer and nearer but does not meet it within a finite distance. (The word means "not falling together".)

-PTYSIS
a spitting (Gk. *ptysis*).

haemo'ptysis—a spitting of blood.

PTYXIS
a fold, a folding, a doubling up (Gk. *ptyx; ptyxis*).

ptyxis—the manner in which a leaf is folded up in a bud.

PULMO-, PULMON-
a lung (L. *pulmo, pulmon-*).

pulmon'ary—pertaining to the lungs, e.g. pulmonary artery.

pulmon'itis — inflammation of the tissue of the lungs, pneumonia.

pulmo'gastric—pertaining to the lungs and the stomach.

pulmo'branchiate—having gills modified for breathing air.

-PULSE See -PEL.

PULVER-
dust (L. *pulvis, pulver-*).

pulver'ize—to make into powder or dust.

pulver'ulent—powdery, dusty, covered with powder, (rock) liable to crumble into powder.

PULVIN-
a cushion, a bolster (L. *pulvinus*).

pulvin'ate—shaped like a cushion.

pulvinus—a swollen base of a leaf (often able to change its form and so move the leaf).

pulvi'llus—a pad on the claw of an insect. (The word is a contraction of *pulvinulus*, a small cushion.)

PUR-
pus (thick yellow matter) from an infected wound (L. *pus, pur-*).

pur'ulent—pertaining to, or full of, or discharging, pus.

sup'pur'ate—to form pus.

-PUS
a foot, footed (Gk. *pous*, a foot).

octo'pus—"eight footed"—a marine cephalopod which has eight tentacles round the mouth.

Platy'pus — "broad footed" — the Duckbill or Duck-mole, an egg-laying mammal of Australia with a duck-like snout and broad, webbed feet.

poly'pus—"many footed"—a kind of soft tumour (e.g. in the nose) with a much branched foot.
The word *polyp* is a contraction of *polypus*. A **polyp** is an animal like the freshwater Hydra, with a tube-like body and tentacles round the mouth. The term is also used for an individual of a compound organism or hydroid colony such as Obelia.

PY- See PYO-.

PYCN-, PYCNO-, PYKNO-
thick, compact, dense (Gk. *pyknos*).

pycno'sis (of a cell)—a shrinking of nucleus material into a hard knot.

pycno'meter, pyknometer—an instrument for measuring the densities of liquids (and especially for investigating the density of water near its freezing point).

pycn'idium—"a small compact mass" —a round spore-container of some fungi. It contains fertile fungal threads and **pycnospores.**

Pycno'gonida — "thick knees" — the class of Sea Spiders which have enormous legs.

PYEL-, PYELO-
Literally, a trough or vessel for feeding animals (Gk. *pyelos*).
Used in medical terms to denote the pelvis of the kidney (that part into which the urine drains and from which the urine passes out).

pyel'itis—inflammation of the pelvis of the kidney.

pyelo'cyst'itis—inflammation of the pelvis of the kidney and of the urinary bladder.

pyelo'graphy—the taking of an X-ray photograph of the pelvis of the kidney.

PYG-, PYGO-, -PYGIUM
the rump, the buttocks, the posterior (back) end of an animal (Gk. *pygē*).

pyg'al—pertaining to the upper, back end of an animal.

pygo'style—"rump pillar"—a bone at the end of the spine of a bird formed by the joining of some of the tail bones.

pygo'podes — an order (class) of aquatic birds, including the Auks, whose legs are set very far back.

uro'pygium—"tail rump"—the short tail-stump of the body of a bird.

PYL(O)-, -PYLE
a gate, an opening (Gk. *pylē*).

pyl'orus—"the keeper of the gate"—the opening of the stomach into the duodenum (upper part of intestine) controlled by the **pyloric** muscle. The combining form of this term is PYLORO-, e.g. **pyloro'plasty**—a widening ("moulding") of the pylorus by surgery.

micro'pyle—"a small gate"—the tiny opening at the top of an ovule through which the pollen tube usually passes.

apo'pyle—"away gate"—an opening in Sponges by which water escapes.

PYO-, PY-
pus (the thick yellow matter from an infected wound) (Gk. *pyon*).

py'aemia—an infection of the blood by bacteria which causes abscesses (infected areas) to develop in various parts of the body.

pyo'genesis—the production of pus.

pyo'rrhoea—"a flow of pus"—an infected inflammation round the base of a tooth.

em'py'ema—"a state of pus in"—a collection of pus in a cavity of the body (especially the pleural cavity).

PYR- See PYRO-.

PYREN-, PYRENO-
a fruit-stone (Gk. *pyrēn*).

pyreno'carp—"a fruit with a stone"—a soft fruit, such as a plum or cherry, with a single stone.

Pyreno'mycetes—a major division of the Ascomycetes (large class of Fungi)

in which the fruit-body is flask-shaped with the reproductive bodies inside.
The root is also used to denote a nucleus, a dense mass.

pyren'oid—a small, dense mass of protein found in the chloroplasts (green bodies) of some Algae and Mosses and concerned with the formation of starch.

PYRET-, PYRETO-
an increase above the normal temperature of the body, a fever (Gk. *pyretos*, burning heat, a fever).

pyrexia—a fever (Gk. *pyrexis*).

pyret'ic—pertaining to, or resembling, a fever.

anti'pyretic—(drug) which acts against a fever.

pyreto'logy — the medical science which deals with fevers.

pyreto'therapy — the treatment of a disease by artificially raising the temperature of the body above normal.

PYRI-, PYR-
a pear (L. *pyrum*, a misspelling of *pirum*).

pyri'form—pear-shaped.

pyr'uline—pertaining to the gastropod genus **Pyrula** ("a little pear") the members of which have pear-shaped shells.

PYRO-, PYR-
fire, heat (Gk. *pyr, pyro-*).
A **pyre** is a heap of material for burning (especially a funeral pile for burning a corpse). **Pyrotechnics** is the art of making, or of displaying, fireworks.

pyro'genesis—the production of heat (especially within the body).

pyro'clastic rocks—"broken by fire"—rocks made up of broken pieces which came from a volcano.

pyro-electricity—positive and negative electric charges which are produced in different parts of some crystals (e.g. tourmaline) when the temperature is changed.

pyro'lysis — "heat loosening" — the breaking down (decomposition) of a substance by heat.

pyro'meter—an instrument for measuring a high temperature (e.g. that of a furnace) from a distance.

pyr'helio'meter — an instrument for measuring the rate at which heat is received from the sun.

pyr'ites — "fire stones" — (vaguely) rocks which make fiery sparks when struck together; (specifically) mineral sulphides of iron or of copper and iron.
In the names of chemical compounds PYRO- indicates that the compound (or the parent of the compound) is obtained by, or was originally obtained by, the action of heat.

pyro'sulphuric acid—$H_2S_2O_7$—Sodium pyrosulphate is made by heating sodium bisulphate to about 400° C.

pyro'gallol, pyrogallic acid — $C_6H_3(OH)_3$—a substance, used as a developer in photography, obtained by heating gallic acid.

pyro'racemic acid (pyruvic acid)—$CH_3.CO.COOH$—a substance obtained by the dry distillation of either racemic acid or tartaric acid.

pyr'idine—a colourless, pungent liquid (C_5H_5N) produced by the dry distillation of bones. (It also occurs in coal-tar.)

pyr'ene—a four-ringed hydrocarbon, $C_{16}H_{10}$, obtained from that part of coal-tar which boils above 360° C.
In the names of minerals PYRO- usually indicates that some property is shown or change produced by the action of fire or heat. Sometimes it indicates a fiery red or yellow colour.

pyro'lusite — native manganese dioxide. So called from its use, when heated, in decolourising ("washing") glass.

pyro'chlore—a niobate of cerium and other metals. The crystals, which are brown, become "yellow-green" when strongly heated.

pyro'morphite—a chloro-phosphate of lead. The globules produced by melting assume a crystalline form on cooling.

pyr'ope — "fiery eyed" — a deep red garnet.

pyr'argyrite—ruby silver ore, a sulphide of silver and antimony.

pyro'xenes—metasilicates of calcium and one or more other metals (e.g. iron, magnesium). So called because they were thought to be "foreign" to igneous ('fire-formed') rocks.

Note. *Pyramid* is not related to this root. The word comes through Latin from Greek and is perhaps of Egyptian origin.

PYRR-

flame-coloured, fiery-red (Gk. *pyrros*).

pyrr'ole—a colourless liquid whose vapour turns a pine-wood shaving (which has been moistened with hydrochloric acid) bright red. The molecule is a ring of four CH groups and one NH group.

Q

QUADRAT-

square, made square, right-angled (L. *quadratus*, square set; *quadro, quadrat-*, to make square).

quadrate—rectangular, e.g. quadrate bone of the skull.

quadrat'ure — (*Mathematics*). The process of finding a square with the same area as that of a figure bounded by curves, i.e. the process of finding the area of the figure. (*Astronomy*.) The Moon (or an outer planet) is in quadrature if the line joining the Moon (or planet) to the Earth is at right-angles to the line joining the Earth to the Sun.

quadrat'ic equation — an algebraic equation which involves the square, but not a higher power, of the unknown quantity, e.g. $x^2-3x+4=0$.

QUADRI-, QUADRU-

four (L. *quattuor, quadri-*).

(The form QUADRU- is generally used before the letter *p*.)

A **quadrangle** is a four-angled (and four-sided) figure, especially a square or rectangle, the shape of a court-yard in a school or college. **Quadruplets ('quads')** are four children born at one birth.

quadri'lateral—having four sides; a four-sided figure.

quadri'ceps—a muscle having "four heads" (points of attachment) as has one of the thigh muscles in Man.

quadri'valent — having a valency (chemical combining-power) of four, e.g. a carbon atom can combine with four hydrogen atoms.

quadru'ple—four-fold.

quadru'ped—an animal with four legs.

QUART-

fourth, one-fourth (L. *quartus*).

A **quart** is one-fourth (a **quarter**) of a gallon.

quart'an fever, malaria—fever which flares up every fourth day (i.e. at intervals of 72 hours).

quart'ic equation—an algebraic equation which involves the fourth power (x^4) of the unknown quantity (x).

quart'iles—In a set of measures showing the distribution of a property (e.g. how many people there are of various heights), the quartiles are the measures which divide the group into four equal parts (e.g. all people below a certain height are in the lowest quartile of the population).

Note. The word *quartz* is from German and has no connection with this root.

QUASI-

as if, seemingly but not really (L. *quasi*).

quasi-optical waves—waves of short wavelength which are not visible but behave as light waves.

QUATER-, QUATERN-

four, four together, by fours (L. *quater*). (This prefix must not be confused with QUART-.)

quater'centenary — a four-hundredth anniversary.

quatern'ary—consisting of four parts.

Quatern'ary era—the geological era, fourth in order, covering the most recent periods.

QUIN-, QUINON-

Quinine is an alkaloid found in Cinchona

bark and used as a febrifuge and a remedy for malaria. The name is derived from the Spanish *quina* from the Peruvian *kina* meaning bark. The substance has a complicated structure. On decomposition it yields a simple derivative of **quinoline.**

The Cinchona barks contain, in addition to alkaloids, certain organic acids and neutral substances. One of these acids (which is also found in coffee beans) has been named **quinic acid.** From this acid **quinone** was first prepared.

quin'one—a compound whose molecule consists of a benzene ring in which two of the hydrogen atoms have been replaced by two oxygen atoms. There are two forms: *ortho-* and *para-* (q.v.). Figure (a) below shows the *para*-form.

quin'ol—a compound whose molecule consists of a benzene ring in which two hydrogen atoms have been replaced by two hydroxyl (-OH) groups. (See fig. (b).) It may be obtained by the reduction (addition of hydrogen) of quinone and so is sometimes called **hydroquinone.** It is used as a developer in photography.

quinon'oid structure—the characteristic structure of *para*-quinone (fig. (c)), present in certain dye-stuffs.

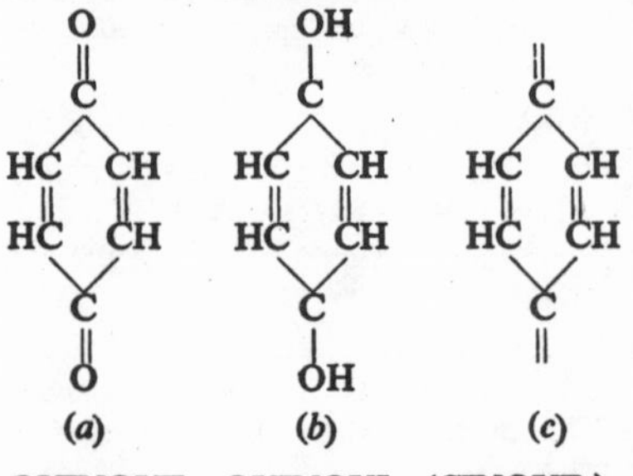

(*a*) (*b*) (*c*)

QUINQUE-, QUINQUI-, (CINQUE-)

five (L. *quinque*).

quinque'foliate — having leaves arranged in fives; having five leaflets to a leaf.

quinque'valent — having a valency (chemical joining-power) of five. (Also spelt **quinquivalent**; also called pentavalent.)

quinqui'fid—cleft into five parts.

The prefix CINQUE- has come through the French.

cinque'foil — a leaf made up of five leaflets; a five-leaf decoration (e.g. in carving).

QUINT-, QUINTU-

The Latin *quintus* means fifth but the prefix is more commonly (though incorrectly) used to mean five.

quintu'ple—five-fold.

quintu'plets (**'quins'**) — five children born at one birth.

R

RACEM-, RACEME

a bunch or cluster of grapes (L. *racemus*). This root has given rise to terms which are rather remote from the original meaning.

raceme—a simple cluster of flowers (e.g. Currant, Lily) in which the separate flowers are on short, nearly equal stalks spaced at equal distances along a short axis.

racem'ic—pertaining to grapes or to grape juice.

racemic acid—a form of tartaric acid found in the mother liquor of fermented grape juice after *d*-tartaric acid (in the form of cream of tartar) has been deposited. It is optically inactive (i.e. does not twist the plane of polarisation of polarised light) and is a combination of *dextro-* (right) and *laevo-* (left) forms of tartaric acid. The salts are called **racem'ates.**

racemic compound — any compound which, like racemic acid, is optically inactive because it is a mixture or combination of two oppositely active forms.

racemisation—the conversion of an active form of a compound into a racemic form.

RACH-, RACHI-, RACHIS, (RHACH-)

spine (Gk. *rhachis*).

rachis—the spine; the axis of a flower

spike; the axis of a leaf composed of rows of leaflets; the shaft of a feather.

rach'itis—"inflammation of the spine"—a disease, now more commonly known as rickets, in which there is failure to form bones properly and a softening of the bones (especially of the spine and leg-bones) because of lack of vitamin D.

rachi'anaesthesia — anaesthesia (unconsciousness) produced by injecting an anaesthetic into the spine.

rachi'odont — "spine teeth" — having some of the spinal vertebrae at the back of the throat modified so as to serve as egg-breaking teeth (as in certain egg-eating snakes).

RADIC-, RADICUL-

a root (L. *radix*, *radic-*, a root; *radicula*, a little root).

A **radish** is a well-known root vegetable.

e'radic'ate—to "root out", to remove completely.

radicle—"a little root"—the root of the embryo within the seed and of the young seedling; any small root. (Also an alternative spelling of radical (chemistry) as below.)

radic'al—(1) relating to the root. (2) (*chemistry*) a group of atoms which remains unchanged in a series of compounds and acts as a single atom, e.g. the nitrate radical $-NO_3$. (The word is also used in other senses in general speech.)

radic'ation—the general character and arrangement of a root system.

radicul'ectomy—the surgical cutting of the roots of spinal nerves.

RADIO-, RAD-

L. *radius*—a staff, a spoke, a ray.

L. *radio*, *radiat-* —to emit rays or beams.

The idea of a shaft or spoke is seen in the English word **radius** which denotes the thicker of the two bones in the forearm and also the straight line from the centre of a circle to the circumference. The idea of the emission of rays is seen in the words **radiate, radiator** and **radiation.**

The combining-elements RADI-, RADIO- have gradually been extended in meaning and application. It is convenient to treat these applications separately.

(1) Pertaining to the radius bone.

radio-carpal — pertaining to the radius and the carpals (wrist-bones).

(2) Pertaining to the radius of a circle, to radii or rays spreading out from a central point.

radi'al—pertaining to a radius, arranged along a radius or radii.

radi'an—a unit for the measurement of an angle. The angle subtended at the centre of a circle by an arc equal in length to the radius (about 57°).

steradian—"solid radian"—a unit for the measurement of a solid (e.g. cone-shaped) angle. It is the solid angle subtended at the centre of a sphere by a piece of the surface equal to the square of the radius.

Radi'olaria—an order of marine Protozoa the members of which have numerous, fine, radial pseudopodia (projections of protoplasm).

(3) Pertaining to the sending out of rays, to radiation in general.

radio'meter — an instrument for measuring the intensity of radiation; also an instrument for showing the conversion of radiant energy into mechanical energy.

radio'therapy — the treatment of disease by radiations, e.g. light, heat, and (especially) X-rays.

(4) Pertaining to X-rays.

radio'logy—the branch of medical science which is concerned with the use of X-rays, e.g. for examining parts of the body, for treatment of disease.

radio'graph — an X-ray photograph.

(5) Pertaining to 'wireless'.

radio'telegraphy—telegraphy (the sending of messages or signals over a distance) by means of wireless waves.

radio'telephony — telephony (the sending of sounds, especially speech, over a distance) by means of wireless waves.

It is from these two words that the modern word **radio** (originally an abbreviation) has been derived.

radio frequencies—those frequencies (of alternating current and of waves) which are suitable for wireless propagation.

radio'location—Radar—a means of finding the position of an object by sending out radio pulses which are reflected back by the object.

(6) Pertaining to radioactivity.

radio'activity — the spontaneous (without external cause) breaking up of the atomic nuclei of certain chemical elements with the sending out of rays in the form of charged particles and/or gamma-rays (like short X-rays). **Radium** is a typical radioactive element.

radio'thorium — a product of the breaking up of thorium.

Radio- is now commonly used to denote an artificial form of an element (an isotope of the normal form) which is radioactive, e.g. **radio-iodine.**

RAM-, RAMI-

a branch (L. *ramus*).

rami'fy—to form branches, to branch out.

ram'ose—with many branches, much branched.

rami'colous—living on twigs.

rami'corn—(insect) having branched antennae.

RAPH-, RAPHID-

Gk. *rhaphis, rhaphid-*, a needle; *rhaphē*, a sewing, a seam.

(The first *h* of this root is often omitted in English derivatives; it is normally retained in the element -RHAPHY (q.v.).)

raphe—a seam-like join between two halves of an organ, etc., e.g. the tongue, the perineum, the brain.

raphide—a needle-shaped crystal (of calcium oxalate) found in some plant cells.

RAS-

to scrape, to rub (L. *rado, ras-*).

e'rase—to rub out, to obliterate.

ab'rasion — a wound formed by the scraping off of skin.

ab'rasive — a substance (e.g. emery) used for rubbing away rough particles on wood, stone, etc.

RE-

This very common prefix (from L. *re-*) which, in general, corresponds to English 'again' and 'against', has many variants of meanings. Most of them are well-known. The more important meanings (in relation to science) are summarised below.

(a) in return, on each other.

re'action—This term is widely used in science with the general idea of the action of B on A when A acts on B. Examples:

(*Chemistry.*) A chemical change in which substances act on each other, e.g. hydrochloric acid and zinc **react** to form hydrogen and zinc chloride. The substances are the **reactants.**

(*Mechanics.*) When a book rests on a table, the weight of the book acts downwards on the table and the table provides an upward reaction (force) to balance it; when a bullet moves forwards from a gun, the gun moves backwards from the bullet (the recoil).

(*Biology.*) The change in the behaviour of an organism in response to a stimulus which acts on it.

re'ciprocal—(*Mathematics*) a number or expression so related to another number or expression that when they are multiplied together the result is 1, e.g. $\frac{1}{4}$ is the reciprocal of 4 (and 4 is the reciprocal of $\frac{1}{4}$).

(b) opposing, acting against (as in **rebel).**

This meaning is seen in **re'sistance** ("a stand against").

(c) back, backwards (as in **retreat, retire).**

re'flect—to "bend back" as (e.g.) when a mirror reflects a ray of light.

re'gression — "a stepping backwards"—the tendency to return from an extreme condition towards the average (e.g. offspring of very tall parents tend to be nearer the average height of Man); a slipping back towards an earlier form or stage of development.

(d) back again, returning to an original state (as in **re-unite**).

re'combine—combine again after having been separated.

re'crystallise—to form into crystals again.

re'gelation — "freezing again" — the process in which ice melts under pressure and freezes again when the pressure is removed (as in squeezing snow to make a snowball).

re'lapse—"to slip back"—to fall back into weakness, illness, etc.

(e) repetition of an action, as in **revolve** ("to turn again and again"), **revise** ("to see again and again"), **research** ("to search repeatedly") and **refine**.

RECT-, RECTI-

right, correct, straight (L. *rectus*).

rect'angle—a figure formed by four straight lines and having four right-angles

recti'fication—putting right, making straight. Examples: (1) of a liquid—purifying and refining (usually by distillation); (2) of an electric current—changing an alternating current into a current which flows only in one direction; (3) of a curve—finding the length of a straight line equal to that of the curve, i.e. finding the length of the curve.

recti'linear — consisting of straight lines, forming straight lines, in straight lines (e.g. the 'rectilinear propagation of light').

recti'rostral—having a straight beak.

rectum — *rectum* (*intestinum*), the "straight intestine"—the final part of the large intestine ending at the anus (outlet).

REN-, RENI-, RENO-

the kidneys (L. *renes*).

ren'al—pertaining to the kidneys.

ad'renal—near to the kidney. **adrenal gland, suprarenal gland**—a gland just above the kidney. **adrenalin**—a substance produced by the inner part of this gland.

reni'form—kidney-shaped.

reno-pericardial — pertaining to the kidneys and the pericardium.

RETI-, RETICUL-

a net; a little net (L. *rete*; *reticulum*).

reti'form — in the form of a net, resembling a net.

reticul'ar, reticul'ate—like a little net, forming a little net.

reticle—a network of fine threads; a network of wires or of lines engraved on glass as in some optical instruments.

RETIN-, RETINO-

Pertaining to the **retina** (L.)—the layer at the back of the eye-ball which is sensitive to light and receives images.

retin'itis—inflammation of the retina.

retino'scopy—a method of examining the refractive qualities of the eye.

RETRO-

(1) backwards (L. *retro*).

retro'grade—moving ("stepping") backwards.

retro'gression—a degeneration, the taking on of features characteristic of lower forms.

(2) behind.

retro'peritoneal — situated or occurring behind the peritoneum.

retro'pharyngeal—situated or occurring in the tissues behind the pharynx.

retro'sternal—behind the sternum (breast-bone).

RH-, -RRH-

An initial ϱ (=*r*) of a Greek word is aspirated. This is represented in English formations by RH-, e.g. in **rheumatism, rhinoceros, rhubarb, rhythm**. (There are a few roots, e.g. *rachi-* (spine), from which it is now more usual to omit the *h*.)

When the combining-form of another

root (or a prefix) is added in front of a root beginning with RH-, a second *r* is inserted (following the Greek practice), so producing the combination -RRH-, e.g. **rhexis**—a bursting, **hepato'r'rhexis**—a bursting of the liver.

RHABD-, RHABDO-
a rod, a stick (Gk. *rhabdos*).

rhabd'ites—rod-like bodies found in the outer cells of Turbellarian worms (a class of Flatworms).

rhabdo'myoma — a tumour composed of striated (striped, rod-like) muscle fibres.

rhabdo'mancy — the art of using a divining rod for discovering underground water.

RHACHI- See RACHI-.

-RHAGE
a breaking (Gk. *-rhagia*, from *rhēgnumi*, to break, to burst).

haemo'r'rhage — "blood breaking" — an escape of blood from a broken blood-vessel, bleeding.

ulo'r'rhage—bleeding ("breaking") of the gums.

RHAMPH-, RHAMPHO-
a curved beak (Gk. *rhamphos*).

rhamph'oid—shaped like a beak.

rhampho'theca — "beak container" — the horny covering of the upper and lower jaws of birds.

-RHAPHY
the sewing up of a wound (Gk. *rhaphē*, a sewing, a seam).

perineo'r'rhaphy—sewing up a tear in the perineum (e.g. after a difficult childbirth).

nephro'r'rhaphy—the sewing in position of a displaced kidney.

salpingo'r'rhaphy—the stitching up of a Fallopian tube (see *salpingo-*) after a part of the ovary has been removed.

RHEO, -RHOEA (-RHEA), RHYO-
a flow, a stream, a current (Gk. *-rhoia*, from *rheō*, to flow).

rheo'logy—the study of the flow and deformation of matter.

rheo'pexy—making "a flow set solid" —the making of certain colloidal solutions quickly set into a jelly or solid by gentle rhythmic shaking.

rheo'stat—an instrument for making "the flow (of electricity) stay still"—a variable resistor for adjusting the electric current in a circuit.

-RHOEA (-RHEA) usually denotes of flow of matter from the body.

dia'r'rhoea — "a through flow" — abnormal looseness of the bowels.

pyo'r'rhoea — "a flow of pus" — The term is now used for a state of inflammation round the socket of a tooth.

gono'r'rhoea—a disease in which there is a mucous discharge from the sex organs.

The root also appears in other forms. In **catarrh** ("a down flow") it is reduced to *rh*. The obsolete word **rheum** ("that which flows") denoted any watery or mucous discharge from the body, e.g. tears, saliva, catarrh. The present word **rheumatism** is derived from this word though there is little (if any) medical connection. **Rhyo'lite** is a general name for certain igneous rocks (similar in chemical composition to granite) formed from laval flows.

-RHEXIS
a breaking, a bursting, a rupture (Gk. *rhēxis*).

rhexis—the rupture of a body structure.

hepato'r'rhexis—the bursting of the liver (e.g. as a result of being run over by a car).

cardio'r'rhexis—rupture of the wall of the heart.

karyo'r'rhexis—the breaking up of the nucleus of a cell.

RHIN-, RHINO-
the nose (Gk. *rhis*, *rhin-*).

A **rhinoceros** has a horn (or in some species two horns) on its nose.

rhin'itis—inflammation of the membrane of the nose.

rhino'scope — an instrument for examining the cavities of the nose.

rhino'phyma—overgrowth and coarseness of the skin of the nose.

rhino'plasty—"nose moulding"—the repair of a damaged nose by plastic surgery.

rhin'encephalon—the part of the brain of a vertebrate animal which is concerned with smelling.

RHIPID(O)-

a fan (Gk. *rhipis, rhipid-*).

rhipid'ate—fan-shaped.

rhipidium—"a little fan"—a fan-shaped head of flowers.

RHIZ-, RHIZO-, -RHIZA

a root (Gk. *rhiza*).

rhiz'oid—root-like; a root-like part of a Fungus or Moss.

rhizo'me—a thick horizontal stem, just below the ground, which sends out roots and new shoots, e.g. of Couch Grass.

rhizo'philous—"liking", i.e. growing on, roots.

rhizo'sphere—the region of the soil around the roots of a plant.

rhizo'podia—"root feet"—root-like pseudopodia (projections of protoplasm) of certain Protozoa (e.g. Foraminifera).

myco'r'rhiza—a beneficial association between a fungus and the root of a higher plant.

RHOD-, RHODO-

rose, rose-red (Gk. *rhodon*, the rose).
A **rhododendron** is a well-known genus of evergreen shrub which bears large flowers of various shades of red.

rhodo'sporous—producing red spores.

Rhodo'phyceae—the group of Red Algae (the red seaweeds).

rhod'ium—a chemical element, similar to platinum, which forms red and rose-coloured salts.

rhodo'crosite—manganese carbonate which occurs as rose-pink crystals.

rhodon'ite—a rose-coloured mineral (metasilicate of manganese).

rhod'ops'in—"red seeing-substance"—a purple pigment found in the rods of the retina of the eye.

-RHOEA See RHEO-.

RHOMB, RHOMBO-

A **rhomb,** or **rhombus,** is a figure with four equal sides but whose angles are not right-angles, i.e. a diamond or lozenge shape. The name is taken from the Greek *rhombos* (a spinning-top, a magic wheel). The term **rhomb** is also used for a **rhombohedron** (see below).

rhomb'oid—(*noun* and *adjective*)—of, or nearly, the shape of a rhomb; also (loosely) a four-sided figure whose opposite sides and opposite angles are equal.

rhombo'hedron—a solid figure with six faces each of which is (strictly) a rhombus or (more generally) a parallelogram, e.g. a crystal of calcite. (The **rhombohedral system** of crystal shapes, to which calcite belongs, is also called the trigonal system.)

ortho'rhomb'ic crystal—"a right-angled rhomb"—a crystal whose shape is based on three axes which are unequal in length but at right-angles to each other. A brick-shape is the simplest example. (The **orthorhombic system** of crystal shapes is also called the trimetric ("three measures") system.)

rhombic sulphur—a form of sulphur whose crystals are orthorhombic. (Also called octahedral sulphur.)

RHYNCH-, RHYNCHO-

a snout, the beak of a bird (Gk. *rhynchos*).

rhyncho'phorous—having a beak. **Rhynchophora**—the family of Weevils (beetle-like insects).

rhynch'odont—having a toothed beak, e.g. a Falcon.

Rhyncho'spora—"snout seed"—a genus of sedge-like plants, the Beak Rush or Beak Sedge.

RHYO- See RHEO-.

RIB-, RIBO-

Ribose is a simple pentose sugar. The name is said to have been formed by the rearrangement of certain of the letters of the name *arabinose* (another sugar, identified earlier, to which ribose is closely related).

ribo'flavin(e)—vitamin B2—a complicated compound whose molecule contains a ribose side-chain. (The flavins form a group of natural yellow pigments.)

ribo'nucleic acid—one of the two nucleic acids, found in the nucleus and especially the surrounding protoplasm of living cells. It is built up from chains consisting essentially of a nitrogenous base, a phosphate group and a ribose molecule.

desoxy'ribo'nucleic acid—the other (related) nucleic acid which is found only in the nuclei of living cells.

ROD-, -RODE, -ROSION

to gnaw (L. *rodo, ros-*).

e'rosion (to **erode**)—"a gnawing away"—the wearing away of a land surface by the weather and the carrying away of the material by wind, water, etc.

cor'rosion (to **corrode**)—"thorough gnawing"—the wearing away and gradual destruction by rust, acids, etc.

rod'ents, Rodentia—animals, such as Rats, Rabbits, Squirrels, which have prominent cutting teeth for gnawing.

rodent ulcer—a sore place on the skin which "gnaws" away the skin round it.

ROS-, ROSA-, ROSE, ROSEO-

a rose (flower), rose (colour) (L. *rosa* (n.), *roseus* (adj.)).

ros'aceous—having the characteristics of a rose.

Rosaceae—the Rose family of plants (including Rose, Cherry, Apple, Strawberry).

rose'ate—rose-coloured.

rose'ola—a rose-coloured rash (e.g. that caused by Measles).

ros'aniline—a substance derived from aniline, the base of magenta-coloured fuchsine dyes.

roseo-cobalt salts—rose-coloured compounds made from cobalt salts, ammonia and water in certain proportions.

Note. The word *rosin* is not derived from this root; it is a variant of *resin*.

ROSTR-

a beak (L. *rostrum*).

A **rostrum** is a platform for public speaking; that in a Roman forum was decorated with the beaks of captured galleys. In zoology the term is used for the beak of a bird or a projecting part of certain other animals.

rostr'ate—having a beak; ending in a long, hard point.

longi'rostral—having a long beak.

denti'rostral—having a toothed or notched beak.

rost'ellum—"a little beak"—a beak-shaped outgrowth, e.g. the abortive third stamen of an orchid flower.

ROT-, ROTA-, ROTI-

L. *rota*, a wheel; *roto, rotat-*, to turn.

Rotate and **rotation** are well-known words derived from this root. A **rota** is a list of persons who are to act, or of duties to be carried out, in turn.

rotor—the part of an electric machine (e.g. a generator) which turns; a set of revolving vanes which (in some types of aircraft) give lift (support). (The word is irregularly formed from *rotator*.)

rota'plane—an aircraft which obtains its lift from vanes which turn freely on a vertical axis.

rota'gravure—photogravure printing (based on plates made photographically) from revolving cylindrical plates.

Roti'fera—the Wheel Animalcules—a class of Protozoan animals which have a ring of cilia (lash-like projections) round the mouth by which they swim.

-RRH-

For the origin of this combination of letters, see RH-. For roots apparently beginning with RRH-, see roots beginning with RH-.

RUB-, RUBE-, RUBI-, RUBID-

red (L. *ruber; rubidus*).

The name **ruby** for the red precious stone is derived from the Old French *rubi(s)* and this was probably derived from the Latin *rubeus* (red).

rube'facient—causing redness (e.g. of the skin by a hot plaster).

Rubella — German Measles (one feature of which is a red rash).

rubi'celle—an orange-red variety of spinel (an aluminate of magnesium).

rubid'ium — a chemical element, similar to potassium and caesium, named from the prominent red line which it gives in its spectrum.

RUMEN, RUMIN-

The **rumen** (L. *rumen*, *rumin-*, throat) is the first division of the 'stomach' of an animal such as a cow. It is an expansion of the lower end of the food-pipe and is used for the storage of food before re-chewing.

rumin'ants — animals such as cows which chew the cud.

rumen'otomy — the surgical cutting into the rumen.

rumin'ate — to chew the cud. (This word, like its Latin parent *ruminare*, has been extended in meaning and, in ordinary speech, means to meditate, to ponder over.)

RUPT-

to break (L. *rumpo*, *rupt-*).

rupt'ure—a forcible breaking or tearing of an organ or structure of the body; the pushing of a part of an organ through the wall of the cavity which contains it (especially the wall of the abdomen); a bursting of a blood vessel.

dis'rupt—to break apart, to shatter.

e'rupt'ion—"a breaking out", e.g. of material from a volcano, of teeth through the gums, of pimples, etc. through the skin.

S

SAC, SACC-, SACCUL-

A **sac** (L. *saccus*) is a bag-like structure, e.g. a cavity in a plant or animal enclosed by a membrane, a membranous covering of a tumour, cyst, fruit, etc.

sacc'ate — expanded into a sac; enclosed in a bag-like cover.

sacc'iform—in the form or shape of a sac.

saccule—"a small sac". **sacculi'form**—shaped like a small sac.

SACCHAR-, SACCHARO-

sugar (Gk. *sakchar*).

sacchar'oses, sacchar'ides—a general name for carbohydrates especially the simpler ones which are sugars. **mono-saccharide**—a sugar consisting of one molecular unit, e.g. glucose; **di'saccharide**—a sugar consisting of two molecular units, e.g. sucrose (cane sugar); **poly'saccharide**—a carbohydrate built up from many simple sugar units, e.g. starch, cellulose.

sacchar'imetry—the measurement of the amount of sugar present in a solution (especially by measuring the power of the solution to twist the plane of polarisation of polarised light). The instrument which is used is a **saccharimeter**. (Note. The name **saccharometer** is sometimes given to a hydrometer used for estimating the amount of sugar in a solution by measuring its density.)

saccharo'lytic — "sugar loosening" — (bacteria) which break down sugar and simple starches.

Saccharo'myces—the "sugar fungus" —yeast, which ferments sugar (and produces alcohol).

sacchar'in—a very sweet, crystalline substance, made from toluene (in coal-tar) and used as a substitute for sugar.

SACR-, SACRO-

Pertaining to the **sacrum** (L. *os sacrum*, the sacred bone, so called from its use in sacrifices)—the triangular wedge of bone (formed by joined vertebrae) where the backbone joins the pelvis.

sacr'algia—pain in the sacrum.

sacro-caudal—pertaining to the sacrum and the tail region.

SAGITT-

an arrow (L. *sagitta*).

Sagitt'arius—the star-group known as The Archer.

sagitt'ate—(leaf, etc.) shaped like an arrow-head.

The root has also given rise to the word **sagittal**—pertaining to, or lying in, a direction from the front of the body to the back and in a central position, e.g. the position of the join (on the top of the skull) between the two parietal (side) bones.

SAL, SAL(I)-

salt (L. *sal*).

The ancient Romans served out rations of salt (*sal*) and other necessities to their soldiers. When, later, money was substituted for the rations it was called *salarium* ("salt money"). Today a regular payment for services is called a **salary.**

sal ammoniac — "salt of Ammon" (see AMIDE)—ammonium chloride.

sal volatile — "the volatile salt" — a sweet-smelling solution, containing some ammonium carbonate, used in cases of faintness.

sal'ine—salty, (natural water, medicine) containing salt or salts.

sali'ferous—(rock strata) containing much salt.

SALI-, SALTAT-, SALT(I)-, SILI-

to leap, to jump (L. *salio* (*-silio*), *salt-*; *salto, saltat-*).

sali'ent—jutting forward, e.g. a salient angle, a salient (=prominent) feature.

saltat'ory—used in, or adapted for, jumping.

saltat'ion — a variation (in an offspring) which occurs suddenly, i.e. it is not inherited, a mutation.

salti'grade — progressing by jumps; with legs adapted for jumping.

re'sili'ent — "leaping back again" — elastic, resuming shape and size after being distorted, able to recover.

SALIC-

Pertaining to the Willow tree (L. *salix, salic-*). The Willow is typical of the order of plants called the **Salicales.** Hence, pertaining to substances derived from the Willow tree.

salic'in — a substance (a glucoside) obtained from the bark of the willow tree. Used medicinally.

sali'genin—a substance formed by the breakdown of salicin. This can be converted into **salicylic acid.**

salic'ylic acid—hydroxy'benzoic acid, $HO.C_6H_4COOH$. Used as an antiseptic and in the manufacture of aspirin. The salts are called **salicylates**; Oil of Wintergreen is **methyl salicylate.**

SALPING-, SALPINGO-

The Greek *salpinx, salpingo-* means a trumpet. The root (abbreviated) is seen in the name of the herbaceous plant **Salpiglossis** ("trumpet tongue"). In medical terms the root always refers to the trumpet-shaped Fallopian tube into which the ovum (egg cell) passes when it leaves the ovary.

salping'itis—inflammation of a Fallopian tube.

salping'ectomy—the surgical removal of a Fallopian tube.

salpingo'stomy—the making (by surgery) of an opening into a Fallopian tube because the natural opening has become closed.

SALTAT-, SALTI- See SALI-.

SANGUI-, SANGUIN-, SANGUINI-

blood, race (L. *sanguis, sanguin-*).

sanguin'eous — of blood; blood-coloured; full-blooded.

con'sanguin'eous—of the same blood, tribe or race.

sangui'vorous, sanguini'vorous—feeding on blood (as do Fleas).

SANE, SANIT-

sound, healthy; health (L. *sanus* (adj.), *sanitas* (n.)).

This root is seen in such well-known words as **sane** (sound of mind), **insane, sanitary** (bringing about healthy conditions) and **sanitation.** A **sanatorium** (L. *sano, sanat-*, to make healthy) is a place for the treatment of invalids (especially those who are convalescing or are suffering from consumptive diseases).

SAPON-, SAPONI-

soap (L. *sapo*, *sapon-*).

sapon'aceous—of, like, or containing, soap.

saponi'fication—the process of making soaps from fats; hence, the conversion of an ester into an alcohol and an acid.

sapon'ins—glucosides which have the characteristic property of forming a soapy lather.

sapon'ite—a soft, soapy mineral (a silicate of magnesium and aluminium).

SAPR-, SAPRO-

rotten, putrid, decaying (Gk. *sapros*).

sapro'phagous — (insect, etc.) which feeds on decomposing matter.

sapro'philous—growing on ("liking") decaying matter.

sapr'aemia—the presence in the blood of poisons produced by decaying materials in the body.

sapro'phyte—a plant (e.g. a Fungus) or bacterium which obtains its food from dead or decaying matter.

SARC-, SARCO-

flesh (Gk. *sarx*, *sark-*).

Sarco'phagous means flesh-eating; a **sarcophagus** was a stone coffin (as used, e.g. by the Greeks) in which the corpse was supposed to be consumed.

sarco'logy—that part of the study of anatomy which deals with the fleshy tissues of the body.

sarc'ode — fleshy; the term is also sometimes used (as a noun) for animal protoplasm, especially in unicellular animals. Hence **Sarcod'ina**—a class of the Protozoa comprising forms which are usually free-living and are able to thrust out protoplasmic projections (pseudopodia).

sarco'lemma—"flesh peel"—the sheath of a muscle fibre.

sarco'ma—a diseased growth of muscle and connective tissues.

sarcos'ine — a crystalline compound obtained by the decomposition of creatine (found in muscles).

sarco'lactic acid—a form of lactic acid (also called *d*-lactic acid) found in muscle tissue.

sarco'lite—"flesh stone"—a silicate of aluminium, sodium and calcium found as flesh-coloured crystals.

SAUR-, SAURO-, -SAUR, -SAURUS

a lizard (Gk. *sauros*).

saur'ian — pertaining to, or like, a lizard.

sauro'phagous—feeding on reptiles.

The suffix -SAUR(US) is used in forming the names of the large extinct reptiles which flourished in the Mesozoic era (about 180 to 80 million years ago), e.g. **Brontosaurus, Ichthyosaurus, Plesiosaurus.** The **Dinosaurs** belong to a group of fossil reptiles related to the Crocodiles.

SCAL-, SCALE

a ladder (L. *scala*).

This root is seen in the word **escalator** and in **scale** (a series of degrees or measurements).

scal'ar quantity—a quantity which can be measured on a scale, i.e. it can be wholly stated in terms of its size, e.g. volume but not force (for which the direction must also be stated).

Note.

(1) The word *scale* (as of a fish) has come through Old French from Old Teutonic *skalā*.

(2) The word *scalene* (descriptive of a triangle with all sides unequal) comes from Gk. *skalēnos*, limping, uneven.

SCALP-, SCALPR-

knife, chisel (L. *scalprum*; *scalpellum* (diminutive)).

scalpel—a small knife (as used by a surgeon) held in the hand like a pen.

scalpr'iform—chisel-shaped (cutting-teeth).

scalp'ella—"little knives"—in Flies, a pair of pointed projections belonging to the mouth parts.

SCAPH-, SCAPHO-

anything hollowed out, like the hull of a boat (Gk. *skaphē*; *skaphos*).

scaph'oid—boat-shaped (as one of the bones in the wrist and one in the ankle).

scapho'cephalic — having a boat-shaped or keel-shaped head.

bathy'scaphe—a kind of submarine, used in deep-sea exploring, consisting of a buoyant chamber filled with petrol supporting a cabin underneath.

SCAPULO-

Pertaining to the **scapula** (L.) — the shoulder-blade.

scapulo-humeral — pertaining to the scapula and the humerus (upper arm-bone).

scapul'odynia—pain in the region of the shoulder-blade.

SCHIST, SCHISTO-

cloven, split, as if split (Gk. *schistos*).

schist—a rock (such as mica or talc) which has a tendency to split and form flakes.

schisto'some—"split body"—a parasitic blood-fluke which causes **schisto'som'iasis** (Bilharzia).

schisto'cyte—a very small, flattened, or fragmentary, red blood cell. Their presence in the blood causes the condition known as **schisto'cyto'sis.**

schisto'c(o)elia—a fissure of the abdomen from birth.

SCHIZO-

split, cleft (Gk. *schizō*, to split).

schizo'carp—a dry fruit (such as that of Mallow, Nasturtium) which splits into two or more parts each with one seed.

schizo'genesis — reproduction by splitting.

Schizo'phyta — the group of minute plant-like organisms, including Bacteria and Blue-green Algae, which only reproduce by splitting.

schizo'gnathous—having a cleft palate.

schizo'phrenia—"split mind"—a disordered state of mind in which thoughts, feelings and actions are not consistent with each other.

There is considerable variation in the pronunciation of this word-element. The first three letters are usually pronounced *sk-* but one occasionally hears them pronounced *sh-*. The biologists seem to prefer a long *i* (as in *prize*) but most people pronounce *schizophrenia* with the beginning *skitso-*.

SCIA-, (SKIA-), SCIO-, (SKIO-)

shade, a shadow (Gk. *skia*).

scia'graphy—"shadow drawing"—the art of shading a drawing; finding the time by shadows (as by a sun-dial); making a drawing to show the inside of a house as if it were cut through vertically; X-ray photography.

scia'gram—an X-ray photograph.

scio'philous—(plant) which flourishes in the shade. Such a plant is a **scio'phyte**.

scio'phyllous — (plant) having leaves which can endure shading.

Note. The word *sciatic* (of the hip) is not derived from this root but is a corruption of *ischiadicus*; see ISCHI-.

SCISS-

to cut (L. *scindo*, *sciss-*).

sciss'ile — capable of being cut or divided; (mineral) which splits into thin sheets.

sciss'ion—a cutting, a division, a cleft.

ab'scission—the "cutting off" of a part of a plant, e.g. the separation of spores, the shedding of leaves.

ab'scissa—the measurement along the X (horizontal) axis of a point which is plotted on a graph. It is the piece of the X axis which is "cut off" by the ordinate (vertical line) from the point.

SCLER-, SCLERO-

hard (Gk. *sklēros*).

sclero'dermatous, sclerodermic — having a hard outer skin or covering.

scler'enchyma—hard, woody tissue in plants which supports and protects the softer tissues.

scler'oma—a hard nodule or tumour (e.g. in the nose).

sclero'sis—a hardening of cell walls (in plants) or of tissues (in animals). **arterio-sclerosis** — a hardening of the arteries.

sclero'meter — an instrument for measuring hardness (e.g. of crystals).

sclera—the tough, outer coat of the eye in vertebrate animals. Also called the **sclerotic** (coat). **scler'itis**—inflammation of the sclera. **sclero-iritis**—inflammation of the sclera and the iris.

SCOLEX, SCOLEC-, SCOLECO-

a worm (Gk. *skōlēx, skōlēk-*).

scolex—the terminal organ of attachment of a Tapeworm.

pro'scolex—the Bladderworm (larval) stage of a Tapeworm.

scolec'ite — a mineral (a kind of zeolite) which occurs in fibrous or needle-shaped crystals.

scoleco'sporous — producing thread-shaped or worm-shaped spores.

SCOLI-, SCOLIO-

bent, crooked, oblique (Gk. *skolios*).

scolio'sis—an abnormal sideways curvature of the spine.

Scoli'odon—the genus which includes the oblique-toothed Sharks.

SCOLO-, SCOLUS

a prickle, a thorn (Gk. *skōlos*).

scolus—a thorn-like outgrowth of the body wall of the larvae of some insects.

scolo'phore—a thorn-like nerve-ending, for hearing purposes, in insects.

SCOPA, SCOPAR-

twigs, a broom, a brush (L. *scopa*).

scopa—a bundle or tuft of bristly hairs on the legs of bees on which pollen is collected.

scop'iform — arranged in bundles, broom-shaped.

Scopar'ium—a pharmaceutical name for the tops of the Common Broom.

scopar'in — a drug (a diuretic) prepared from Broom.

-SCOPE, -SCOPIC, -SCOPY

The Greek *skopos* means "one who watches". In scientific terms the suffix -SCOPE usually indicates an instrument which is used for observing and showing an effect but not for measuring it. Thus a **microscope** is used for observing small things; a *micrometer* is used for measuring small things.

Note: *telescope* comes directly from the Greek *tēleskopos*—far-seeing.

peri'scope—an instrument which enables one to see round a corner, over an obstacle, etc.

thermo'scope — an instrument for showing changes in temperature but not for measuring them in degrees.

electro'scope — an instrument for showing the presence of electric charges.

spectro'scope — an instrument by which a spectrum may be observed and studied.

sclero'scope—an instrument for observing the hardness of metals. (This instrument permits measurements and so would more appropriately be called a sclerometer.)

The suffix is much used in the names of medical instruments for the inspection of the interior of parts of the body.

oto'scope—an instrument for inspecting the inside of the ear.

laryngo'scope—an instrument for inspecting the inside of the larynx.

stetho'scope—an instrument for observing the inside of the chest (by listening to the movements of the heart, lungs, etc.).

As is usual when a suffix is convenient and becomes well-known, -(o)SCOPE has been freely added to all sorts of words.

The adjectival form -SCOPIC usually means 'pertaining to the use of the instrument -SCOPE', e.g. a **spectroscopic examination** of a substance is an examination by means of a spectroscope. Sometimes the suffix is better interpreted 'as would be revealed by the instrument'. Thus **microscopic detail** is the detail which would (only) be seen by the use of the microscope. Similarly, a **hygroscopic substance** is one which tends to absorb water—as would be revealed by a hygroscope.

-SCOPY, e.g. **microscopy**, indicates the process or the art of using the instrument.

SCOTO-
darkness (Gk. *skotos*).

scoto'graph—a machine for writing in darkness.

scoto'ma—a blind (or partially blind) area in the field of view.

scoto'phyte—a plant which flourishes in the dark.

-SCRIBE, -SCRIPTION
to write, to draw (L. *scribo*, *script-*).
This root is well-known in **describe** ("to write down") and **scribble** ("little writing").

circum'scribe—to draw a line round another figure, e.g. to draw a circle round a triangle passing through the three corners.

in'scribe — to draw a figure inside another, e.g. to draw a circle inside a triangle touching the three sides.

e'scribe — to draw a figure outside another, e.g. to draw a circle outside a triangle touching one side and the extensions of the other two sides.

SCUT-, SCUTI-, SCUTUM
a shield (L. *scutum*).

scutum—a shield-like plate or scale.

scuti'form—shield-shaped.

scut'ellum — "a little shield" — any small shield-shaped thing, e.g. a horny scale on a bird's foot, a flattened structure in a grass seed.

SCYPH-, SCYPHO-, SCYPHUS
a drinking-cup (Gk. *skyphos*).

scyphus—a cup-shaped part of a plant (as in some Lichens).

scyph'iform—cup-shaped.

Scypho'zoa—a large class of coelenterate animals which includes the common Jellyfish. So called from the cup-like shapes of its members.

SEB-, SEBI-, SEBO-
fat (tallow), oil (L. *sebum*).

seb'aceous—looking like a lump of fat; producing or containing fatty material (as **sebaceous glands** which produce oil to lubricate the hair, skin, feathers, etc.).

sebi'ferous, sebi'parous — bearing or producing fat or oil.

sebo'rrhoea—"a flow of fat (oil)"—a disorder of sebaceous glands resulting in a greasy skin.

SECT-, SECTION
cut, cutting (L. *seco*, *sect-*).
A **section** is a separation by cutting or, more usually, a piece which is cut off; a **cross-section** is what is revealed when a thing is cut across; **conic sections** are the curves which are formed when a plane cuts a cone at various angles. An **insect** is an animal which is "cut into", i.e. it has a notched body; compare the Greek ENTOMO-.

sect'or—a piece of a circle, ellipse, etc. which is "cut" out by two radii.

bi'sect — to cut into two (usually equal) parts.

dis'sect—to cut apart into pieces, e.g. to dissect a plant or animal for examination. (This word is sometimes wrongly pronounced with a long *i*, by confusion with *bisect*, and hence wrongly spelt.)

inter'sect — "to cut between" — to divide a thing into two by lying across or passing through it, e.g. two lines may intersect.

re'section—the "cutting back" (away) of a part of the body, especially of a diseased bone.

vivi'section — the cutting of living animals for the purposes of scientific study.

SED-, SEDAT-, SEDI-, SESS-
to sit, to be seated, to settle (L. *sedeo*, *sess-*, to sit, to be seated; *sedo*, *sedat-*, to settle, to allay).
This root is seen in **sedentary** (e.g. a sedentary occupation in which one does much sitting). A **session** is basically an assembly at which people are seated. A **sedate** person is one who is "settled", calm, not easily disturbed by passion or excitement.

sedi'ment—matter that settles to the bottom of a liquid. **sedimentary rocks**—rocks formed from fragments of earlier rocks, hard parts of organisms, etc.

which have been deposited, e.g. on the sea bottom.

sess'ile — "seated" — (flowers, leaves, etc.) having no stalk; (animals) which are fixed in one place, e.g. to a rock.

sedat'ive — (medicine, etc.) which settles, soothes or allays.

SEISM-, SEISMO-

an earthquake (Gk. *seismos*, a shaking, a shock, an earthquake).

seism'ic—pertaining to earthquakes.

seismo'logy — the study of earthquakes.

seismo'graph — an instrument which records the occurrence of earthquakes.

micro'seism—a small earth tremor.

SELEN-, SELENO-

the Moon (Gk. *selēnē*).

seleno'graphy—the study of the surface features of the Moon. (Compare geography—of the Earth.)

selen'ite — a colourless, transparent variety of gypsum (calcium sulphate). This is apparently the mineral which the ancients believed waxed and waned with the Moon; hence the name *selēnitēs* (*lithos*), Moon stone. (A selenite is also a salt of selenious acid H_2SeO_3.)

selen'odont — (mammal) which has crescent-shaped ridges on the grinding surfaces of the teeth.

The chemical element **selenium** was discovered by Berzelius in 1817. It was named after the Moon because of its association with a similar element tellurium (named after the Earth, L. *tellus*) discovered some years previously. The root SELEN-, especially in chemical names, usually refers to selenium.

seleni'ferous — containing or yielding selenium.

selenide — a compound of selenium with another element.

selenate — a salt of selenic acid H_2SeO_4.

SEMI-

half (L. *semi-*; *semis*, a half).

This prefix is used in its original sense in the word **semicircle** (half a circle) but more often it means 'partly, rather less than, imperfectly'.

semi-conductor — a substance, e.g. germanium, which is part way between a conductor (of electricity) and an insulator.

semi'palmate—having the toes partly webbed.

semi-permeable — (membrane) which allows a liquid to pass through it but not substances dissolved in the liquid.

semi-fluid, semi-liquid — part way between a solid and a liquid, a thick sticky liquid or paste.

SENS-, SENSIT-

sense, feeling (L. *sentio*, *sens-*, to feel).

A number of well-known words, e.g. **sense, sensation, sensitive, insensible,** are derived from this root.

sens'ory nerves—nerves which carry impulses from sense organs, indicating feelings such as pain, taste, etc.

sensit'ize—to make sensitive, e.g. to treat paper with chemicals so that it is sensitive to light.

sensit'ometry — the measurement of the sensitivity to light of chemically treated materials (e.g. photographic films).

SEPSIS, SEPT-

rotting, putrefaction. Used especially to denote a poisoning of a part of the body by bacteria (Gk. *sēpsis*).

Care is needed to distinguish this root from each of the two following roots.

sepsis—festering, poisoning (e.g. of a wound). The wound becomes **septic**.

anti'septic—(substance) used "against sepsis", i.e. to prevent the growth of and to destroy harmful bacteria.

a'septic—free from poisoning bacteria, i.e. surgically clean.

septic'aemia — the invasion of the blood by bacteria, blood-poisoning.

SEPT-, SEPTEM-, SEPTI-

seven (L. *septem*).

This prefix, which is seen in **September** (the seventh month of the Roman calendar), is not much used in scientific terms. The Greek HEPTA- (q.v.) is more common.

sept'angular—having seven angles (and sides).

septi'lateral—seven-sided.

septem'fid — deeply cleft into seven parts.

SEPT-, SEPTO-, SEPTUM

a fence, a partition (L. *s(a)eptum*).

septum — a cross-wall, e.g. a cross-wall in an algal or fungal thread, a partition separating two cavities (as in the nose).

sept'ate—having septa. **a'septate**—not having septa.

septo'nasal—pertaining to the septum of the nose.

SER-, SERO-

Pertaining to **serum**—a watery liquid, especially the watery liquid which separates from blood when it clots (L. *serum*, whey, watery liquid).

ser'ous—of, or pertaining to, serum; of the nature of serum; consisting of, or containing, serum.

sero'purulent—containing serum and pus.

sero'logy — the scientific study of serums.

sero'therapy—the prevention or treatment of disease by injecting into the body a serum which contains antibodies (opposing substances) to the bacteria (etc.) causing the disease.

sero'phyte — a micro-organism (e.g. Streptococcus) which grows in the fresh serum oozing into a wound.

SERR-, SERRI-

like a saw-edge, notched (L. *serra*, a saw; *serratus* (adj.)).

serrate—(leaf, etc.) with a notched edge.

serri'rostrate — (bird) having a serrated bill.

SESQUI-

one and a half (L. *sesqui-*).

This unusual prefix, which is seen in **sesquicentenary** (a hundred and fiftieth anniversary), is mainly used in the names of chemical compounds which contain three parts of one element (group, etc.) to two of another.

nickel sesqui'oxide—an oxide of nickel (also called nickelic oxide) composed of three parts of oxygen to two of nickel, Ni_2O_3.

sodium sesqui'carbonate — a form of soda (used for washing wool) which contains three parts of sodium to two of the carbonate group, $Na_2CO_3.NaHCO_3.2H_2O$.

SESS- See SED-.

SET-, SETI-

a bristle (L. *s(a)eta*).

set'aceous — bristly, having bristles, shaped like a bristle.

seti'form, seti'gerous—bearing or producing bristles.

seti'form — having the form of a bristle.

SEX-

six (L. *sex*).

This prefix is far less common than the corresponding Greek prefix HEX-.

sex'fid—deeply divided into six parts.

sex-digitate—having six fingers.

sextant—a sixth part of a circle; an instrument of this shape used for measuring angular distances (e.g. of the height of the Sun above the horizon).

Note. This prefix must be distinguished from the word *sex* (L. *sexus*) and its derivatives (e.g. *unisexual*).

SIAL-, SIALO-

saliva (Gk. *sialon*).

sial'agogue—a drug which stimulates a flow of saliva. (Also spelt **sialogogue**.)

sialo'rrhoea — an excessive flow of saliva.

sialo'adenitis — inflammation of a salivary gland.

Note. *Sial*, the name given to the rock-shell forming the foundation of the continental masses, has no connection with this root. The name is taken from *Si*licon and *Al*uminium of whose compounds the rocks are mainly composed.

SICC-
dry (L. *siccus*)

sicc'ative—(substance) which has drying properties (especially a substance mixed with oil-paint to dry it).

de'sicc'ated—(food, milk, etc.) which has been dried. (This word is supposed to be a spelling-catch. Possibly errors occur because it is the first syllable which is stressed—and because many people are unaware of the derivation.)

ex'sicc'ate—to dry, to drain away, e.g. to drain water from swampy ground.

SIDER-
a star (L. *sidus, sider-*).
This root must be distinguished from that which follows.
The familiar word **consider**, meaning to contemplate, to weigh the merits, has come (through French) from the Latin *considero*. It is probable that this was derived from *sidus*, a star, for in considering a matter one might examine the stars to see what they foretell.

sider'al—pertaining to the stars, e.g. **sideral day** — the time between two successive crossings of the meridian by a star (especially by the first point in Aries), about four minutes shorter than a day as measured by the Sun.

sider'ostat — an instrument attached to a telescope which causes the stars to appear motionless, i.e. always to be in view in spite of the rotation of the Earth.

SIDER-, SIDERO-
iron (Gk. *sidēros*).

sider'ite—an iron ore (chiefly iron carbonate).

sidero'lite—"iron stone"—a meteorite consisting of iron and stone.

sidero'sis—(1) a lung disease caused by breathing in particles of metal (e.g. by metal-workers). (2) excessive deposits of iron in the tissues of the body.

SILI- See SALI-.

SILIC-, SILICI-, SILICO-
sand (L. *silex, silic-*, flint); hence, relating to the element silicon.

silica — the dioxide of **silicon**, SiO_2, occurring as quartz, sand, etc., and as a part of many minerals.

silic'ates—compounds of silica with metal oxides (equivalent to salts of the **silic'ic acids**). They form a large part of the Earth's crust.

silic'eous—sandy, containing silica.

silici'fication — the process in which silica is worked into rocks, as a kind of cement, after they have been deposited.

silic'osis — a lung disease caused by breathing in particles of sand (as by stone-workers).

silico-methane — a compound of silicon and hydrogen SiH_4, corresponding to methane CH_4. (Also called **silicane** and **silane**.)

silic'ones — plastic-like substances built up from rings or chains of silicon and oxygen atoms together with hydrocarbon groups. (See -ONE.)

Note. *Silicle* and *silicula* (special types of pod-like fruits) are derived from L. *siliqua*, a pod.

SILV- See SYLV-

SINAP-
mustard, mustard seed (L. *sinapis*).

sinap'ism—a plaster or poultice consisting wholly or partly of mustard flour (put on the skin to bring blood to the part).

sinap'ine — an organic compound found in white mustard seed. **sinapic**—derived from or related to sinapine.

sinapis'ine—a white crystalline substance obtained from black mustard seed by extraction with alcohol and ether.

SINISTRO-
left, left-handed (L. *sinister*).
The word **sinister** is still used in heraldry to denote the left side of the shield. To the Greeks a bird appearing to the left (the west) was considered to be an unlucky sign. Hence the left has come to be regarded as the side from which evil comes. **Sinister** (e.g. a sinister person) thus also means unfavourable, suggestive of evil.

sinistro-cerebral — pertaining to the left hemisphere of the brain.

sinistrorse—twisting in a left-handed spiral (as the stems of some climbing plants).

SINU-, SINUS

a curve, a hollow, a fold, a bay (L. *sinus*).

To **insinuate** ("to curve in") means to introduce oneself, a thing, a statement, into a place especially by subtle or indirect methods.

sinu'ous—wavy, with many curves.

sinu'ate—(leaf) with a wavy margin.

sinuato-dentate — (leaf) with a wavy, toothed margin.

sinus—a bend or hollow between two lobes of a leaf; a cavity or depression in an organ, tissue or bone. **sinus'itis**—inflammation of any of the air-containing cavities of the skull connected with the nose.

sinus'oidal—(quantity, e.g. an alternating current) which, when plotted against time, on a graph, gives a wave-like curve (especially a sine curve).

SIPHONO-

a tube, a pipe (Gk. *siphōn*).

This root is readily recognised in the English word **siphon.**

siphono'gam — "tube marriage" — a plant in which the contents of the pollen grain pass down a pollen tube into the embryo sac (cavity of the ovule).

siphono'stele—a hollow tube of conducting tissue enclosing the pith in a plant stem.

siphono'stomatous—having a tubular mouth; (of Gastropods) having the front edge of the shell aperture prolonged into a spout.

-SIS

The addition of this Greek suffix to the stem of a Greek verb forms a noun (denoting an action or, sometimes, a state). Thus:

lyō—to loosen; *lysis*—a loosening, a setting free (whence **ana'lysis** — "a loosening up again").

klaō—to break; *klasis*—a breaking (whence **osteo'clasis**—a breaking of a bone).

Such nouns were also formed by the addition of the suffix to the stem or combining-form of a noun or adjective.

osteon — a bone; **exosto'sis** — an outgrowth of bone.

sklēros—hard; **sclero'sis**—a hardening.

Many words of this type have passed into English or have been constructed on the Greek pattern.

haemat'eme'sis — the vomiting of blood (*emeō*, to vomit).

sym'bio'sis—a state of living together (as an Alga and a Fungus in a Lichen) (*bioō*, to live).

meta'morpho'sis — "a changing of form"—a marked change of form and structure taking place in an animal within a fairly short time, e.g. the change of a caterpillar into a butterfly (*morphē*, form; *morphoō*, to form).

ec'dy'sis—the casting off of a coat (as by Snakes and many Arthropods) (*dyō*, to put clothes on or off).

Both -ASIS and (especially) -OSIS, in which the suffix has taken with itself the terminal vowel of the preceding stem, have become living suffixes and are much used in medical terms to denote a diseased, unhealthy or damaged condition. As such they are added to stems of various origins.

filari'asis—a diseased state caused by thread-like parasitic worms (Filaria).

silic'osis—a lung disease caused by breathing in particles of sand (as by stone-workers).

For further examples, see -ASIS and -OSIS.

-SIST, -SISTOR, (-ISTOR)

to set, to stand, to stay, to stop (L. *sisto*).

This root is seen in **consist** ("to stand together"), **persist** ("to stand through") and **resist** ("to stand against", to stop).

re'sistor—an electric component or apparatus used to provide resistance to the flow of the current.

The element -(S)ISTOR has been used in forming the names of electrical components and devices which have peculiar resisting properties.

therm'istor — thermally sensitive resistor—a component whose resistance decreases markedly with temperature.

var'istor—variable resistor—a component whose resistance varies markedly with the voltage which is applied.

trans'istor—the name given by Bell Laboratories in 1948 to a device which may be simply described as a crystal triode; extended to apply to more complicated devices embodying a semiconductor and electrode junctions. (Probably from 'transfer resistor'.)

SKELET-, SKELETO-, SKELETON

the skeleton (Gk. *skeletos*, dried up; *skeleton* (*sōma*), a dried up body).

skelet'al—pertaining to the skeleton, e.g. **skeletal muscles** — those muscles which move parts of the skeleton.

endo'skeleton — a skeleton which is within the body of the animal (e.g. the bones of Man). **exo'skeleton** — hard, supporting or protective structures which are on the outside of the body of the animal (e.g. finger nails in Man, the shell of a Crab, the horny coat of an insect).

skeleto'genous—(parts) which help to form the skeleton.

skeleto'graphy—the drawing and description of the skeleton.

SKIA-, SKIO- See SCIA-.

SOL-

the Sun (L. *sol*).

A **parasol** is to "ward against the sun", i.e. a sunshade.

sol'ar—pertaining to the Sun, e.g. a solar eclipse, a solar flare, solar power. For solar plexus, see PLEX-.

in'sol'ation—the radiation received at a place from the Sun. (In medicine, a sunstroke.)

sol'stice—"sun standing still"—times of the year (about June 21st and December 22nd) when the midday Sun is overhead at places the maximum distance from the equator. The Sun in its apparent movement north and south of the equator seems to stop and then turn back at these times.

-SOL

A **sol** is a colloidal solution, i.e. a 'solution' in which particles of solid, too small to be seen or to fall as a sediment, are dispersed (spread out and held up) in a liquid. (Some sols set to form gels—jelly-like substances.)

hydro'sol — a colloidal solution in which the particles are dispersed in water.

organo'sol — a colloidal solution in which the particles are dispersed in an organic liquid.

aero'sol—a colloidal system in which particles (e.g. of a liquid) are dispersed in a gas, e.g. a mist.

SOLEN, SOLENO-

a tube, a pipe; also a mollusc (Gk. *solen*).

solen'oid—a long, closely wound coil of wire, especially that used for magnetising purposes.

soleno'stele—a cylindrical bundle of conducting tissue as in some Ferns.

soleno'cyte — "a tube cell" — a tube-like excretory organ of some lower animals (especially the Polychaeta).

soleno'stomatous—(fish) having a tube-like mouth.

solen'ite—a fossil **Solen** (Razor shell).

SOLUT-, SOLV-, -SOLVE

The basic meaning of the Latin verb *solvo*, *solut-*, is "to loosen, to separate, to disengage". To **solve** a problem (or difficulty) is to unravel the complications and so find a **solution** (answer or way out). A person is **absolved** from blame, sin, etc., when he is "loosened away from" it.

In science the root is usually met with reference to the act of **dissolving** a substance (e.g. sugar) in (usually) a liquid (e.g. water) to make a **solution**.

solute—the substance which is dissolved to make a solution.

solvent — the substance (usually a liquid) in which a solute is dissolved.

soluble—able to dissolve. **solubility**—a measure of the ability to dissolve (e.g. the weight of the substance which can dissolve, at a given temperature, in a standard weight of solvent).

SOM-, SOMAT-, SOMATO-, -SOME

the body of an animal (or of a plant) as distinct from the germ cells (Gk. *sōma, sōmat-*).

somat'ic—pertaining to the body, e.g. **somatic cell** — a body cell as distinct from a germ cell.

som'ite — one of the divisions (segments) of the body of a metameric animal (such as an Earthworm).

somato'genesis—arising or developing in the body or from the body cells.

somato'blast—a cell in a developing embryo which will give rise to somatic cells.

-SOME usually denotes "a body", i.e. a small organism or a particle.

trypano'some—"a boring body"—one of several similar kinds of Protozoa some of which cause diseases, e.g. sleeping-sickness.

centro'some—a very small body near the nucleus of a cell which divides when the nucleus divides.

chromo'some — "a coloured body" — one of the rod-like bodies, which can be deeply stained, seen in a cell nucleus which is about to divide. It carries hereditary qualities.

SOMN-, SOMNI-

sleep (L. *somnus*).

somn'ambulation — walking in one's sleep.

in'somnia—sleeplessness.

somni'ferous—(drug, etc.) which induces sleep.

SON-, SONO-, SONOR-

sound (L. *sonus* (n.); *sonorus* (adj.)). **Sonorous** means resounding, loud-sounding; a **sonata** is a musical composition (of a certain form) for one or two instruments.

re'son'ance—the resounding of a wire (or other vibrating body) when sound waves or impulses of the correct frequency fall on it. (Also similar effects with other forms of vibrations.)

sub'sonic—less than the normal speed of sound. **super'sonic** — at a greater speed than that of sound.

ultra'sonics—the study of vibrations which have a greater frequency than those corresponding to audible sound.

sono'meter—an instrument, consisting of a wire stretched between two supports, used for testing and measuring frequencies of sounds.

SOPOR-

sleep (L. *sopor*).

sopor'ific — (drug) which tends to produce sleep.

SORB-

Pertaining to, or derived from, the Service tree or the Mountain Ash tree (L. *sorbus*, the Service tree; *sorbum*, the Service berry).

Sorbus—the genus of plants (now incorporated with *Pyrus*) which includes the Service tree and the Mountain Ash.

sorb—the berry of the Service tree; the Service tree or the Mountain Ash.

sorb'ic acid—an acid contained in the berries of the Mountain Ash, $CH_3.CH:CH.CH:CH.COOH$. The salts are **sorb'ates.**

sorb'itol—a related alcohol, $CH_2OH.(CHOH)_4.CH_2OH$, containing six hydroxyl groups (-OH), isomeric with mannitol.

sorb'ose—a related carbohydrate, isomeric with fructose.

SORB-, SORBE-, SORPTION

to suck in, to drink, to absorb (L. *sorbeo, sorpt-*).

ab'sorb — "to suck away" — to take liquid or gas into itself (as blotting-paper takes in ink). So also, to take in light, etc. **ab'sorption** (n.).

ad'sorb—"to suck to"—to take up a liquid or gas on to the surface (but not into itself). **ad'sorption** (n.).

sorbe'facient—(substance) causing or producing absorption.

SORO-, SORUS

a heap (Gk. *sōros*).

sorus—a cluster of spores on the underside of a leaf of a Fern.

soro'sphere—a hollow ball of spores.

soro'sis—a fleshy fruit formed from a number of crowded flowers (e.g. a pineapple).

soredium — "a little heap" — one or more algal cells enclosed in fungal threads forming a little mass which separates from a Lichen and can give rise to a new plant.

SPAR

a general name for a non-metallic, crystalline mineral (e.g. calcite) which splits fairly readily. (From Middle Low German.)

heavy spar — barytes — the mineral barium sulphate.

fluor'spar—fluorite—the mineral calcium fluoride.

feld'spar — a general name for any one of a group of minerals consisting of silicates of aluminium and sodium, potassium or other metal, occurring as crystals or crystalline masses in many rocks. (The more common spelling **felspar** is strictly incorrect. The word is derived from *feld*, a field, not *fels*, a rock.)

SPECI-, SPECIES

The Latin word *species* has a range of meanings (e.g. an appearance, a kind, a figure, a likeness) but it has come to denote "a particular kind of thing, a thing with particular qualities". This meaning is seen in the common words **special** (of a particular kind) and **specimen** (a particular thing taken as an example of a class or whole).

The word **species** itself denotes, in biology, the smallest group of organisms (plants or animals) which possess distinctive and invariable characteristics. The members of the group are usually able to breed among themselves but not with members of other groups. A number of species make up a genus.

speci'fy — "to make a species" — to name expressly, to indicate exact items, details, etc.

speci'fic—definite, distinctly pertaining to a named thing.

specific gravity—the precise statement of the 'weight' of a particular kind of substance, namely the weight of a piece of the substance compared with the weight of the same volume of water.

SPECTR-, SPECTRO-

The Latin word *spectrum* means "an image, a vision". The English word **spectre** reflects this meaning. In science the word **spectrum** is used to denote the band of rainbow colours into which white light can be spread out (e.g. by a prism); hence, any band of radiations which are spread out according to their wavelengths.

spectr'al—pertaining to a spectrum.

spectro'meter—an instrument for producing a spectrum and with which measurements can be made (e.g. of the wavelengths of the light of the different colours). A **spectro'scope** is a simpler instrument by which a spectrum can be viewed but which does not necessarily permit measurements to be made.

spectro'scopic analysis—the analysis of a substance (finding out what it is composed of) by examining the spectrum of the light which it emits.

spectro'gram — a photograph of a spectrum.

spectro'helio'graph—an instrument for taking photographs of the Sun using only a part of the spectrum (only one colour) of the light which it emits.

SPECUL-

like a mirror (L. *speculum*, a mirror).

specul'ar reflection—regular reflection as by a mirror.

speculum—a curved instrument for widening and viewing a passage or cavity in the body.

speculum metal—an alloy of copper and tin as used in making the mirror of a reflecting telescope.

SPELAE-, SPELAEO-, (SPELEO-)
a cave (Gk. *spēlaion*).

spelae'an—pertaining to, or living in, a cave; of the nature of a cave.

spelaeo'logy—the scientific study of caves.

SPERM, SPERMA-, SPERMATO-, SPERMI-, SPERMO-, -SPERM
a seed (Gk. *sperma, spermat-*).

(1) the seed of a plant.

spermato'phyte, spermophyte — a seed-bearing plant. **Sperma(to)phyta** —the great class (sub-kingdom) of plants which bear seeds.

Angio'sperms—plants which bear their seeds in "containers" (ovaries) as do flowering plants.

Gymno'sperms — plants with "naked seeds" (not in an ovary)—the cone-bearing plants.

endo'sperm—the tissue within a seed which serves as food for the embryo.

spermo'derm—the outer covering of a seed.

(2) the male 'seed' of an animal.

sperm—a male germ-cell.

spermato'zoon—the characteristic male germ-cell of an animal consisting of a head (with a nucleus) and a tail.

spermato'genesis—the formation of sperms.

spermi'duct—the duct (pipe) by which sperms are carried from the testis of an animal to the external opening.

spermo'toxin—a substance which destroys spermatozoa.

sperma'theca — "sperm-container" —the cavity for receiving and storing sperms in many invertebrate animals (e.g. Earthworm).

sperma'ceti—white wax obtained from the head of the sperm whale. (It was once supposed to be whale spawn.)

SPHAERO- See SPHER-.

SPHEN-, SPHENO-
a wedge (Gk. *sphēn*).

sphen'oid — wedge-shaped. Used especially of the **sphenoid bone** at the base of the skull.

sphenoid'itis—inflammation of the air-cavity in the sphenoid bone.

Sphen'odon—"wedge tooth"—a South American fossil sloth; a primitive New Zealand reptile.

SPHER-, SPHERO-, SPHAERO-, -SPHERE
a ball (Gk. *sphaira*).

This root is recognised in the words **sphere** and **spherical**.

spher'oid — sphere-like but not perfectly spherical.

sphero'meter—an instrument for finding the radius of a sphere or of a spherically curved surface (such as the face of a lens).

The spelling SPHAERO- occurs in some Natural History terms.

sphaero'carpous — having globe-shaped fruit.

sphaero'cephalous—having the flowers crowded into a rounded head.

-SPHERE often denotes a hollow sphere, i.e. a spherical shell or layer.

atmo'sphere — "vapour sphere"—the layer of gases surrounding a heavenly body (especially the Earth).

litho'sphere — "stone sphere" — the outer rocky layer of the Earth which surrounds the **centrosphere** (central core).

tropo'sphere — "turning (changing) sphere" — the lower layer of the atmosphere in which temperature decreases with height and in which variations of weather occur.

iono'sphere—a layer of the upper atmosphere which is ionised and reflects radio waves.

SPHYGMO-, -SPHYXIA
the pulse (of the blood) (Gk. *sphygmos*).

sphygmo'graph—an instrument which draws a line to show the pulse.

sphygmo'manometer — (manometer — an instrument for measuring pressure)— an instrument for measuring blood pressure.

a'sphyxia — "without pulse" — suffocation and unconsciousness due to lack of oxygen in the blood.

SPIR-, -SPIRE, -SPIRATION
to breathe (L. *spiro, spirat-*).
To **inspire** is to "breathe in" (and hence, to infuse thought or feeling into another person); to **expire** is to "breathe out" (often used, in ordinary speech, to mean breathing out for the last time); to **perspire** is to "breathe through" (to pass liquid or vapour through pores in the skin). The word **spirit** (which is used in a variety of senses) is also derived from this root.

re'spiration — (loosely) breathing air in and out; (more exactly) the taking in of oxygen and the giving out of carbon dioxide produced by the oxidation of body materials (internal respiration).

trans'pire—to "breathe across"—used chiefly in science to refer to the losing of water vapour by a plant (especially by the leaves).

spir'acle—a tube-like breathing hole in some animals, e.g. in insects and in whales.

spir'ometer — an instrument for measuring the amount of air breathed in and out.

a'spir'ator—(*ad-*, to + *spir-*, breathe) —an apparatus for drawing a stream of air or gas through a liquid, etc.

SPIRO-
a coil, a spiral (Gk. *speira*; L. *spira*).

spiro'chaetes — "spiral bristles" — spiral-shaped bacteria including those which cause syphilis and relapsing fever (**spirochaetosis**).

spiro'lobous — (plant embryo) which has spirally rolled cotyledons (seed leaves).

SPLANCHN-, SPLANCHNO-
"the inward parts", especially the bowels, etc. (Gk. *splanchnon*).

splanchn'ic — pertaining to the entrails.

splanchno'ptosis—a dropping of the organs of the abdomen.

splanchno'megaly—abnormal enlargement of the organs of the body.

micro'splanchnic—having a very small body in comparison with the legs (as some Spiders).

SPLEN-, SPLENO-
the spleen (Gk. *splēn*).
The **spleen** is an organ near the stomach which acts as a blood reservoir and is concerned with the formation of new blood cells and the destruction of old red cells.

splen'itis—inflammation of the spleen.

splen'algia—pain in the region of the spleen.

spleno'tomy—the surgical cutting into the spleen.

splen'ectomy—the surgical cutting out of the spleen.

SPOR-, SPORO-, -SPORE
a spore (Gk. *spora*, a sowing, a seed).

sporo'genesis — the formation of spores.

spor'angium—"spore container"—an organ in which spores are formed (especially in a Fungus or Moss).

sporo'phore—the spore-bearing structure of a Fungus.

sporo'phyte—a form of a plant which reproduces by means of spores (especially a form which, as in a Fern, alternates with a sexually-reproducing form).

sporo'gonium — "spore generator" — the sporophytic form of a Moss which develops almost as a parasite on the sexually-reproducing form.

oo'spore — "egg spore" — a thick-walled spore which normally germinates only after a period of rest.

SQUAM-, SQUAMI-
a scale (of a fish, insect, reptile, etc.) (L. *squama*).

squama—(in general) a scale or scaly part of a plant or animal; a scaly portion of bone, especially of the temporal bone of the skull.

squamose, squamous, squamate — scaly.

squami'form—having the form of a scale.

squam'ule—a small scale.

de'squam'ation—the casting off of the surface layer of the skin.

-STALSIS, -STALTIC, -STOLE

The parent word of these elements is the Greek verb *stellō*—to send, to set, to place; whence

peristellō—to "place round", to wrap up.

diastellō—to "place out", to separate, to expand.

systellō—to "draw together", to contract.

peri'stalsis — "a wrapping round" — the wave-like contraction in successive circles by which the contents of the alimentary canal (food-pipe, intestine, etc.) are pushed along.

dia'stole—the regular expansion of the heart and the arteries. **systole** — the regular contraction of the heart and arteries. Together they form the pulse (the regular beating).

sy'staltic—expanding and contracting in turn, pulsating.

STANN-

tin (L. *stannum*).

The early Romans appear not to have clearly distinguished tin from lead and the word *stannum* was sometimes applied to certain lead alloys. Later the use of the word was restricted to tin proper. The chemical symbol for tin, **Sn**, is derived from the Latin name.

stann'ous chloride — the tin chloride $SnCl_2$; **stann'ic** chloride — the tin chloride $SnCl_4$.

stann'ite—a complex sulphide of tin and other metals.

STAPHYLO-

a bunch of grapes (Gk. *staphylē*).

staphylo'coccus — a coccus (round bacterium) which tends to associate with others in clusters. (Staphylococci cause inflammation and the formation of pus.)

staphylo'ma—"a grape-like state"—a bulging of any part of the ball of the eye. The root also denotes the uvula—the hanging fleshy part of the soft palate (soft roof of the mouth at the back)—which, apparently, the Greeks likened to a bunch of grapes. (It is interesting to note that the Latin *uvula* means "a small bunch of grapes".)

staphylo'rrhaphy—the surgical operation of "sewing up" a cleft in the soft palate.

staphylo'plasty — the operation of closing a cleft palate by plastic surgery.

-STASIS, -STASY

a standing, a standing still, a state or position (Gk. *stasis*).

This root is seen in **ecstasy** which denotes a state in which a person "stands outside" his senses.

stasis—a "standing still"—the stoppage of the flow of blood in a part of the body or of the movement of the contents of the bowel.

haemo'stasis—"blood standing still" —the stopping of bleeding.

epi'stasis—a "standing over"—a state in which one hereditary character is dominant to another although the two characters do not (as brown and blue colour of eyes) form a contrasting pair.

meta'stasis—a "changed position"—a transfer of diseased tissue (e.g. of a tumour) from one part of the body to another.

dia'stase — a ferment (group of enzymes) which changes starch into sugar, e.g. in a germinating seed. (The term comes from *diastasis*, a standing apart, a separation.) The suffix -ASE, denoting an enzyme, is taken from the parent word diastase.

iso'stasy—"equal standing, balancing"—the state in the Earth's crust in which large heavy masses above the surface (e.g. a mountain range) are assumed to be balanced by similar masses below the surface.

-STAT, -STATIC, -STATICS, STATO-

standing, standing still (Gk. *statos*).

The Latin verb *stare, statum*, to stand, is

clearly related to the Greek root. The Latin root gives rise to such well-known words as **state, station, stationary,** and **statue.** A few of the terms given below are derived from the Latin rather than from the Greek root.

pro'state—"one which stands before" —a gland associated with the male reproductive organs.

stator — the part of an electrical machine (e.g. a generator) which remains still (i.e. does not turn).

stato'lith — "position stone" — (1) (*Botany*)—a solid grain (e.g. of starch) in a cell (a **statocyst**) which comes to rest at the bottom of the cell and possibly plays some part in indicating the position of the tissue in relation to the direction of gravity. (2) (*Zoology*) — a chalky particle in an organ lined with sensory cells (a statocyst); the movement of the particle (or particles) helps the animal to perceive its position in space.

statics — the branch of mechanics which deals with forces that are balanced so that the bodies on which they are acting remain at rest (e.g. the forces on a balanced lever, the forces in the parts of a bridge).

hydro'statics—the study of the pressure, etc. of water and other liquids which are at rest.

electro'statics—the study of electric charges which are at rest (e.g. the charge in an electrified ebonite rod).

-STATIC is used to form adjectives, e.g. **hydrostatic pressures, electrostatic instruments, epistatic characters** (see -STASIS).

(**Statistics** may be broadly described as that branch of study in which numerical information (e.g. heights of people, weights of crops, etc.) is collected, classified, and examined mathematically. (The term is also used for the numerical information itself.) The term originally denoted that branch of political science which deals with the collection, classification and discussion of facts bearing on the affairs of a state or community.)

-STAT denotes an instrument or device which makes a quality or thing "stand still", i.e. remain at rest or unchanged. The first term of this type appears to have been *heliostat*.

helio'stat—an instrument attached to a fixed telescope which keeps the Sun always in view in spite of its apparent motion across the sky.

thermo'stat—an instrument which controls the heating of a tank of water (oven, etc.) so that the temperature remains (almost) the same.

rheo'stat — (literally) an instrument which makes a flow (of electricity) stay still; (actually) a variable resistor by which the current in a circuit can be varied and adjusted.

STEAR-, STEAT-, STEATO-

fat, tallow (Gk. *stear, steat-*).

stear'ic acid—the acid which, when combined with glycerine, forms the hard fat of animals.

stear'in—the hard fat so formed.

stear'one — a ketone (q.v.) derived from stearic acid.

steat'ite—a soft, greasy form of talc. (Also called soapstone.)

steato'rrhoea—the presence of excess fat in the matter which flows from the bowels.

STEG-, STEGO-, -STEGE

a covering, a roof (Gk. *stego*, to cover, to shelter; *stegē*, a covering).

stego'carpous—"with roofed fruit"—(a Moss) whose spore-containers have covers.

Stego'saurus—a fossil Dinosaur which was covered with large bony plates and spines.

stego'gnathous — with the jaw composed of overlapping plates.

branchio'steg'al — pertaining to the gill-covers.

uro'stege—"a tail cover" on the under side of a snake.

STELE

a pillar, a post (Gk. *stēlē*).

In botany a **stele** is the bundle of conducting tissue in a stem or root.

proto'stele—"first (primitive) stele"—

a stele in which the conducting tissue forms a solid central core.

soleno'stele — a tube-shaped stele (as in the fern Dennstaedtia).

dictyo'stele—a stele in the form of a network (as in most common ferns).

STELL-

a star (L. *stella*).

stell'ar—pertaining to stars.

stell'ate—star-shaped; radiating from the centre like rays from a star.

con'stell'ation—a group of stars which (it has been imagined) form a picture and so has been given a name, e.g. Taurus (the Bull).

Stellaria—the hedgerow plant known in English as Stitchwort. The expanded flowers are star-shaped.

STENO-

narrow (Gk. *stenos*).

Stenography is commonly called shorthand; it means "narrow writing".

steno'sis—a narrowing or constricting of a passage in the body, e.g. of the pipe through which urine passes out.

steno'phyllous—having narrow leaves.

steno'thermous — able to live within only a narrow range of temperatures.

STERCO-, STERCOR-

dung, faeces (waste matter from the bowel) (L. *stercus*, *stercor-*).

stercor'aceous — consisting of, containing, or pertaining to, faeces.

sterco'bilin — the brown colouring matter of faeces.

stercor'ate—to manure with dung.

sterco'lith — a hard mass of faeces (containing calcium salts) in the bowel.

STEREO-

stiff, rigid, solid, cubic; hence, relating to space, to three dimensions (Gk. *stereos*).

stereo'scope—an instrument by which two pictures of the same thing are seen at once and combined to give a solid, three-dimensional effect.

stereo'chemistry — the study of the ways in which the atoms of a molecule are arranged in space (not as if drawn on a flat surface).

stereo'isomers—two substances whose molecules are identical in structure (*isomer*="equal parts") but with the atoms differently arranged in space, e.g. the different forms of tartaric acid.

stereo'type—a printing plate made by forming a mould (often of papier mâché) of the type and then taking a cast of this in metal.

stereo'phonic reproduction—a way of reproducing sound so that the sound seems to come, not from a loudspeaker, but from various directions in space.

STERN-, STERNO-, STERNUM

The **sternum** is the breast-bone (Gk. *sternon*).

xiphi'sternum — the "sword-shaped" lower end of the sternum.

stern'algia—pain in the region of the sternum.

sterno'mastoid muscle — a muscle running from the mastoid process (just behind the ear) to the sternum.

-STEROL, -STERONE

Chole'ster'ol ("bile-solid-alcohol") is a complex alcohol found in nerve tissues and gall stones. It is typical of the **sterols**—complex alcohols, built up from rings of carbon atoms, found in nature combined with fatty acids. The **sterones** are related compounds but containing a ketone (q.v.) group.

ergo'sterol—a sterol found in ergot (diseased rye), yeast and, in traces, in animal tissues. With the help of sunlight it forms vitamin D.

cortico'sterone—a substance produced by the outer shell of the suprarenal glands.

testo'sterone—a male hormone (stimulating agent) which stimulates the development of male characteristics.

STETHO-

the breast, the chest (Gk. *stēthos*).

stetho'scope—an instrument for "inspecting the chest" by enabling the movements of the heart, lungs, etc. to be heard.

stetho'graph — an instrument which automatically records the movements of the chest in breathing.

STICH-

in a line, in a row (Gk. *stichos*).

mono'stichous — in one line, forming one line.

di'stichous — in two opposite rows (e.g. as fruits on a vertical stem).

di'stich'iasis—"a state of having two rows"—the state of having two complete rows of eye-lashes on either or both eye-lids.

rhipido'stichous—in fan-like rows (e.g. the stiffening rods at the base of the fins of some fish).

stich'idium—"a little row"—a special branch of the body of a Red Alga on which the spores are formed.

STIGMA, STIGMAT-

a prick, a brand-mark; (more generally) a mark, a spot (Gk. *stigma, stigmat-*). In non-scientific language a **stigma** is a mark or stain on one's good name. Certain saints are said to have developed marks (**stigmata**) on their bodies resembling the wounds on the crucified body of Christ.

stigma—a distinctive mark on an animal (e.g. the eye-spot of a Protozoon, a mark on a butterfly's wing); the top end of the pistil of a flower on which the pollen alights. (The Greek plural **stigmata** is normally used.)

ptero'stigma — "wing mark" — an opaque cell on the wing of a butterfly.

a'stigmat'ism—"not (coming to) a point"—a defect of a lens in which light from a point is not brought to a focus at a point (but on to two short lines).

an'astigmatic—(lens) which has been corrected so that it is not astigmatic. (In this curious word the prefix *a-/an-* (=not) has been used twice.)

-STOLE See -STALSIS.

STOM-, STOMA, STOMAT-, -STOMY

a mouth, an opening (Gk. *stoma, stomat-*).

The word **stomach** is derived, through French and Latin, from the Greek *stomachos*, "a little mouth". The Greek word originally denoted the throat or gullet, hence it denoted the mouth or orifice of any organ especially the stomach, and hence the stomach itself.

stoma — an opening (pore) in the surface of a leaf by which gases can pass in and out. (Plural **stomata**.)

siphono'stomat'ous—having a tubular mouth.

gnatho'stomat'ous—having the mouth provided with jaws.

ana'stom'osis — "state of being mouthed up"—a join between two blood vessels; an artificial joining of two parts of the intestine.

peri'stome—(in general) the area surrounding the mouth of an animal; a fringe round the opening of a moss capsule.

-STOMY denotes a surgical operation in which an artificial opening is made into an organ, etc. (The element must be distinguished from -TOMY, and -ECTOMY.)

colo'stomy—the making of a hole in the colon (large intestine) for the escape of faeces when the bowel is blocked further on.

gastro'stomy—the making of an artificial opening into the stomach through which food may be passed when food cannot be passed from the mouth.

tracheo'stomy — the making of an artificial opening into the wind-pipe when normal breathing is not possible.

STRATI-, STRATO-, STRATUM, -STRATE

spread out, in a layer (L. *sterno, strat-*, to spread out; *stratum*, a coverlet, a bed).

stratum—a layer, e.g. of a sedimentary rock, of cells.

strati'fied—occurring, or formed, in layers, e.g. **stratified epithelium**—a skin made up of layers of different cells.

strati'graphy—the historical study of rock layers of the Earth's crust.

stratus — a level, unbroken sheet of cloud. (The termination *-us* has been given to this name to bring it into line with the names for other types of cloud, e.g. *cirrus*, *cumulus*.)

strato'cumulus—a stratus cloud which has begun to break up into rounded humps.

strato'sphere — the layer of the atmosphere (upwards from 7 miles above the Earth's surface) in which there are no weather changes and no variation of temperature with height.

sub'strate — "the layer below" — the substance which is acted on by an enzyme; the substance (e.g. meat juice, gelatin) on which bacteria, fungi, etc., are grown.

STREPTO-

twisted (Gk. *streptos*).

strepto'neural — having "twisted nerves" — said especially of certain Gastropods in which the loop of the visceral nerves is twisted into a figure of eight.

The root is frequently taken to mean a chain (especially a chain with a beaded or twisted appearance).

strepto'bacilli — bacilli (rod-like bacteria) linked in chains.

Strepto'coccus — a coccus (rounded bacterium) which tends to associate with others in chaplets or chains.

strepto'cyte—a shapeless body occurring in bead-like strings from the blisters of foot-and-mouth disease.

STROBIL-, STROBILI-

a cone (of Pine tree, etc.) (Gk. *strobilos*).

strobilus—a cone.

strobili'form—cone-shaped.

strobil'ate—of the nature of a cone.

STROBO-

a twisting, a whirling round (Gk. *strobos*).

strobo'scope—an instrument consisting essentially of a slotted disc which is made to rotate. From the appearance of a rotating or vibrating thing as seen through the slot, the speed of rotation or vibration of the thing can be determined.

STROMBULI-, STROMBUS

a top, anything whirled round, a spirally shaped thing (Gk. *strombos*).

strombus—a spirally coiled pod.

strombuli'form — (fruit) of a spiral form.

strombuli'ferous — having organs spirally coiled; bearing spirally coiled structures.

STROPH-

a turning, a twisting (Gk. *strophē*).

Both **apostrophe** ("a turning away") and **catastrophe** ("a down-turning") are derived from this root but their modern English meanings are somewhat remote from the original Greek meanings. A *katastrophe* was originally the change which produced the final event of a drama, and hence a conclusion (usually unhappy).

Stroph'anthus — "twisted flower" — a genus of tropical trees and shrubs whose flowers have a tubular corolla with five twisted lobes.

strophanthin — a poisonous drug (a glucoside) extracted from some species of *Strophanthus*, used as a heart stimulant.

geo'strophic—(winds, ocean currents) whose turning and direction are caused by the rotation of the Earth.

STYLE, STYLI-, STYLO-

a pillar, a post (Gk. *stylos*).

style — the part of the pistil of a flower between the ovary and the stigma.

hetero'stylous—having styles of one length in some flowers and of a different length in other flowers (as in Primrose).

styl'et—a small, pointed, bristle-like projection.

styl'oid process — a spiny projection from the base of the temporal bone of the skull. **stylo'glossus**—a muscle connecting the styloid process with the tongue.

pygo'style—"rump pillar"—a bone at the end of the spine of a bird formed by the joining of some of the tail bones.

SUB-

This Latin prefix has been much used in the formation of English words, both in association with Latin roots and, as a living prefix, with words of other kinds. In some formations the *b* is changed to match the initial letter of the root, e.g. **suffer** ("to bear under"), **support** ("to carry under") and occasionally the prefix takes the form *sus-* (a contraction of *subs-*), e.g. **suspend** ("to hang under").

The prefix has acquired a range of related meanings which, for the present purpose, may be considered under two headings.

(1) under, underlying, below (as in **subway, submarine, submerge**).

sub'costal—below the ribs.

sub'cutaneous—just below the skin.

sub'soil — the layer of coarsely broken rock lying between the true soil and the bed-rock below.

sub'strate—"the underlying layer" —the substance on which an enzyme acts; the substance (e.g. meat juice) on which bacteria, fungi, etc. are grown.

(2) less than, of a less degree, not quite, (as in **subnormal, sub-tropical, sub-conscious**).

subacute illness—an illness which is apparently less serious than the acute form.

sub-standard film — cinema film which is of a narrower width than the standard 35 mm. film.

sub'sonic — less than the normal speed of sound.

sub-group—a division of a group.

carbon sub'oxide — an oxide of carbon, C_3O_2, which contains a less proportion of oxygen than the more common oxides CO and CO_2.

SUBER-

cork (L. *suber*).

suber'eous, suber'ic, suber'ose—of, or like, cork.

suber'ized—changed into cork.

suber'in — a complex mixture of waxy substances present in the cell walls of corky tissues.

SUCCIN-

amber (L. *succinum*).

succin'ite — a variety of amber; a granular garnet of amber colour.

succin'ol—purified amber tar oil.

succin'ic acid—an acid, $HOOC.CH_2.CH_2.COOH$, occurring widely in nature, originally obtained by the dry distillation of amber.

In the names of chemical compounds SUCCIN- indicates a derivative of succinic acid.

succin'amide — the amide (q.v.) of succinic acid, $H_2N.CO.CH_2.CH_2.CO.NH_2$.

succin'amic acid—the acid, $H_2N.CO.CH_2.CH_2.COOH$, in which one of the acid groups of succinic acid has been replaced by an amide group.

succin'yl—the 'stem' of succinic acid, $-CO.CH_2.CH_2.CO-$.

SUCCUL-, SUCCUS

juice (L. *succus*).

succus—a juice secreted by a gland, e.g. **succus entericus**—the digestive juices secreted by the intestinal glands.

succul'ent—juicy, soft and thick. (The word is often used when *fleshy* would be more appropriate.)

SUDAT-, SUDOR(I)-, SUDAMIN-

sweat (L. *sudor*; *sudamen*, *sudamin-* (nns.); *sudo*, *sudat-* (v.)).

sudat'ory—stimulating sweating.

sudori'fic—(drug) which causes sweating.

sudori'ferous — sweat-producing (gland).

sudamin'a—white blisters on the skin due to the holding back of sweat in the sweat glands.

Note. **Exudate** (to ooze out, to give out liquid) is derived from this root but the *s* is dropped after the *x*.

SULPH-, SULPHO-, (SULF-)

sulphur (L. *sulfur*, *sulphur*, brimstone).

In England the root is always spelt *sulph-* but in the U.S.A. the spelling *sulf-* is normal.

sulph'ide — a compound of sulphur with another element (or group of elements), e.g. zinc sulphide ZnS.

sulph'ite — a salt of **sulphurous acid** H_2SO_3, e.g. sodium sulphite Na_2SO_3.

sulph'ate — a salt of **sulphuric acid** H_2SO_4, e.g. sodium sulphate Na_2SO_4.

sulph'ones—compounds in which an SO_2 group is joined to two hydrocarbon groups, e.g. di'ethyl sulphone $(C_2H_5)_2.SO_2$.

sulphon'ic acids—acids which contain the group of atoms $-SO_2.OH$, e.g. benzene sulphonic acid $C_6H_5SO_2.OH$.
The prefix SULPHO- is now less commonly used than its Greek counterpart THIO- (q.v.).

potassium sulpho'cyanide (potassium thiocyanate)—the compound KCNS in which a sulphur atom takes the place of an oxygen atom.

SUPER-

This Latin prefix (*super-*) primarily means "above, over, on the top of" as in **supervisor** (who watches from above). Extensions of meaning include "beyond", "more than usual" and "of a higher degree" (as, e.g. in **supernatural**).

super'natant liquid — the clear liquid which is "swimming above" a sediment which has settled out.

super'ficial — of, or on, the surface only. (In this word the prefix really means 'not under'.)

super'numerary — in excess of the normal number.

super'sonic—above (greater than) the normal speed of sound.

super'conductivity — the property of some substances at very low temperatures of having practically no electrical resistance, i.e. they conduct extremely well.

super'charger—a kind of pump used to supply a motor-car or aeroplane engine with air and fuel at a pressure greater than that of the atmosphere.

super'heated steam—steam heated (out of contact with water) to a higher temperature than that at which it would be formed from water.

super'phosphate — a fertilizer, made from calcium phosphate, which contains a higher proportion of the phosphate group than normal calcium phosphate.
The word **superior** (L. *superior*, comparative of *superus*, above, high) may similarly be interpreted as "higher" in position or in rank or degree.

superior ovary (of a flower) — an ovary which is attached to the receptacle (top of flower stem) above the points of attachment of other parts such as petals, etc.

superior planet—a planet whose orbit lies beyond (outside) that of the Earth.

SUPRA-

above, on the upper side (L. *supra*).
This prefix, similar in meaning to SUPER-, is freely used in forming anatomical terms.

supra'orbital—above the orbit (bony hollow) of the eye.

supra'renal—situated above the kidneys, e.g. the suprarenal gland. (Hence, **supra'renal'ectomy** — the surgical removal of the suprarenal gland.)

supra'clavicular — above the clavicle (collar bone).

SYLV-, SYLVI-, SYLVESTR-, (SILV-, etc.)

a wood, a forest (L. *silva* (n.); *silvestris* (adj.)).
(Although the basic Latin root is spelt with an *i*, a *y* is more usual in English spellings.)

sylva—the forest trees of a region.

sylvi'culture, silviculture—the growing and tending of trees in a wood.

sylvestr'al—growing in woods and in shaded hedgerows.
The root also occurs in the names of chemical compounds which are obtained from trees, especially from turpentine oils and resins. For details of these, e.g. **sylvic acid, sylvestrene,** a textbook of chemistry should be consulted.

Note. *Sylvanite* (an ore of tellurium, gold and silver) is named after Transylvania where it is found. *Sylvite* (or *sylvine*) is mineral potassium chloride. The early chemists called potassium chloride *Sal digestivus Sylvii*, the digestive salt of Sylvius.

SYM-, SYN-, SYL-, SYS-, SY-
with, together (Gk. *syn*, *syn-*, *sym-*).
This important prefix takes various forms according to the initial letter of the stem to which it is attached:

SYM- before *b*, *m*, and *p*;
SYL- before *l*;
SYS- before *s* (not followed by a consonant);
SY- before *s* (followed by a consonant) and *z*;
SYN- before other letters.

It is seen in such familiar words as **sympathy** ("a feeling with (another person)"), **synagogue** (a place in which Jews are "drawn together"), and **system** ("a setting (of things) together").

sym'biosis—a "living together" of two organisms for their mutual advantage (as an Alga and a Fungus in a Lichen).

sym'physis—a "growing together" of two bones to form a fixed joint (as in the pelvis and in the skull).

sym'ptoms — the various pains, feelings, etc. which "fall together", and so help to indicate the nature of a disease.

syn'apse — a "knotting together" of nerve cells.

syn'chronise—to cause to happen (act, etc.) at the same time.

syn'optic chart—a weather map on which the details of temperature, atmospheric pressure, amount of cloud, etc. are shown so that they may be "seen together".

syn'thesis—"a putting together"—the building up of complicated substances from simpler substances, e.g. the synthesis of sugar by a plant, of indigo by a chemist.

sy'zygy — "a yoking together" — the Moon is in syzygy when it is in line with the Earth and the Sun, i.e. when it is New and when it is Full.

The prefix may sometimes be better interpreted as "joined, united".

sym'petalous — (flower) in which the petals are joined together.

syn'dactyl—having two (or more) of the digits joined (as in some birds).

SYNCHRO-, (SYNCRO-)
This modern element is an abbreviation (or corruption) of *synchronised* (or *synchronous*). The *n* of *chron-* (time) has been dropped and sometimes the *h* is also dropped.

synchro'mesh gear (syncromesh) — "synchronised mesh gear"—a system of gear wheels in which the two wheels which are to engage each other are automatically made to turn at the same speed before doing so.

synchro'tron—a machine for producing very high speed atomic particles. The accelerating forces keep in time with the movement of the particles in spite of the change of mass of the particles with speed.

SYRING-, SYRINGO-, SYRINX
a pipe, a tube (Gk. *syrinx*, *syring-*).
This root is seen in the English word **syringe**.

syrinx—(1) the song-producing organ of a bird. (2) the Eustachian tube which runs from behind the ear-drum to the throat.

syring'itis — inflammation of the Eustachian tube.

syringo'myelia—a disease of the spinal cord in which irregular, pipe-like cavities are formed.

Syringa—the Mock Orange. So called because of the use of the stems, when cleared of pith, as pipe-sticks.

T

TACH-, TACHEO-, TACHO-, TACHY-
swift, quick; swiftness, speed (Gk. *tachys*, *tache-* (adj.); *tachos* (n.)).

tacheo'meter, tachymeter—a surveying instrument adapted to the quick determination of distances.

tacho'meter — an instrument which indicates the number of revolutions per minute of a turning axle.

tachy'cardia — unusually fast heartbeat.

tachy'pnoea—unusually fast breathing.

tachy'glossate — having a tongue capable of being quickly thrust in and out (as has an Ant-eater).

tachy'lyte — a black, glassy form of basalt so called because it readily melts ("quick loosening") under the blowpipe. (The alternative spelling **tachylite** probably arose because of confusion with -LITE, a stone, a rock.)

TACT-, TAG-, TANG-

to touch (L. *tango*, *tact-*, and its derivatives).

tact'ile — pertaining to the sense of touch, e.g. tactile bristles.

con'tact — a state of "touching together".

con'tag'ious—(disease) which is passed on from one person to another when they "touch together".

tang'ent — a straight line which touches but does not cut (even if extended) a circle or other curve.

TALI-, TALO-

the ankle (L. *talus*).

talus — the main ankle bone (the astragalus).

talo'calcaneal—pertaining to the ankle bone and the heel.

tali'pes — "ankle foot" — club-foot (misshapen foot usually from birth).

TANG- See TACT-.

TARS-, TARSO-

Pertaining to the **tarsus**—the region at the lower end of the hind leg of a vertebrate animal, e.g. the ankle in Man, the shank (upright part) of a bird's leg (Gk. *tarsos*, the sole of the foot).

tarsals—the ankle bones.

meta'tarsals — the bones "after (beyond) the tarsus", i.e. the bones of the foot (before the toes).

tars'algia — pain in the instep of the foot.

TAUTO-

the same (Gk. *tautos*).

Tautology is the saying of the same thing twice over in different words, e.g. 'He goes occasionally but not often'.

tauto'merism—the existence of a substance (e.g. acetoacetic ester) as a balanced mixture of two forms (**tautomeric forms, tautomerides**) which are composed of the same atoms ("same parts") and are readily converted into each other. (The substance has two sets of chemical properties corresponding to the two forms.)

TAXI-, TAXO-, -TAXIS, -TAXY

arrangement, a setting up in order (Gk. *taxis*).

taxi'dermy — "arrangement of the skins"—the art of preparing and mounting animals' skins in a life-like form.

taxo'nomy — "the management (law) of arranging"—the science of classifying (e.g.) in botany and zoology.

phyllo'taxis, phyllotaxy—the arrangement of the leaves on a shoot.

chaeto'taxy—the arrangement of the bristles on an animal.

a'taxy—"lacking arrangement, not set in order"—a condition in which the muscles do not work together and so movements are irregular.

-TAXIS (-TAXY) is also used to denote the response, reaction or disposition of an organism to a stimulus. (In the case of a plant it refers to the response of the plant as a whole (contrast -TROPISM) and so only to microscopic plants and reproductive units which are capable of movement.)

photo'taxis—the response of an organism to light.

geo'taxis—the response of an organism to the stimulus of gravity.

pneumo'taxis — the response of an organism to gases (e.g. to dissolved carbon dioxide).

chemo'taxis — the movement of an organism (or a reproductive unit) in relation to a concentration of a chemical substance.

Note. *Taxin(e)*, an alkaloid obtained from the Yew Tree, derives its name from L. *Taxus*, the Yew.

TECHNIC-, TECHNO-
art, skill (Gk. *technē*).
This root, which is seen in **technical** (pertaining to a particular art or skill), generally refers, in modern words, to industrial and mechanical arts and skills.

techno'logy — the science of the industrial arts and skills.

pyro'technics—"fire arts"—the art of making, or of displaying, fireworks.

Techni'color — (=technical colour) — the trade name of a process for producing coloured cinematograph films.

TECT-, TECTORI-, TEGUL-, TEGUMENT
These word-elements are derived from the Latin verb *tego, tect-*, to cover. (But see certain similar Greek elements below.) To **protect** means to hold out a covering.

tectum—a roof-like or covering structure in an animal's body.

tectori'al—like a roof or cover, e.g. the tectorial membrane in the inner ear.

tectrices—(feminine plural of *tector*, a cover)—small feathers which cover the bases of the main wing and tail feathers of a bird and fill up the spaces between them.

tegul'ar—of or like roof-tiles.

teguli'colous—(plant) living on roof-tiles.

tegmen — the natural covering of an animal's body (or part of it).

in'tegument—a covering layer of tissue, e.g. that of an ovule or a seed.

Note.

(1) The Greek word *tektōn* means a builder. An *architect* (*architektōn*) is a chief builder; *tectonic* means relating to building.

(2) A *eutectic mixture* (e.g. of two metals) is one of such proportions that it solidifies (and melts) at one temperature like a pure substance (Gk. *eu-*, well, *tēkō*, to melt).

TELE-
far off, at a distance (Gk. *tēle*).
This prefix is well-known in **telescope** (by which one sees things far away) and **telephone** (an instrument for sending sounds, especially speech, over a distance). As a living prefix it is freely added to words of various kinds and origins. In the word **television** the Greek prefix has been added to a word of Latin origin.

tele'pathy—"feeling at a distance"—the (supposed) communication between two people's minds without the use of sight, hearing or other senses.

tele'communication — the sending of information (e.g. speech, pictures, signals) over a distance by electrical methods.

tele'photo lens—a camera lens used for obtaining close-up photographs of distant things.

tele'therapy — medical treatment of diseases by the use of X-rays or gamma-rays. The rays are directed in a narrow beam from a distance on to the diseased part.

tele'meter—an instrument by which measurements of a physical (especially electrical) quantity are indicated at a distance away. (Also, an instrument used in surveying for determining distances.) **Telemetry** — the science of indicating measurements at a distance (e.g. from a machine, aircraft, rocket, etc.).

TELE-, TELEO-, TELEUTO-
complete, perfect (Gk. *teleos*, complete, perfect; *telos*, an end, a completion; *teleutē*, a bringing to an end, a completion).

Tele'ostei—the Bony Fishes (in which the skeleton has completely turned into bone).

Teleo'saurus—"a complete lizard"—a genus of fossil Mesozoic crocodiles.

teleuto'spore — a thick-walled spore, formed by a Rust Fungus towards the end of the season, which rests for some time before germinating.

TELO-
an end (Gk. *telos*, an end, fulfilment, completion).

telo'taxis — the movement of an animal in order to attain some end.

telo'dendra—the small "end branches" into which the stem of an efferent (outward carrying) nerve divides where it joins with another nerve cell.

telo'syn'apsis—"end-together-joining" —the end-to-end joining of the elements of a pair of chromosomes (hereditary bodies) in a dividing nucleus.

tel'angi'ectasis — "end-vessel-extension"—a state of enlargement of the small arteries and capillaries.

Note. ATELO- (q.v.) means lacking completion, not fully developed, imperfect.

TEMPOR-, TEMPORO-
Pertaining to the temple (the flat side of the head between the forehead and the top of the ear (L. *tempora* (plural)).

tempor'al—pertaining to the temple, e.g. the temporal bone.

temporo-maxillary—pertaining to the temporal and maxillary (upper jaw) regions.

TEND-, TENS-, TENT-
to stretch (L. *tendo*, *tens-*, or *tent-*).
This root is seen in **tend** ("to stretch towards"), **extend** ("to stretch out") and **extent** (the amount of stretching out).

tens'ion—the force within a bar, wire, etc. which is being stretched. (This idea of internal forces is seen in the word **hyper'tension**—high blood pressure—but the blood is not being stretched.)

tens'ile strength — the strength of a metal to withstand forces which stretch it.

tens'or—a muscle which stretches or tightens a part of the body but does not move it.

ex'tens'or—a muscle which extends (straightens) a limb or other part which has been bent.

ex'tens'ometer — an instrument used for testing the stretching of metals.

tend'on—a band of fibres by which a muscle is joined to a bone or to another muscle.

tendo — a tendon (L.), e.g. **tendo calcaneus** — the tendon which connects the calf muscle with the heel (the Tendon of Achilles).

Note. The word *tentacle* ("a little feeler") is not derived from this root, but from L. *tento*, to touch, to feel.

TENO-
a tendon, a sinew (Gk. *tenōn*).

teno'tomy—the cutting of a tendon.

TER-, TERN-, TERTI-
thrice, third (L. *ter*; *tertius*).

tern'ate—(leaves, etc.) attached in groups of three.

ter'valent—having a valency (chemical joining-power) of three. (Also called trivalent.)

terti'ary—third, e.g. the Tertiary era is the third geological era in order of time (from about 75 million to 1 million years ago).

terti'an fever—a form of malaria with intense effects every third day (at intervals of 48 hours).

TERAT-, TERATO-
a monster (Gk. *teras*, *terat-*).

terato'logy — the study of plant and animal monstrosities.

terato'geny—the production of monstrosities.

terato'ma—a tumour-like mass in the body, often containing several different tissues, caused by abnormal development.

The prefix TERA- is occasionally used in the metric system of units to denote a million million (10^{12}).

TERE-, TEREB-, TEREBENTH-, TEREBINTH-, TERP-
Terebinth (or *Pistacia terebinthus*) is the turpentine tree, the source of Chian turpentine. The name comes from the Greek *terebinthos*. The English word **turpentine** may be traced back through various spellings (e.g. *terbentyne* and *terebentyne* in the fourteenth and fifteenth centuries) to the Latin *terbentina*

or *terebenthina* (*resina*) and thus to the same root.

terebinthine—relating to terebinth or to turpentine.

terebenth'ene—a form of the hydrocarbon pinene, the chief constituent of French turpentine.

tereb'ene—a mixture of hydrocarbons formed by the action of sulphuric acid on pinene, used as a disinfectant.

The chemical term **terpene** has been formed from *terp*(*entine*)+*-ene*. The **terpenes** include oil of turpentine, oil of citron, orange-oil, and similar plant oils. The terpenes proper have the formula $C_{10}H_{16}$.

hemi'terpenes—hydrocarbons with the formula C_5H_8.

poly'terpenes — hydrocarbons whose molecules are multiples of C_5H_8.

terp'ane — the hydrocarbon corresponding to terpene but without any double joins between carbon atoms, $C_{10}H_{20}$.

terp'ineol — a terpene alcohol, $C_{10}H_{17}OH$, used as the basis of certain perfumes.

tere'phthalic acid—a form of phthalic acid, $C_6H_4(COOH)_2$, obtained by the oxidation of turpentine.

TERG-, TERGUM

the back, the hide of the back (L. *tergum*).

terg'al—pertaining to the back.

tergum—the back part of a segment of an arthropod animal; one of the plates of the shell of a Barnacle.

terg'ite—the horny back-covering of some arthropods.

TERN- See TER-.

TERP- See TERE-.

TERR-, TERRESTRI-, TERRI-

the Earth, land, earth (L. *terra*; *terrestris* (adj.)).
This root is seen in *terra firma* (dry land), **territory**, and **terra-cotta** ("baked earth").

terrestri'al—relating to the Earth (as distinct from the sky or heavens), e.g. **terrestrial telescope**—a telescope suitable for viewing things on the Earth because (unlike an astronomical telescope) it gives an image the right way up.

terri'colous—living in or on earth (as do Earthworms).

terri'genous — produced by the Earth (the land), e.g. **terrigenous sediments**—sediments, deposited in the shallow parts of the seas, formed from material taken from the land.

sub'terr'anean—under the (surface of the) Earth.

TEST-, TESTA

an earthen pot, hence a shell, a crust (L. *testa*).
In botany the term **testa** denotes the seed-coat.

test'aceous—made of baked earth; of the colour of old, red bricks; pertaining to shells or to shellfish.

Testacea—a name sometimes given to various groups of invertebrate animals with shells (other than Crustaceans), especially to (1) a sub-order of pteropod molluscs which have chalky shells. (2) an order of the Protozoa the members of which have a kind of shell (usually one chamber) through apertures in which the pseudopodia (projections of protoplasm) are pushed.

TESTUD-, TESTUDIN-

a tortoise, tortoise-shell (L. *testudo*, *testudin-*).

testudin'eous—like, or relating to, the shell of a tortoise.

testudin'ate—arched like the shell of a tortoise.

TETR-, TETRA-

four (Gk. *tettares*, *tetra-*).

tetr'ad—a group of four things, e.g. of spores.

tetr'ode — a radio valve with four electrodes.

tetra'hedron—a solid figure with four faces (a triangular pyramid).

tetra'dactyl — (animal) having four fingers or toes.

tetra'merous—having "four parts" or

parts arranged in fours, e.g. a flower with sepals, petals, etc. in fours.

tetra'valent — having a valency (chemical joining-power) of four. (Also called quadrivalent.)

carbon tetra'chloride — a compound whose molecule consists of a carbon atom joined to four chlorine atoms, CCl_4.

lead tetra'ethyl—$Pb(C_2H_5)_4$, a colourless liquid used for improving motor-spirit.

THALAM-, THALAMI-, THALAMO-, THALAMUS

The Latin *thalamus* and the Greek *thalamos* denote a bed-chamber, an inner private room. The meanings of the root as used in biological terms seem rather remote from the original meaning.

(1) *Botany*. The **thalamus** is the receptacle of a flower (the enlarged top of the flower-stalk on which the petals and other parts are situated).

thalami'floral—having all the parts of the flower joined separately to the receptacle with the ovary above the other parts.

(2) *Zoology*. The **thalam'encephalon** ("inner room brain") is the back part of the forebrain of a vertebrate animal. Its sides become thickened to form the **thalami** from which nerves (especially the optic nerves) originate (or appear to originate).

THALASS-, THALASSO-

the sea (Gk. *thalassa*).

thalass'ic—pertaining to the seas (as distinct from the greater oceans).

thalasso'phyte — "a sea plant", a seaweed, a marine Alga.

thalass'in — a poisonous substance produced by some Sea Anemones.

THALL-, THALLO-, THALLUS

A **thallus** is a plant body which is not clearly separated into leaves, stem and root. (This Latin word is derived from the Greek *thallos*, a young green shoot.)

Thallo'phyta — the great group of plants which comprises the Algae, Fungi and the Lichens.

thall'iform—in the form of a thallus.

pro'thallus — a small, thallus-like plant which grows from a fern spore and which reproduces sexually to give rise to the familiar form of fern.

hetero'thall'ic — "of different plant-bodies"—said of some Fungi in which there are two forms of mycelia (body threads). Spores are formed only when different forms of mycelia come together.

Thallium, the rather uncommon chemical element, was named after the brilliant green ("young shoot" colour) line which it gives in its spectrum.

THANAT(O)-, -THANASIA

death (Gk. *thanatos*).

thanat'oid—deadly, poisonous; death-like.

thanat'ophidia — poisonous (deadly) snakes.

eu'thanasia—easy painless death; the bringing about of this in cases of people with incurable diseases.

THECA, THEC(O)-, -THECA

a container, a receptacle (Gk. *thēkē*).

theca — a spore-container of some Fungi and Mosses; a pollen-container; a case or sheath round an organ.

thec'odont—having the teeth set in "containers" (sockets) in the bone.

sperma'theca—a cavity or sac used by many lower animals (e.g. an Earthworm) for receiving and storing sperm-cells.

gnatho'theca — "jaw container" — the horny part of the lower beak of a bird.

-THELIUM

Primarily the root denotes a teat or nipple (Gk. *thēlē*).

poly'thelia — the occurrence of more than the necessary (or normal) number of nipples.

The meaning of the root has been extended in the word **epi'thelium** which denotes a cellular tissue forming the outer layer of a mucous membrane in animals or, more generally, a covering or lining of a free surface. Hence the root is often more conveniently interpreted as 'a thin layer of cells'.

meso'thelium—the middle of the three tissues forming the wall of a developing embryo.

-THERAPY

healing, medical treatment (Gk. *therapeia* (n.); *therapeuō* (v.)).

The basic meaning of the root is "waiting upon, serving"; it is now used to denote "waiting upon, attending upon" in a medical sense. **Therapeutics** is the branch of medical science which is concerned with the treatment of disease and the use of curative and remedial agents. -THERAPY denotes medical treatment; the first part of the word indicates the method of treatment.

chemo'therapy—the treatment of disease by chemicals (drugs).

radio'therapy — treatment by radiations (e.g. heat); **electro'therapy** — by electricity; **organo'therapy**—by extracts from the organs of animals, etc.

Therapy is also used as a noun, e.g. **occupational therapy.**

-THERIA, THERI(O)-, THERO-

a beast, a wild animal (Gk. *thēr*; *thērion*).

Eu'theria—"good (higher developed) animals — a sub-class of mammals in which the young are born in an advanced stage of development.

theri'odont—"animal teeth"—(*n.* and *adj.*) a fossil reptile with teeth of a mammalian type.

Thero'poda—an order of carnivorous Dinosaurs having feet like those of mammals.

Note. The name of the disease *diphtheria* is derived from Gk. *diphthera*, a skin, a hide. So named because of the tough membrane which is formed on the affected parts.

THERM-, THERMO-, -THERMY

heat (Gk. *thermē*).

This root is seen in the familiar word **thermometer** — an instrument for measuring how hot things are, i.e. their temperatures. A **therm** is a unit (equal to 100,000 British Thermal Units) for expressing a quantity of heat.

therm'al — relating to heat or to temperature, e.g. thermal expansion of a bar, thermal capacity of a block of substance.

thermo'stat — a container which is automatically kept at a steady temperature.

thermo'dynamics — the study of the relation between heat energy and mechanical energy.

thermo'setting plastics — plastics in which a chemical change takes place while they are being moulded (under heat and pressure) so that the finished article is not affected by heat.

therm'ionics—the study of the giving out of electrons by hot bodies (e.g. a heated wire). Extended to include the control and use of such electrons (as in radio valves).

exo'thermic — (chemical reaction) in which heat is given out.

dia'therm'anous — (substance) which lets heat rays pass through it. (Compare a *transparent* substance which lets light rays pass through it.)

dia'thermy—the use of electric currents to produce heat in the deeper tissues of the body.

steno'thermy — "narrow heats" — the ability to live within only a small range of temperatures.

hypo'thermy — medical treatment by keeping the patient at a temperature a few degrees below normal body temperature.

-THESIS, -THETIC

a setting, a placing, a putting (Gk. *thesis*).

A **thesis**, in academic circles, is essentially a proposition which is laid down or stated (and if necessary defended) before an examiner.

hypo'thesis — "a placing under, a foundation"—a reasonable explanation of observed happenings, used as a starting point for reasoning and for designing tests which will establish its truth (or otherwise).

pros'thesis — "a putting to" — the supplying of an artificial part of a body (e.g. a wooden leg, false teeth) to make up for a part which is ineffective or has been lost.

syn'thesis—"a putting together"—the building up of complex chemical substances from simpler substances, e.g. the synthesis of sugar by a plant, of a rubber-like substance by a chemist.

A **synthetic substance** is, strictly, one which has been **synthesised,** i.e. built up by synthesis, either within a living thing or in a chemist's laboratory. But nowadays the term is used more loosely for almost any artificial (or substitute) substance as manufactured by a chemist, whether by synthesis or not, e.g. synthetic rubber, fibres, and diamonds.

THI-, THIO-, THION-

sulphur (Gk. *theion*).

thio'phene — a liquid similar to benzene. The molecule consists of a ring of four CH groups and one sulphur atom.

thion'yl chloride—a colourless liquid with the formula $SOCl_2$.

The prefix often denotes the presence of a sulphur atom (in a compound) in a place normally occupied by an oxygen atom.

potassium thio'cyanate — the compound with the formula KCNS. (Compare potassium cyanate KCNO.) Also called potassium sulphocyanide.

sodium thio'sulphate—a white crystalline compound (known to the photographer as 'hypo') with the formula $Na_2S_2O_3$. (Compare sodium sulphate Na_2SO_4.)

thi'amides — compounds similar to amides but with a sulphur atom in place of the oxygen atom, e.g. $CH_3.CS.NH_2$. (Compare acetamide $CH_3.CO.NH_2$.)

thio'ethers — compounds similar to ethers (q.v.) but with a sulphur atom in place of the oxygen atom, e.g. $CH_3.S.CH_3$

thi'ols—compounds similar to alcohols but with a sulphur atom in place of the oxygen atom, e.g. ethane-thiol C_2H_5SH. (Compare ethyl alcohol C_2H_5OH.)

THIXO-

touch, touching (Gk. *thixis*).

thixo'tropy—"a turning by touching" —the property of some jelly-like substances of becoming liquid when shaken and of re-forming a jelly when left to stand.

THORAC-, THORACO-, (THORACI-, THORACICO-)

The **thorax** (Gk. *thōrax*, breast-plate, breast) is (in land vertebrates) the part of the body between the neck and the abdomen, i.e. the chest; (in insects) the middle of the three main parts of the body; in general, a similarly situated part of any animal.

thorac'ic—pertaining to the thorax.

thoraco'tomy—the surgical cutting of the wall of the chest.

thoraco'plasty—"thorax moulding"—the taking out of parts of ribs and the collapsing of a lung so that the lung may rest and recover after a disease.

thoraco'scope — an instrument for viewing the inside of the chest.

THROMB-, THROMBO-

the clotting of blood (Gk. *thrombos*, a clot of blood).

thrombus—a clot of blood formed in a blood vessel.

thromb'in — a substance formed in blood which helps the blood to clot (e.g. in a wound).

thrombo'cyte — "a clotting cell" — a small, plate-like cell in the blood which plays an important part in clotting.

thrombo'sis—the formation of a blood clot in a blood vessel or organ.

THYRO-, THYROID-

The Greek word *thyreos* means an oblong shield, hence **thyroid** means shield-like. The **thyroid cartilage** is the largest of the cartilages of the larynx (voice-box) and is formed from two plates like a shield. It produces the lump known as the Adam's Apple on the front of the throat. Hence THYRO- and THYROID are used to denote this part of the body.

thyro-glossal—pertaining to the thyroid region and the tongue.

thyroid gland—a large ductless gland situated near the larynx.

More commonly the elements refer to the thyroid gland itself.

thyroid'itis—inflammation of the thyroid gland.

thyroid'ectomy—the surgical cutting out of the thyroid gland.

thyr'oxine—a complex substance (containing iodine) produced by the thyroid gland and secreted into the blood stream.

thyro'toxic'osis — "a state of thyroid poisoning" — the condition which is caused by over-activity of the thyroid gland (as in Grave's disease).

TINCT-

dye, stain (L. *tingo*, *tinct-*, to colour, to dye).

The English word **tinge** is derived from this root.

tinct'orial — of dyes, e.g. **tinctorial power**—a measure of the depth of colour produced by a dye.

tinct'ure—a solution (especially of a drug) in alcohol, e.g. tincture of iodine.

TOC-, TOCO-, TOK(O)-

childbirth (Gk. *tokos*).

toco'logy—the medical science which deals with childbirth.

dys'tocia—difficulty in giving birth to a child.

mono'tocous—producing one offspring at each birth.

poly'tocous, polytokous — producing many offspring at one birth (as by some animals).

-TOM-, -TOME, -TOMY

to cut; a cutting (Gk. *temnō* (v.); *tomē* (n.)).

The word **a'tom** means "not cut", i.e. indivisible. The idea that matter consists of ultimate, indivisible particles goes back to the time of the Greeks but it was not until the beginning of the nineteenth century that Dalton formulated the atomic theory in terms and concepts which were scientifically acceptable. An atom is the smallest particle of an element which can take part in a chemical reaction.

It is now known that an atom is not indivisible. The central nucleus may be separated from the electrons which travel round it and the nucleus itself may be split into simpler parts. The original name **atom** might therefore seem to be misleading, or even incorrect, but it remains true that an atom cannot be split into smaller parts which still have the qualities of the element.

En'tomo'logy is another interesting word. It means "the study of insects". It is derived from the Greek *entomos*, meaning "cut into", a reference to the characteristic notched body of an insect.

The root is most frequently met in the form of the suffix -TOMY ("a cutting"). A **dichotomy** is a cutting, i.e. division, into two separate parts, e.g. the division of the growing point of a branch into two branches. The suffix may be added to the appropriate form of the name of almost any part of the body to indicate the operation of cutting into that part by surgery.

gastro'tomy — the surgical cutting of the wall of the stomach.

tracheo'tomy—the surgical operation of cutting into the trachea (wind-pipe).

hepato'tomy—the surgical cutting of the liver.

litho'tomy—the cutting of the urinary bladder to remove a stone.

(**Anatomy** strictly means the process of cutting up a plant or animal body for the purposes of examination of the parts. But the word has been extended in meaning, both by literary writers and by scientists themselves, to denote the parts which are revealed by the process of anatomy. Hence one speaks of an animal's anatomy and of the science of Comparative Anatomy.)

Whereas -TOMY denotes a "cutting into" a body part (technically called an incision), -ECTOMY (q.v.) means a "cutting out", i.e. a removal, of a body part.

gastr'ectomy—the surgical cutting out of the whole or a part of the stomach.

hepat'ectomy—the surgical removal of a part of the liver.

(-STOMY (q.v.) must be distinguished from these suffixes; it means the making of a "mouth" (artificial opening) by surgery.)

The suffix -TOME is used in two senses.

(1) A section or segment (as if cut off). This sense is seen in the ordinary word **epitome** (a summary, an abstract—a part cut off from the original). A **myotome** is a segment of a muscle (especially of an animal with a segmented body).

(2) An instrument used for cutting.

pharyngo'tome—a surgical instrument used for pharyngotomy (cutting into the pharynx). So also **tracheto'tome** (for cutting the trachea), **litho'tome** (for cutting the bladder in lithotomy).

micro'tome — an instrument for cutting thin sections for examination under a microscpe.

TONIC

The Greek word *tonos* means that which strains or tightens, a stretched or tightened thing. This meaning is seen in the word **peri'toneum** — the membrane which lines the body cavity containing the viscera (soft organs) and stretches over the viscera themselves.

The English word **tone** (which has been derived through Latin from the Greek) has acquired a range of meanings. It may be broadly interpreted as meaning strength, intensity or quality. **A tonic** is a medicine which invigorates and gives strength.

iso'tonic—(solution) having an equal (equivalent) strength to that of another solution and so having the same osmotic pressure (q.v.).

hypo'tonic—having a lower strength (and so lower osmotic pressure); **hyper'tonic**—having a greater strength.

TOP-, TOPO-, -TOPE

a place (Gk. *topos*).

topo'graphy — the description of the natural and artificial features of a place (town, district, etc.) and the representation of these features on a map; the features themselves.

ec'topic pregnancy—the fertilization of the egg and the growth of the embryo "out of place", i.e. in some other place than the ovary (e.g. in the Fallopian tube).

iso'topes — forms of a chemical element which have identical chemical properties but different atomic weights. So called because they are put in the "same place" in the Periodic System (classification of elements).

TORQU-, TORS-, TORT-, TORTI-

to turn, to twist (L. *torqueo, tort-*).

This root is seen in **torture, distort** ("to twist away"), **extort** ("to twist out") and **contortions** ("complete twistings"). A chemist's **retort** has a long neck which "twists back" from the bowl which holds the liquid.

torque—a set of forces which causes an axle to turn; the size of this turning-power.

tors'ion—twisting, e.g. of a rod, of a plant stem, of the cut end of an artery.

torti'collis—"twisted neck"—a turning of the head or neck to one side because of diseased neck-bones or (more usually) of stiff and contracted muscles.

torti'cone—a spirally twisted shell.

TORUS, TOR-, TORUL-

a bulge, a swelling, a protuberance (L. *torus*).

The term **torus** is used in the various sciences in different senses. It denotes, for example, a large outwardly curved moulding, the rounded top of a flower stem (the receptacle), a rounded protuberance of a part of an organ, a smooth ridge as of a muscle. In mathematics it denotes an anchor-ring, i.e. a solid ring with a circular cross-section.

tor'ose, tor'ous — bulging, swollen. Used especially to describe a roughly cylindrical thing with bulges at intervals.

tor'oid, toroidal—like an anchor-ring.

tor'ic lens—a spectacle lens of which one surface is shaped like a piece of the surface of a torus (anchor-ring).

torul'oid—having a series of small swellings, e.g. like a necklace.

Torula—a kind of yeast-like fungus which forms chains of round cells. Infection by such a fungus causes **torul'osis.**

torulus—a socket for an antenna of an insect.

TOX-, TOXICO-, TOXO-

(1) a bow (for arrows) (Gk. *toxon*, a bow; *toxa*, bow and arrows).
A **toxophilite** is "a lover of the bow", an archer.

Tox'odon—a genus of large extinct quadruped animals which had strongly curved molar teeth.

(2) The Greek *toxikon* (*pharmakon*) was originally "the arrow drug", i.e. poison used for smearing on arrows. Hence the root TOX(ICO)- has come to denote a poison.
Intoxication is, etymologically, a state of poison in the body.

tox'ic—poisonous; pertaining to a poison.

toxico'logy—the branch of medical science which deals with the nature and effects of poisons.

toxico'dermat'itis — inflammation of the skin caused by irritating poisons.

toxic'osis—a disease or unhealthy condition caused by a poison.

tox'in—a specific poison, especially one produced by an organism (e.g. a bacterium), which causes a disease. **anti'toxin**—a substance produced by the body to counteract a toxin.

tox'aemia — an unhealthy condition caused by the absorption of toxins into the blood and tissues.

TRACHE-, TRACHEO-

the trachea (the wind-pipe).
The name **trachea** is derived through Latin from the Greek *tracheia* (*artēria*) which means "the rough artery" (*trachys*, rough).

trache'itis—inflammation of the membrane of the trachea.

tracheo'bronchial — pertaining to the trachea and the bronchi (the two main branches of the trachea).

tracheo'cele—an air-containing swelling in the neck due to a bulging of the wall of the trachea.

tracheo'scope—an instrument for inspecting the inside of the trachea.

tracheo'tomy—the surgical operation of cutting into the trachea.

tracheid(e)—a long, pointed cell, in the wood-tissue of a plant, which has lost its living contents and serves to conduct water. (It was once thought to be an air-pipe; hence the name.)

TRACHEL-, TRACHELO-

the neck (Gk. *trachēlos*). Used also to denote the neck of the uterus (womb).

trachel'ate—neck-like.

trachel'ectomy — the surgical cutting out of the neck of the uterus.

trachelo'rrhaphy—the sewing up of a torn neck of the uterus.

TRACHY-, TRACH-

rough (Gk. *trachys*).

trachy'glossate — having a rough tongue (suitable for scraping and rasping).

trachy'spermous—having seeds with a rough surface.

trach'oma—a disease of the eye in which there are rough, granular growths on the inner surfaces of the eye-lids.

TRACT-

to draw, to drag (L. *traho*, *tract-*).
This root occurs in a number of well-known words, e.g. **contract** ("to draw together"), **extract** ("to draw out"),

abstract ("to draw away (or out) from"). A **tractor** is an engine for pulling.

re'tract'ile — able to be drawn back (e.g. the claws of most members of the Cat family).

con'tract'ile — capable of, or producing, contraction, e.g. **contractile root**—a form of fleshy root which, as it ages, wrinkles and shortens and so drags the plant deeper into the ground; **contractile vacuole** — a cavity, in some Protozoa (e.g. Amoeba), filled with liquid, which periodically collapses and expels its contents.

tractrix — a mathematical curve. Imagine a heavy particle attached to a piece of string, lying on a rough surface (e.g. a table) with the string lying out straight. If the free end of the string is moved along a straight line (other than the original direction of the string) the particle is dragged along a path which is a tractrix.

TRANS-

across, through, to or on the other side (L. *trans*, *trans-*).

This common prefix needs little explanation. It is seen, for example, in **transfer** ("to carry across"), **transmit** ("to send across"), **transparent** ("appearing through") and **transverse** ("turned crosswise").

trans'fusion — "a pouring across" — a passing of a liquid from one container to another, e.g. of blood from the veins of one person to those of another.

trans'lucent—(material) through which light can pass (but through which one cannot see things clearly).

trans'ducer—that which "leads across" —a device or apparatus which takes power from one system (e.g. an electrical system) and supplies it to another system (e.g. an acoustic system). A loudspeaker is an example of a transducer.

trans'uranic elements — chemical elements (produced artificially) which have higher atomic numbers than that of uranium.

In chemical names the prefix *trans-* indicates that two atom groups of a molecule are situated on opposite sides of the main axis of the molecule. Below are shown two forms of an acid $C_2H_2(COOH)_2$; (a) is the *trans*-form (in this case fumaric acid) and (b) is the *cis*-form (in this case maleic acid).

(a)	(b)
HC.COOH	HC.COOH
‖	‖
HOOC.CH	HC.COOH

TRAUMA, TRAUMAT-, TRAUMATO-, TRAUMO-

a wound, an injury (Gk. *trauma*, *traumat-*).

trauma — a wound, a body injury. (Also an emotional shock.)

traumat'ic—pertaining to, or caused by, wounds.

traumato'logy—the scientific description of wounds.

traumo'tropism—"wound turning"— the curved growth of a part of a plant which results from a wound.

TRI-, TRIPLO-

three; threefold. (L. *tres*, *tri-*; *triplex*. Gk. *treis*, *tri-*; *triplous*.)

A **triangle** has three angles and three sides, a **trident** has three teeth, a **tripod** has three feet. **Triple** means threefold; **triplets** are three offspring born at one birth.

tri'cuspid — having three cusps or points, e.g. the tricuspid valve of the heart.

tri'foliate—having leaves or leaflets in sets of three.

tri'gon'ometry—"three angle measurement"—the branch of mathematics which is based on calculations about the angles and sides of triangles.

Tri'lobite—an extinct aquatic arthropod, living in Palaeozoic times, which had three main lobes to its body.

tri'merous — having "three parts" or parts arranged in threes, e.g. (flower) having petals, stamens, etc., arranged in threes.

tri'ode — a radio valve with three electrodes.

tri'valent—having a valency (chemical joining-power) of three. (Also called tervalent.)

sulphur tri'oxide—an oxide of sulphur, SO_3, in which three oxygen atoms are joined with one sulphur atom.

tri'nitro'toluene — an explosive (T.N.T.), $CH_3.C_6H_2(NO_2)_3$. In the molecule, three nitro groups ($-NO_2$) replace three hydrogen atoms of toluene $C_6H_5CH_3$.

triplo'blastic — having three primary layers in the developing embryo.

TRIBO-

rubbing (Gk. *tribos* (n.), *tribō* (v.)).

tribo-electricity—electricity produced by the friction between two bodies.

tribo-luminescence—the production of light by friction, e.g. by the grinding of certain solids (notably ordinary sugar).

TRICH-, TRICHO-

a hair, the hair (Gk. *thrix*; *tricho-*).

trich'oid—like a hair.

trich'ite—a hair-like crystal occurring in groups in certain volcanic rocks.

Trich'ina—a hair-like parasitic worm which causes the disease **trichiniasis** or **trichiniosis**. It may enter the body as a result of eating bad pork.

tricho'gyne—a hair-like growth on the female organ of some lower plants.

trich'oma—a disease of the hair.

trich'ome—a plant-hair, an outgrowth of one or more cells on the surface.

tricho'pathy—the medical treatment of diseases of the hair.

tricho'phyllous—(plant) having leaves and young stems protected from excessive drying by a thick coating of hairs.

dermo'trichia — "skin hairs" — the horny rays (supporting elements) of the unpaired fins of fish.

TRIPLO- See TRI-.

TROCH-, TROCHO-

a wheel, a disc (Gk. *trochos*).

troch'al—wheel-shaped.

troches—small discs of medicine to be taken by the mouth.

troch'oid—a curve traced out by any point fixed relative to a circle (e.g. a point on a radius or on an extension of a radius) which rolls along a straight line. (In the special case when the point is on the circumference of the circle the curve is called a cycloid.)

trocho'phore, trocho'sphere — a larval form of many Molluscs and certain Annelid worms characterised by a spherical body and a ring of cilia (hair-like outgrowths).

meso'trochal — having a ring of cilia round the middle of the organism.

trochlea—any pulley-shaped structure in the body, e.g. the grooved lower end of the upper arm bone. (Gk. *trochilia*, L. *trochlea*, a pulley.)

Note. A *trochanter* is a bony protuberance on the upper part of the thigh bone. The word is derived from Gk. *trochanter*, from *trechō*, to run.

-TRON See -ON.

TROPH-, TROPHO-, -TROPHY

food, nourishment, nutrition (Gk. *trophē*).

troph'ic—pertaining to nutrition.

tropho'neurosis — faulty, defective nutrition due to a nervous disorder.

tropho'plasm — protoplasm which is mainly concerned with nutrition.

a'trophy — "lacking nutrition" — the wasting away (of a cell or body member) due to under-nourishment or disease.

hyper'trophy — "over feeding" — abnormal enlargement of a cell or body member.

-TROPIC, -TROPISM, TROPO-, -TROPE

a turning, direction (Gk. *tropos*).

The **Tropics** is the name given to the region between the **Tropic of Cancer** and the **Tropic of Capricorn.** The midday Sun is overhead at the Tropic of Cancer on about June 21st, at the equator on about September 22nd, at the Tropic of Capricorn on about December 22nd and at the equator again on about March

21st. The two tropics mark the "turning points" in the Sun's apparent motion.

laevo'tropic — twisting the plane of polarisation of polarised light to the left.

Helio'trope—a well-known flowering plant whose flowers turn to the Sun.

tropo'sphere—the lower layer of the atmosphere in which the temperature falls with height and within which weather changes occur. The **tropo'pause,** which divides the troposphere from the higher stratosphere, marks the end of the region of change.

A **tropism** is a response, in the nature of a turning or directed growth, of a part of a plant to an outside stimulus.

helio'tropism—the turning and growth of a part of a plant towards the Sun. So also **photo'tropism** — towards (or away from) light; **geo'tropism** — towards (or away from) the stimulus of gravity; **hydro'tropism** — towards water. Roots are positively **geotropic** (they grow in the direction of the stimulus) and shoots are negatively geotropic.

The Greek word *tropos* was extended metaphorically to mean "a manner, a mode".

allo'tropy — "other modes" — the existence of a chemical element in two or more different forms, e.g. diamond, graphite and lamp-black are **allotropic** forms (or **allotropes**) of carbon.

-TRUDE, -TRUSION

to push, to thrust (L. *trudo, trus-*).

To **intrude** is to push one's way in.

in'trus'ive rocks—rocks formed from melted material which has been forced into earlier rocks. Also called **intrusions.**

ex'trus'ive rocks—rocks formed from melted material which has been forced out on to the surface of other rocks. (Also called lava flows.)

ex'trusion — a method of producing rods, etc., by forcing hot metal through a hole.

TRYPANO-

a borer (Gk. *trypanon*).

Trypano'somes — a group of disease-carrying parasites (Protozoa) with long, slender bodies. They cause Sleeping-sickness and Chagas disease.

trypanosom'iasis — Sleeping-sickness, caused by Trypanosomes.

A **trepan** is a tool used by a surgeon for cutting the bone of the skull. The name comes from the same root.

TUBER, TUBERCUL-

a bump, a swelling, a tumour (L. *tuber, tuberculum* (dim.)).

tuber—a short, thick part of an underground stem, e.g. a potato.

tubercle—a general name for a "small swelling", e.g. on the roots of plants, on a bone or other part of the body; a small mass of cells caused by infection with the tubercle bacillus.

tubercul'osis — a disease caused by infection of the body, especially the lungs, lymph glands and joints, by the tubercle bacillus. Tubercles develop in the body tissues.

tubercul'in — a substance, prepared from a culture of the tubercle bacillus, used for telling whether a person has (or has had) tuberculosis.

TUM-, TUME-

to swell (L. *tumeo*).

tume'fy—to cause to swell. **tumid**—swollen.

tume'facient — causing or producing swelling.

tum'our—a diseased swelling or enlargement formed by the growth of new cells.

in'tumescence—the process of swelling, of becoming swollen; the swelling of certain crystals when they are heated.

de'tumescence — the reduction of a swelling.

TURB-

a disorder, a confused crowd (L. *turba*).

This root occurs in the word **disturbance.**

turb'id—(liquid) which has particles of solid floating in it and is not clear.

turb'ulent—disturbed, disordered, e.g. a turbulent flow of liquid in a pipe (as opposed to a smooth, straight-line flow).

per'turb'ations—variations from the regular movement of a planet (etc.) in its orbit round the Sun caused by the disturbing gravitational forces of another planet. (The existence of Neptune was deduced from the perturbations of Uranus.)

Turb'ellaria — a class of Flatworms, comprising the Planarians. So called because the cilia (hair-like lashes) cause tiny currents ("little disturbances") in the water.

TURBIN-, TURBO-

that which turns or spins, a top, a whirl (L. *turbo*, *turbin-*).

turbin'ate—shaped like a spinning top (roughly conical).

turbine — an engine (for producing power) in which water, gas or steam turns a set of blades on an axle.

Turbo—a genus of gastropod molluscs with turbinate shells.

TURBO- usually means 'relating to a turbine'.

turbo-alternator—a steam turbine and an alternating electricity generator directly joined together.

turbo-jet—a jet engine in which hot gases from the burning fuel turn a turbine and this drives a compressor.

turbo-prop—a form of aeroplane in which hot gases from the burning fuel turn a turbine and this turns the propellers.

TYLO-

a knob, a pad (Gk. *tylos*).

tylo'pod—(animal) with pads on its digits instead of hoofs, e.g. a camel.

tylo'sis—a diseased inflammation of the eye-lids in which there is a thickening and hardening at the edges.

-TYPE, TYPO-

The Greek word *typos* has a range of (related) meanings. It means a blow, the mark of a blow (i.e. an impression), an impressed or moulded figure, and hence a model. By metaphor a model or figure is a **type**.

The idea of striking and making a mark is reflected in words connected with printing. The words of this book were first written with a **typewriter**; the printer set the words in **type**.

typo'graphy—the art of printing; the method of printing in which the letters, etc. are raised up from the surface of the printing plate and the ink applied to the top surface of the letters.

'Mono'type' — a machine which puts together type (for printing) a letter at a time.

'Lino'type'—a machine which constructs a complete line of type in one piece.

The idea of a model is seen in the following words.

arche'type, proto'type—the first type or model (e.g. of an aeroplane) from which the copies or improved forms may be made.

TYRO-, TYROS-

cheese (Gk. *tyros*).

tyros'ine — a complex substance, formed from the breakdown of certain proteins, found in the pancreas and in cheese.

tyro'toxicon—a poisonous substance produced in bad cheese and milk.

U

UL- See ULO-.

-ULA, -ULE, -ULUM, -ULUS, (-UL-)

The Latin suffix -ULUS (-A, -UM) is a diminutive, i.e. it indicates a small specimen of the thing denoted by the main part of the word.

Examples are **Convolvulus** ("a small rolled or twisted thing"), **blastula** ("a little bud"), **gastrula** ("a little stomach"), **Campanula** ("a little bell"), **capitulum** ("a little head").

-ULE is the English equivalent of this suffix. It is seen, for example, in **globule** ("a little ball"), **granule** ("a little grain"), **capsule** ("a little case"), and **ovule** ("a little egg").

-UL- is used in forming adjectives (e.g. **globular**). It does not, however, necessarily indicate a diminutive (e.g. **glandular**).
Also see -CULUS and -UNCULUS.

ULNO-
Pertaining to the **ulna** (L.), the inner (not thumb-side) bone of the forearm.
ulno-carpal — pertaining to the ulna and the carpals (wrist bones).

ULO-, UL-
the gums (in which the teeth are set) (Gk. *oulon*, pl. *oula*).
ul'itis—inflammation of the gums.
ulo-glossitis — inflammation of the gums and the tongue.
ulo'rrhage—bleeding ("breaking") of the gums.

ULTRA-
(1) beyond, on the other side (L. *ultra*).
ultra-violet rays—rays (e.g. in sunlight) which, in a spectrum, lie beyond the violet (i.e. are of shorter wavelength).
ultra'marine — a blue pigment originally obtained from Lapis lazuli which was brought from "beyond the sea".
ultra'sonics — the science which deals with rapid mechanical vibrations having frequencies above those which correspond to sound.
(2) very, exceedingly.
ultra-short waves — radio waves with a wavelength less than 10 metres.
ultra-high frequencies—radio frequencies greater than 300 megacycles (300 millions of cycles) per second.
ultra'microbe—a particle (of doubtful nature) which causes disease but is too small to be seen with a microscope.

-UM
(1) The ending of many Latin neuter nouns (and of nouns constructed on the Latin pattern), e.g. **septum, momentum, basidium, sporangium, quantum**. The plural is normally formed by replacing -UM by -A.
(2) An ending of the names of many metallic chemical elements. See -IUM.

UMBEL, UMBELL-, UMBELLI-, UMBRA
a shade, a shadow (L. *umbra*; *umbella*, "a little shade, a sunshade, a fan).
This root is seen in **umbrella** (a word which has come through Italian).
umbel — a group of flowers (like a miniature sunshade) borne on stalks of nearly equal length springing from one point on the main axis.
umbell'ate — having the character of an umbel; (plant) producing umbels.
Umbelli'ferae—the family of flowering plants, including Parsley, Parsnip and Hemlock, which "bear umbels".
umbra — a true shade, a region in which no light falls because of the presence of an obstacle.
pen'umbra — "almost a shadow" — a part-shadow, a region which bounds the umbra when an obstacle obstructs the light which is coming from an extended (not point-like) source; some, but not all, of the light is prevented from reaching the region.

UNC-, UNCI-, UNCIN-
a hook (L. *uncus; uncinus*).
unc'ate, uncin'ate—hooked, bearing a hook or hooks.
unci'form — in the form of a hook, hook-like.

-UNCLE, -UNCULUS
A variant of the suffix -CULE (q.v.), a diminutive which indicates a small specimen of the thing denoted by the main part of the word.
The English word **uncle** is derived from L. *avunculus*, "a little grandfather".
carb'uncle—"a small coal"—(1) a red gem, especially garnet. (2) a pimple, a tumour.
ped'uncle—"a small foot"—a flower-stalk.
Ran'unculus — "a small frog" — the

genus of plants which includes the Buttercup and Crowfoot. (Named from the frogs which live in the kind of places where these plants grow.)

UNDUL-
wavy (ultimately from L. *unda*, a wave).

undul'ating — moving up and down like a wave, having a wavy surface.

undul'ant fever — a fever whose intensity increases and decreases in a regular way.

UNGUI-, UNGUL-
a finger nail, a claw, a talon, a hoof (L. *unguis*; *ungula*).

ungula—a nail, claw or talon.

ungu'al—of, like, or bearing, a nail or claw.

ungui'late—provided with claws.

Ungul'ates—the large group of plant-eating animals, including Sheep, Goat, Horse, Ox, which have hoofs.

mult'ungulate — having the hoof divided into three or more parts.

UNI-
one, single (L. *unus*).
This prefix is seen in **uniform** (of one form or kind) and **universe** (all existing things which "turn as one"). To **unite** means to become, or make as, one.

uni'axial—having one axis.

uni'cellular—composed of one cell.

uni'lateral—on one side (only).

uni'directional current — an electric current which always flows in the same direction. (Commonly called direct current.)

uni'parous—giving birth to one off-spring at a time.

uni'sexual—showing the features of only one sex, e.g. a flower which has stamens (male) or a pistil (female) but not both.

uni'valent—having a valency (chemical joining-power) of one. (Also called monovalent.)

UR- See -URIA, URINO-.

URAN-, URANO-
the heavens (Gk. *ouranos*).

urano'graphy—descriptive astronomy, especially the description of the constellations.

Uranus — the planet, discovered by Herschel in 1781, which is seventh in distance out from the Sun.

uran'ium — a radioactive metallic element which has the highest atomic weight of the natural elements. (Named after Uranus.)
In medical terms the root refers to the vault or roof of the mouth.

urano'plasty—"moulding the roof of the mouth"—the surgical closing of a cleft palate (gap in the roof of the mouth).

UREDIN(O), UREDO-
The Latin *uredo*, *uredin-* means a burning itch, a blight. The root is used especially with reference to certain parasitic Fungi.

Uredin'ales—a large group of parasitic Fungi, commonly called the Rust Fungi, including that which attacks wheat.

uredo'spores — spores (appearing as orange spots on the wheat) produced in the summer. They are formed in the **uredo-stage** of the life-history. (Later, winter spores (teleutospores) are formed which infect Barberry.)

-URET, -URIZE
The suffix -URET (adapted from L. *-uretum*) was formerly used in the names of chemical compounds formed by the combination of two elements; now replaced by -IDE.

sulph'uret — a sulphide. **sulphuretted hydrogen**—hydrogen sulphide, H_2S.

carb'uret — a carbide. Hence **carburetted** — combined with, or impregnated with, carbon.

carburet'tor, carburetter—an apparatus in which air is mixed with a volatile fuel (e.g. petrol) to form a combustible mixture for use in an internal combustion engine.

carb'urize—to impregnate with, or cause to combine with, carbon. Used especially to describe the addition of carbon to iron in the making of cement steel.

URETER, URETERO-
The **ureter** is the duct (pipe) by which urine passes from the kidney to the bladder (Gk. *ourētēr*).

ureter'itis — inflammation of the ureter.

uretero'tomy—the surgical cutting of the ureter.

uretero'colo'stomy—the surgical operation of making the ureter drain into the colon (large intestine).

URETHR-, URETHRO-
The **urethra** is the duct (pipe) by which urine passes from the bladder to outside the body (L. *urethra*; Gk. *ourēthra*).

urethr'itis — inflammation of the urethra.

urethro'tomy—the surgical cutting of the urethra.

urethro'scope—an instrument for inspecting the inside of the urethra.

urethro'cyst'itis — inflammation of both the urethra and the bladder.

-URGY
working, the working (Gk. *-ourgos*; *-ourgia*).

metall'urgy—the science of extracting, refining and working metals.

chem'urgy—"working by chemicals" —the application of chemistry to agriculture and of agriculture to chemistry. The word **surgery** has been derived through Old French *surgerie*, *cirurgerie*, Latin *chirurgia*, from Gk. *cheirourgia*, a working by hand.

-URIA, URINO-, URO-, UR-
Pertaining to **urine**, the 'water' passed out of the body as an excretion of the kidney (L. *urina*; Gk. *ouron*).
(Also see URO- =tail.)

urea — a soluble crystalline compound, $CO(NH_2)_2$, found in the urine of mammals.

ur'ic acid—a compound which occurs in the urine of flesh-eating animals and in the excreta of birds and reptiles. Deposits of uric acid in the body cause gout and rheumatism. The urine of plant-eating animals contains **hipp'uric acid.**

di'uretic—(drug) which stimulates the discharge of urine.

ur'aemia—"urine in the blood"—a state of the blood which is caused by the failure of a diseased kidney to remove urea and similar urinary substances.

haemat'uria—the presence of blood in the urine.

glycos'uria—the presence of sugar in the urine.

URINO- and URO- are usually interchangeable; the latter is now the more common.

uro'logy (urinology) — the branch of medicine which deals with disorders of the urinary organs.

uro'scopy—the scientific examination of urine (e.g. as a means of testing for diseases).

uro-genital system—the organs concerned with the discharge of urine and with reproduction when, as in vertebrate animals, there is a connection between them.

-URIZE See -URET.

URO-, (-URA, -URUS)
a tail (Gk. *oura*).
(Also see URO- (URINO-) =urine.)

uro'delous — having a "clear (plainly seen) tail", e.g. an amphibian such as the Salamander in which the tail persists.

uro'pod — "tail foot" — any limb-like appendage on the abdomen of a Crustacean (e.g. Lobster) especially one which is towards the back and is larger than the rest.

uro'pygium—"tail rump"—the short tail-stump of birds.

uro'stege—"tail covering"—one of the special scales on the under side of a snake's tail.

An'ura—animals "without tails"—an order of tailless amphibians including Frogs and Toads.

Dasy'urus—an animal with a "hairy

tail" — a genus of small carnivorous marsupial animals of Australia and Tasmania.

-US
The ending of a large number of Latin masculine nouns (and of nouns formed on the Latin pattern), e.g. **focus, locus, nucleus, strobilus, Gladiolus**. The plural is properly formed by replacing -US by -I. (Note that a few nouns of another type, e.g. **corpus, genus,** form plurals in other ways.)

UTER(O)-
The **uterus** (L.) is the womb.

uter'itis—inflammation of the womb.

uter'ine — pertaining to the womb; born of the same mother but not the same father (e.g. 'his uterine brother').

intra-uterine — situated within, or developing within, the womb.

UTRICLE, UTRICUL-
a small leather bag, a bladder (L. *utriculus*).

utricle, utriculus—a small sac (bag) in the body, especially that in the inner ear.

utricul'ar—like a bladder; pertaining to a utricle.

Utricul'aria—the Bladderwort, which has little bladders growing among the leaves.

V

VACCIN-
Pertaining to a cow, and hence to cowpox (L. *vaccinus*; *vacca*, a cow).

vaccinia—cowpox, a disease of cows in which there are small blisters on the milk-producing organs, caused by a virus.

vaccine—(1) Pertaining to cows. (2) Liquid (containing the virus) taken from a cowpox blister and injected into the human body as a protection against smallpox. (3) Any similar preparation of micro-organisms or viruses injected into the body to stimulate the production of disease-resisting substances.

vaccin'ation—the injection of a vaccine into the body.

VACU-
void, empty (L. *vacuus*).
The English words **vacate** and **vacant** are derived from the related verb-form of this root.

vacuum—(strictly) a space which contains no gas or other form of matter; (in practice) a space in which the air (or gas) pressure has been much reduced.

vacu'ole—"a small empty space"—a small space or cavity (usually containing a liquid) in the protoplasm of a cell.

e'vacu'ate—to empty, to remove the contents of, e.g. to evacuate a flask, the stomach, the bowels, a building. (In recent years one has heard about 'evacuating people'. This process is not, as the phrase suggests, a medical operation but is the emptying of a building (town, etc.) by the removal of the people.)

-VALENCY, -VALENT
having strength or worth (L. *valens*).
This root is recognised in **equivalent** (of equal worth). To be **ambivalent** is to have either or both of two contrasting values or qualities.
The root has special significance in chemistry.

valency—the combining-power of an atom (or group of atoms) expressed, in its simplest form, in terms of the number of hydrogen atoms with which the atom can combine or which it can replace.

mono'valent (univalent) — having a valency of one, e.g. chlorine (whose atom can combine with one hydrogen atom to produce HCl), sodium (whose atom can replace the one hydrogen atom in HCl to give $NaCl$).

di'valent (bivalent)—having a valency of two, e.g. oxygen (whose atom can combine with two hydrogen atoms to produce H_2O), zinc (whose atom can replace the two hydrogen atoms of H_2SO_4 to give $ZnSO_4$).

tri'valent (tervalent)—having a valency of three, e.g. nitrogen (whose atom can combine with three hydrogen atoms to give NH_3).

tetra'valent (quadrivalent) — having a valency of four, e.g. carbon (whose atom can combine with four hydrogen atoms to give CH_4).

(Both Latin and Greek prefixes of number are used but, in spite of the Latin origin of *-valent*, Greek prefixes are more usual.)

By extension of meaning, -VALENT also refers to the nature or process of chemical combination.

electro'valency — the joining of two atoms by the giving of one (or more) electrons by one atom (e.g. sodium) to another atom (e.g. chlorine) and the subsequent attraction between the two ions (charged particles) so formed.

co'valency—the joining of two atoms by the sharing of a pair of electrons.

VARI-, VARIO-

various, changing, changeable, different (L. *varius*).

Several closely related Latin roots convey the meaning of vary, alter, to be variegated. These roots appear in a number of well-known words, e.g. **vary, variable, variation, various, variety.**

vari'ola—smallpox, a disease in which the body is speckled with pustules.

vario'lite — "variegated stone" — a kind of igneous rock containing many small rounded bodies (feldspar) embedded in it.

vario'meter—an instrument used to show or measure variations in a physical quantity (e.g. atmospheric pressure, magnetic force); (in radio) a tuning coil which consists of a variable inductance.

var'istor — "variable resistor" — an electrical device or component whose resistance varies markedly with the voltage which is applied.

VARIC-, VARICO-, VARIX

A **varix** is an abnormal enlargement of a vein which then becomes lengthened and tortuous (L. *varix, varic-*).

varic'ose vein — an enlarged vein as described above.

varico'cele — a varicose condition of the veins which leave the testis and form the spermatic vein.

VARIO- See VARI-.

VAS, VASO-, VASCUL-

a vessel (L. *vas*; *vasculum*, small vessel).

vas deferens—"a carrying-off vessel" —the duct (pipe) by which the semen (male fluid) is led from the testis to the organ which discharges it.

vas'ectomy — the cutting away, by surgery, of the vas deferens (or part of it).

vaso'constrictor — a nerve or a substance which causes the constriction of an artery.

vaso'motor—(nerve) which causes constriction or expansion of an artery.

extra'vas'ate—to force fluid out from its container.

vasculum — "a small vessel", e.g. a botanist's collecting-case.

vascul'ar bundle—the bundle of tissues in a plant which conducts liquids. It consists essentially of xylem (wood-vessels) and phloem.

vascular system (of an animal)—the vessels and organs responsible for the circulation of blood and lymph.

VEN-, VENE-

a vein (L. *vena*).

ven'ation — the arrangement of the veins in a leaf, in an insect's wing, etc.

vene'section—the cutting of a vein to let out blood.

intra'venous—within, or put within, a vein.

Note. *Venom* is derived from L. *venenum*, a poison. *Venereal* is derived from the name Venus.

VENT-, VENTI-, VENTIL-

air movement, the wind (L. *ventus*, the wind; *ventilo, ventilat-*, to fan).

vent—a hole, opening or passage out of (or into) a closed space and through which gases (or liquids) can pass.

ventil'ate—to cause air to circulate (e.g. in a room), to replace 'used' air by fresh air.

venti'facts—small stones which have been shaped by the wind (usually under desert conditions).

Note. This root must not be confused with VENT- (L. *venio*) meaning "to come" (as in *prevent*—"to come before", *invent*—"to come on to").

VENTR-, VENTRI-, VENTRO-
the belly (L. *venter*).

ventr'al — pertaining to the front, under, belly-side of an animal. (When applied to a leaf it denotes the upper (not back) side.)

dorsi'ventral — (leaf, etc.) which has the front and back sides differently constructed.

ventr'icose, ventricous—having a belly which sticks out; (generally) swollen, especially in the middle.

ventro'lateral — at the side of the ventral region.

ventri'cle—"a little belly"—a small chamber or cavity, especially the cavities in the brain and the two lower chambers of the heart.

ventri'fixation, ventrofixation — the fixing of the womb (by surgery) to the front wall of the abdomen.

VERD-, VERDI-
green (L. *viridis*, hence French *vert*).

verd'ure—the greenness of grass and vegetation.

verdi'gris — "green of Greece" — a form of green copper carbonate as seen on copper which has been exposed to moist air; also green copper acetate.

verd'ite — a green rock consisting chiefly of green mica and clay.

VERM-, VERMI-
a worm (L. *vermis*).

vermi'an—of worms, worm-like.

vermi'form—shaped like a worm, e.g. the vermiform appendix.

vermi'cide — a substance which kills worms.

vermi'fuge — (drug) which expels worms from the intestine.

vermi'cular—like a "little worm" in form or movement.

vermicul'ites — a group of hydrous silicates occurring as decomposition products of the micas. When slowly heated they form flakes and open into worm-like threads.

Vermilion was originally an orange-red dye-stuff obtained from the Cochineal insect ("a little worm"); it is now the red pigment mercury sulphide.

VERS-, -VERSE, VERT-, VERTIC-
to turn (L. *verto*, *vers-*).

This well-known root is seen in **reverse** ("to turn back"), **revert, invert,** etc. The **universe** comprises everything that exists and "turns as one". To be **versatile** is to be able to turn readily from one skill, duty, subject, etc., to another.

extra'vert—(person) who is "turned outwards", who is interested in things outside himself and readily makes contact with other people.

intro'vert—(person) who is "turned to the inside", whose interests are mainly about his own thoughts and attitudes and how he is regarded by others.

vertebra — one of the bones which compose the backbone. **vertebrate**—(animal) which has a backbone (L. *vertebratus*, jointed, able to turn). **invertebrate**—(animal) without a backbone (e.g. an insect, a shellfish, a worm).

vertigo—a feeling of spinning round, giddiness, dizziness.

The primary meaning of the Latin *vertex* (*vertic-*) is a whirl, a whirlpool.

vertic'il—"a little whirl"—a whorl, a set of parts (e.g. leaves) arranged in a ring round an axis.

verticillate—arranged in whorls.

The Latin *vertex* also means the crown of the head, the peak, the top of anything. This meaning is reflected in the English words **vertex** (e.g. of a triangle) and **vertical.**

VESI-, VESIC-, VESICO-
a bladder, a blister (L. *vesica*).

vesica — a bladder (especially the urinary bladder).

vesic'ant—(substance or other agent) which causes blisters.

vesico'tomy—the surgical cutting of the urinary bladder.

vesi'cle—"a little bladder"—any small bag-like cavity, bubble, blister, containing a liquid or gas.

vesi'culose—swollen like a bladder; appearing to be made up of small bladders.

VIBR-, VIBRO-

to shake, to swing (L. *vibro*, *vibrat-*).
This root is immediately recognised in the word **vibrate.**

vibro'tropism — the response of an animal to shaking or vibrations in its surroundings.

vibro'scope—an instrument for observing or counting vibrations (e.g. of a tuning fork).

VINYL

(*Chemistry.*) The hydrocarbon group of atoms CH_2:CH-. (From L. *vinum*, wine + *-yl* (q.v.).)

vinyl chloride—CH_2:CHCl.

poly'vinyl plastics (and **resins**)—polymers built up from vinyl compounds such as the chloride and acetate.

VIR-, VIRID-

green (L. *viridis*).

vir'escence—an unusually green condition, e.g. of flower petals which are normally brightly coloured.

virid'escent—tending to become green.

viridian green—a transparent green pigment made from an oxide of chromium.

VIR(O)-

Pertaining to a **virus**—a particle (probably living), smaller than a bacterium, which causes a number of diseases (L. *virus*, slime, poison).

vir'ology—the study of viruses and the diseases they cause.

vir'osis—a disease due to a virus, e.g. smallpox.

VISCER-, VISCERO-

Pertaining to the **viscera**—the internal organs of the body, especially those of the abdomen (L. *viscera*, entrails).

viscer'al—pertaining to, situated in, or affecting, a **viscus** (an internal organ).

e'viscer'ate—to remove the viscera.

viscero'ptosis—a dropping of the intestines due (e.g.) to weakness of the abdominal muscles.

viscero-motor — (nerves) which bring motor (movement) impulses to the viscera.

VISC(O)-

sticky, semi-liquid.
This root comes from the Latin *viscum* —Mistletoe or bird-lime (sticky substance spread on twigs to catch birds).

visc'in—a sticky substance obtained from Mistletoe berries.

visc'id, visc'ous — sticky, unable to flow easily.

viscosity — the stickiness of a liquid, (strictly) the internal friction between various layers of a liquid which prevents easy flow.

visc'ometer — an instrument for measuring the viscosity of a liquid.

visc'ose — a sticky substance (the sodium salt of cellulose xanthate) used in the making of rayon.

VI-, VIT-, -VIVE, VIVI-

life; to live (L. *vita* (n.); *vivo* (v.)).
This root is seen in the well-known words **vital** (essential to life), **vivid** (life-like, bright and clear) and **revive** (to bring back to life).

vi'able—(embryo, new-born animal) able to live and develop.

vit'amin—"amine of life" (but see the note under AMINE) — a substance (of which there are a number of kinds) which is necessary, usually in only a small amount, for the health and proper working of an animal organism. Hence **a'vitamin'osis**—a state of deficiency of vitamins; a disease due to such a deficiency. (This word is an interesting example of a mixture of Latin and Greek elements.)

vivi'parous—bringing forth the young alive (not hatching them from eggs).

vivi'section — the cutting of living animals as a means of scientific study.

VITELL(O)-
the yolk of an egg (L. *vitellus*).

vitell'ine—pertaining to egg-yolk; of the yellow colour of an egg-yolk.

vitell'in — the chief protein of egg-yolk.

vitell'arium—a gland which forms egg-yolk.

VITR-, VITRI-, VITRO-
glass (L. *vitrum* (n); *vitreus* (adj.)).

vitreous—pertaining to glass; of the nature of glass.

vitri'fy—to change, or cause to be changed, into glass or a glass-like substance.

de'vitri'fication—"the un-making of glass"—the gradual formation of very small crystals in glass and glassy minerals so that the substances lose their clear glassy nature.

blue vitriol — "blue glass" — the old name for crystalline copper sulphate. So also **green vitriol** — ferrous (iron) sulphate; **white vitriol**—zinc sulphate; **oil of vitriol**—concentrated sulphuric acid.

vitro-dentine—the hard outside layer of dentine in a tooth.

VIVI- See VI-, VIT-.

VOLT, VOLTA-
The Italian physicist Volta (1745–1827) was the first to construct a chemical battery which could produce a continuous electric current. Electricity so produced was called **voltaic electricity** and the chemical cell a **voltaic cell**. These terms are now out of date.

volt — the practical unit for the measurement of electric pressure.

volt'meter — an instrument for measuring electric pressure in volts.

volta'meter — (originally) an instrument, based on electrolysis, for the measurement of an electric (voltaic) current; now (more commonly) any cell in which electrolysis takes place.

-VOLUTE, -VOLUTION, VOLV-, -VOLVE
to roll, to turn about or round (L. *volvo, volut-*).
This root is seen in **involve** ("to roll in") and in **revolve, revolution** (in which the prefix *re-* is an intensive).

volute—rolled up (as a leaf, petal, etc.).

volv'ulus — "a small twisting" — a twisting of an abdominal organ, e.g. of a loop of the intestine.

e'volve (evolution)—"to roll out"—to give out heat, gas, etc. (*Biology*) **evolution** is the gradual development of more complex organisms from simpler forms.

con'volution — "a rolling with" — a coil, a twist. **Convolvulus**—a well-known plant, also called Bindweed, which climbs by coiling itself round a support.

in'volute, e'volute — mathematical curves. Imagine a piece of string fixed at one end to a point on a curve A (e.g. a circle). If the string is held tightly by its other end and gradually wrapped round the curve, any point on the string marks out another curve B. Curve A is the **evolute** of curve B; curve B is the **involute** of curve A.

in'volucre — a ring of bracts (scales) round the base of a closely packed flower-head.

-VORE, -VOROUS
eating (L. *voro, vorat-*, to devour).
voracious—greedy in eating.

carni'vore — an animal which eats flesh. It is **carnivorous.**

herbi'vorous — plant-eating. So also **frugivorous** (fruit), **graminivorous** (grass), **piscivorous** (fish), **omnivorous** (all kinds of food), etc.

VULCAN-
Vulcanus was the Roman god of fire, furnaces and metal-working. The name has given rise to the word **volcano.**

vulcan'ite — a preparation of rubber and sulphur hardened by heat, ebonite.

vulcan'ites—a general name for fine-grained igneous rocks normally occurring as lava flows.

vulcan'ization of rubber—the process of treating crude rubber with sulphur and subjecting it to heat as a means of producing rubber of a more durable and useful form.

vulcan'ology—the scientific study of volcanoes.

-VULSION
a plucking, a pulling (L. *vello*, *vuls-*, to pluck).

e'vulsion—the forcible pulling out of a part in surgery.

con'vulsion—violent, irregular motion of a limb or other part of the body due to involuntary contraction of the muscles as in a seizure.

W

The letter W does not occur in the Latin alphabet and has no counterpart in the Greek alphabet. No classical roots, therefore, begin with this letter.

A few scientific terms, the meanings of which are usually inferable, have been formed from ordinary English words such as *water*, *wave* and *wind*. A few other terms (e.g. **watt**) and, more especially, the names of some minerals (e.g. **websterite, willemite**) have been formed from proper names.

Wolfram, or **wolframite,** is a mixture of iron and manganese tungstates, an important source of the element tungsten. The origin of the name is obscure. The name wolfram is sometimes used as an alternative to tungsten, the chemical symbol for which is W.

Wort is derived from an Old English word meaning plant or herb. It occurs in the English names of some plants, e.g. **Bladderwort, Stitchwort.**

X

XANTH-, XANTHO-
yellow (Gk. *xanthos*).

xantho'carpous—having yellow fruit.

xantho'derm(i)a — yellowness of the skin.

xanth'odont—having yellow teeth (as the cutting-teeth of some rodents).

xantho'phyll—"yellow leaf"—one of the two yellow pigments associated with chlorophyll in green plants.

xanth'opsia—"yellow vision"—an unhealthy condition in which objects appear yellow to the observer.

xantho'pterin — a yellow pigment found in the wings of the Lemon Butterfly and in the bodies of Wasps.

xanth'ic acid — an organic acid, $C_2H_5O.CS.SH$, whose salts (**xanthates**) are yellow.

xanthene dyes — dye-stuffs (chiefly yellow or brown) based on a ring of five carbon atoms and one oxygen atom.

XEN-, XENO-
a guest (or a host), a stranger, a foreigner (Gk. *xenos*).

xeno'lith—a piece of rock of one kind in a rock of another kind.

xeno'gamy — "stranger marrying" — cross-fertilization.

xeno'phobia — fear (i.e. dislike) of foreigners.

mono'xen'ous — (fungus) which is restricted to one species only of host plant.

xen'on — a heavy, inert gaseous element discovered, with other rare ("strange") gases, at the end of the last century.

XER-, XERO-
dry (Gk. *xēros*).

xero'derm(i)a—a disease causing dry plates on the surface of the skin.

xero'philous — "dry loving" — (plant) adapted to, and flourishing in, dry conditions.

xero'phyte — a plant able to live in places where there is little supply of water, e.g. in deserts.

xer'ophthalmia—"dry" ophthalmia—an inflammation of parts of the eye, especially of the conjunctiva, in which there is no discharge of liquid. Also called **xerosis.**

XIPH-, XIPHI-, XIPHOS-
a sword (Gk. *xiphos*).

xiph'oid—shaped like a sword.

xiphi'sternum — the lower, pointed part of the sternum (breast-bone).

Xiphos'ura — "the sword tails" — a

class (order) of arthropod animals, including the King Crabs, which have a telson (last section of abdomen) in the form of a long spine.

XYL-, XYLO-
wood (Gk. *xylon*).
A **xylophone** is a musical instrument in which wooden rods are struck and made to vibrate and so emit sound.

xyl'em—the woody part of the conducting system of plant stems, branches, etc.

xyl'ene—a hydrocarbon, $C_6H_4(CH_3)_2$, obtained from coal-tar.

xyl'ose — a sugar-substance obtained by boiling wood-gum, straw or jute with dilute sulphuric acid.

xylo'phagous — (insects) which eat wood.

xylo'tomous — (insects) which cut or bore into wood.

Y

-YL
A suffix used in forming the names of chemical radicals (groups of atoms which act as single units), e.g. **ethyl**—the radical C_2H_5-. The suffix is derived from the Greek *hylē* which, in addition to meaning wood, also means the stuff or raw material of which a thing is made. Thus ethyl means "the ether stuff".

In organic chemistry the suffix is mainly used for two types of radicals.

(a) hydrocarbon groups having a valency (chemical joining-power) of one:

meth'yl — CH_3- (methane CH_4; methyl chloride CH_3Cl).

prop'yl — C_3H_7- (propane C_3H_8; propyl chloride C_3H_7Cl).

phen'yl—C_6H_5- (phenol C_6H_5OH; phenylamine $C_6H_5NH_2$).

benz'yl — $C_6H_5.CH_2$- (benzyl alcohol $C_6H_5CH_2OH$).

(Note that the meaning may be modified by the addition of -ENE (q.v.), e.g. propane $CH_3CH_2CH_3$, propyl $CH_3CH_2CH_2$-, propylene $CH_3.CH:CH_2$. Acetylene is irregularly named.)

(b) the 'stems' (residues) of organic acids:

acet'yl — CH_3CO- (acetic acid $CH_3.CO.OH$; acetyl chloride $CH_3CO.Cl$).

benzo'yl—C_6H_5CO- (benzoic acid $C_6H_5CO.OH$; benzoyl chloride $C_6H_5CO.Cl$).

succin'yl — $-OC.CH_2.CH_2.CO$- (succinic acid $HO.OC.CH_2.CH_2.CO.OH$).

Other important radicals are:

hydrox'yl—the group -OH, e.g. hydroxyl'amine $NH_2.OH$.

carbox'yl — the group -COOH, the characteristic group of organic acids.

In inorganic chemistry the suffix is normally used for a radical which contains oxygen.

thion'yl — $SO=$, e.g. thionyl chloride $SOCl_2$.

sulphur'yl — $SO_2=$, e.g. sulphuryl chloride SO_2Cl_2.

nitros'yl—NO-, e.g. nitrosyl chloride NOCl.

nickel carbon'yl—$Ni(CO)_4$.

Z

ZE-, ZEO-
to boil (Gk. *zeō*).

ec'ze'ma—"an out-boiling"—a general name for various eruptions on the skin.

zeo'lites—"boiling stones"—a group of alumino-silicates of sodium, potassium, calcium, etc., containing loosely held water, occurring in cavities in igneous rocks. So called because they swell, melt and give off water when heated.

ZOO-, -ZOA, -ZOON, -ZOIC, -ZOID
an animal (Gk. *zōon*, a living being, an animal; *zōa* (plural)).
Zoo is now accepted as a word; it is an abbreviation of **Zoological Gardens. Zoology** is the study of animals.

zoo'geography—the study of the distribution of animals over the Earth.

zoo'plankton—plankton (floating and

drifting organisms) which consists of animals. (Contrast phytoplankton — of plants.)

zoo'lite — "animal stone" — a fossil animal; a fossilised animal substance.

zoo'phyte — a plant-like coelenterate animal, e.g. Hydra, Obelia. (A better term would have been *phytozoon.*) These animals form the class **Hydrozoa**; most of them are polyps (often in colonies) with alternating generations of medusae (free-swimming forms).

Proto'zoa—"the first animals"—the sub-kingdom of very simple, one-celled or non-cellular animals, including Amoeba and Paramoecium.

Sporo'zoa—a large class of the Protozoa; the animals are parasitic and normally without means of moving by themselves. (It includes the malaria parasite. Sexual reproduction in the body of the mosquito results in the formation of **sporozoites** ("spore animals"); these are injected into the blood when the mosquito bites and there asexual reproduction occurs.)

Meta'zoa — the sub-kingdom of animals (showing some degree of complexity) which comes "after" the sub-kingdoms Protozoa and Parazoa.

spermato'zoon—"a seed animal"—the typical male germ-cell, consisting usually of a head (with a nucleus), a thin body and a tail.

spermato'zoid—a motile male cell (of various kinds) of lower plants. Also called an **anthero'zoid.**

-ZOIC (Gk. *zōikos*) is adjectival. It occurs, for example, in the names of geological eras.

Palaeo'zoic—"ancient life"—the era from about 500–400 million to 200 million years ago.

Ceno'zoic — "recent life" — the era from about 60 million years ago to the present day in which there was development of life as it is now.

Note. The term **azote** ("without life") was originally proposed for nitrogen. The prefix AZO- (q.v.) indicates nitrogen.

ZYG-, ZYGO-, -ZYGY

a yoke (Gk. *zygon*).

zygo'ma—the bridge of bone in front of the ear and below the eye.

zygote—the cell formed by the union of two sex cells.

hetero'zygous — having "different (other) yoking"—having inherited both of two contrasting factors (e.g. blue eye-colour from one parent, brown from the other). One factor is often dominant (e.g. brown eye-colour) and the other factor remains hidden.

Zygo'mycetes — a group of Fungi, including many common moulds, which form resting spores (**zygospores**) by the fusion of the ends of two fungal threads.

Zyg'nema—"yoked thread"—a genus of fresh-water Algae which reproduce by conjugation so that the algal threads are yoked together.

zygo'morphous—"shaped like a yoke" —(flower) built on an H-plan, i.e. symmetrical only about a vertical central line, e.g. White Deadnettle.

sy'zygy — "a yoking together" — The Moon is in syzygy when it is in line with the Sun and the Earth, i.e. when it is New or Full.

ZYM-, ZYMO-, -ZYME

fermentation (Gk. *zymē*, leaven, yeast).

zym'urgy — the art or practice of fermentation as in brewing, wine-making, etc.

zymo'sis—fermentation.

zymotic diseases—an old name for diseases caused by the multiplication of germs, once regarded as similar to fermentation.

zym'ase—a substance produced (e.g.) by yeast which causes fermentation of sugar.

en'zyme — a substance formed by living cells which brings about (or speeds up) chemical actions in the body. They were formerly regarded as, and known as, ferments.

zymo'gen — a substance formed by plants or animals as a stage in the formation of an enzyme.